BURN

Paul Collins

BURN

THE EPIC STORY OF BUSHFIRE IN AUSTRALIA

In memory of
Kathleen Ann O'Hehir
Winifred Mary O'Hehir
and
James O'Hehir

First published in Australia and New Zealand in 2006
Copyright © Paul Collins 2006

All rights reserved. No part of this book may be reproduced or transmitted in any form or by any means, electronic or mechanical, including photocopying, recording or by any information storage and retrieval system, without prior permission in writing from the publisher. The *Australian Copyright Act 1968* (the Act) allows a maximum of one chapter or 10 per cent of this book, whichever is the greater, to be photocopied by any educational institution for its educational purposes provided that the educational institution (or body that administers it) has given a remuneration notice to Copyright Agency Limited (CAL) under the Act.

This project has been assisted by the Commonwealth Government through the Australia Council, its arts funding and advisory board.

Allen & Unwin
83 Alexander Street
Crows Nest NSW 2065
Australia
Phone: (61 2) 8425 0100
Fax: (61 2) 9906 2218
Email: info@allenandunwin.com
Web: www.allenandunwin.com

National Library of Australia
Cataloguing-in-Publication entry:

Collins, Paul, 1940- .
   Burn: the epic story of bushfire in Australia.

   Bibliography.
   Includes index.
   ISBN 9781741750539.

   ISBN 1 74175 053 9.

   1. Forest fires – Australia. I. Title.

363.3790994

Judith Wright, lines from 'The Two Fires' as published in *A Human Pattern: Selected Poems* (Sydney, 1996) reproduced by permission of ETT Imprint

Maps by Ian Faulkner
Set in 12/15 pt Bembo by Midland Typesetters, Australia
Printed and bound in Australia by Griffin Press
10 9 8 7 6 5 4 3 2 1

CONTENTS

List of abbreviations | vi
Maps | viii
Introduction | xvii

PART 1 A NATION OF FIRE-LIGHTERS | 1

1 Black Friday, 13 January 1939 | 3
2 Over a century of fires, 1788–1938 | 51
3 'Burn, burn, burn', 1939 | 108

PART 2 AFTER STRETTON | 137

4 Black days, 1939–66 | 139
5 Ablaze, southern Tasmania, 1967 | 178
6 On the urban frontier, 1968–2002 | 205

PART 3 THE GREAT FIRES OF 2003 | 245

7 'Stinking hot and windy', the Snowy Mountains and Victoria, 2003 | 247
8 A perfect fire day, Canberra, 17 January 2003 | 287

PART 4 THE GREAT FIRE DEBATES | 327

9 To burn or not to burn? | 329
10 Fire thugs | 349
11 Fireproofing Australia? | 361

Notes | 380
Bibliography | 402
Index | 409

ABBREVIATIONS

ABS	Australian Bureau of Statistics
ACT	Australian Capital Territory
ACT ESB	ACT Emergency Services Bureau
ADB	*Australian Dictionary of Biography*
ADF	Australian Defence Force
ANP	Alpine National Park [Victoria]
ANU	Australian National University [Canberra]
ANZAAS	Australia and New Zealand Association for the Advancement of Science
BFBA	Bush Fire Brigades Association [Victoria]
BOM	Bureau of Meteorology
bp	(Years) Before Present
CFA	Country Fire Authority [Victoria]
CFS	Country Fire Service [South Australia]
COAG	Council of Australian Governments
COR INQ Brin	Coronial Inquiry into the Circumstances of the Fire(s) in the Brindabella Range in January 2003
COR INQ Kos	Coronial Inquiry into the Fires Referred to as 'Jagungal Wilderness Area, Kosciuszko National Park Complex of Fires'
CSIRO	Commonwealth Scientific and Industrial Research Organization
DEC	Department of Environment and Conservation, New South Wales
DEC SUB	'Submission to the Coronial Inquiry into December 2002–March 2003 Bushfires', DEC, New South Wales
DPI	Department of Primary Industry, Victoria
DSE	Department of Sustainability and Environment, Victoria
DSEVAF	Kevin Wareing and David Flinn, *The Victorian Alpine Fires January–March 2003*
EMA	Emergency Management Australia

ABBREVIATIONS

ESPLIN	Bruce Esplin, Malcolm Gill and Neal Enright, *Report of the Inquiry into the 2002–2003 Victorian Bushfires*
FFDI	Forest Fire Danger Index [compiled by A.G. McArthur]
HRNSW	*Historical Records of New South Wales*
ICC	Incident Control Centre
IMC	Incident Management Controller [New South Wales]
IMT	Incident Management Team [New South Wales]
KC	King's Counsel
km/h	kilometres per hour
KNP	Kosciuszko National Park [New South Wales]
MCLR	Ron McLeod, *Inquiry into the Operational Response to the January 2003 Bushfires in the ACT*
MMBW	Melbourne and Metropolitan Board of Works
NAIRN	Garry Nairn (Chairman), *A Nation Charred: Inquiry into the Recent Australian Bushfires*
NLA	National Library of Australia, Canberra.
NPWS	National Parks and Wildlife Service [New South Wales]
PMG	Postmaster General's Department [handled postage and telecommunications]
POL BRIEF	New South Wales Police, 'Coronial Brief', 2003 Fires
PROV	Public Record Office [Victoria]
QFRS	Queensland Fire and Rescue Service
RCME	[Stretton] Royal Commission, Minutes of Evidence
RCR	[Stretton] Royal Commission Report
RCYF	[Stretton] *Report of the Royal Commission to inquire into The Place of Origin and the Causes of the Fires which commenced at Yallourn on the 14th Day of February 1944.*
RFS	Rural Fire Service [New South Wales]
SDI	Soil Dryness Index
SEC	State Electricity Commission [Victoria]
TB	tuberculosis
TFS	Tasmanian Fire Service
UAP	United Australia Party [forerunner of Liberals]
UHF	ultra high frequency [radio]

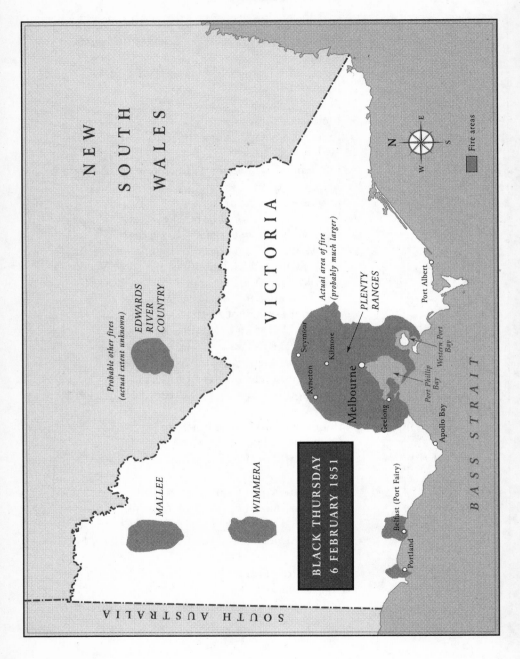

MAPS

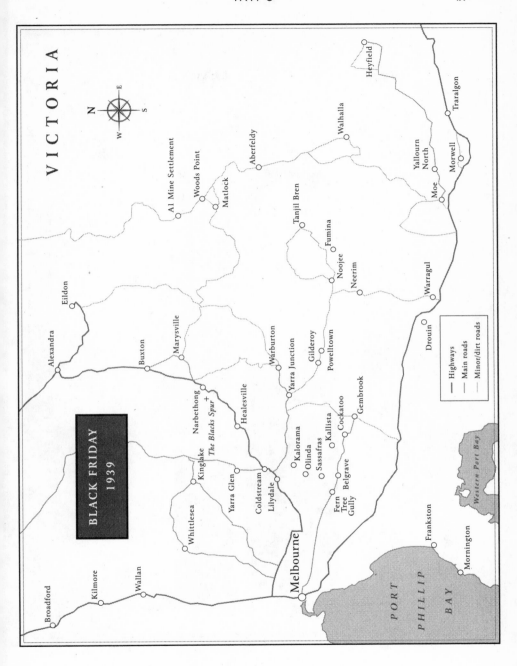

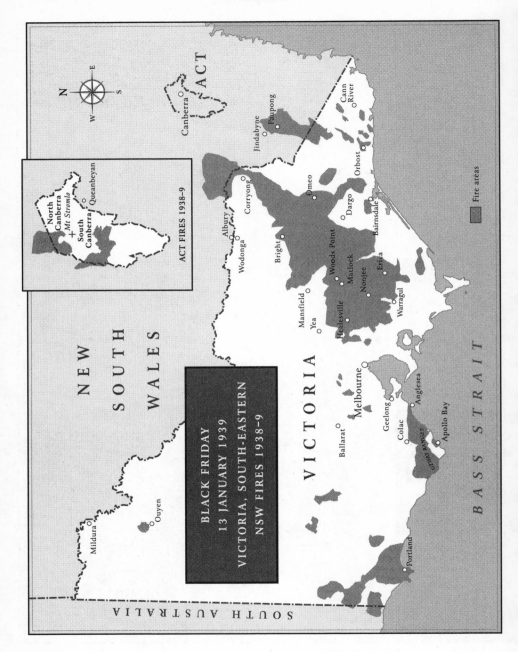

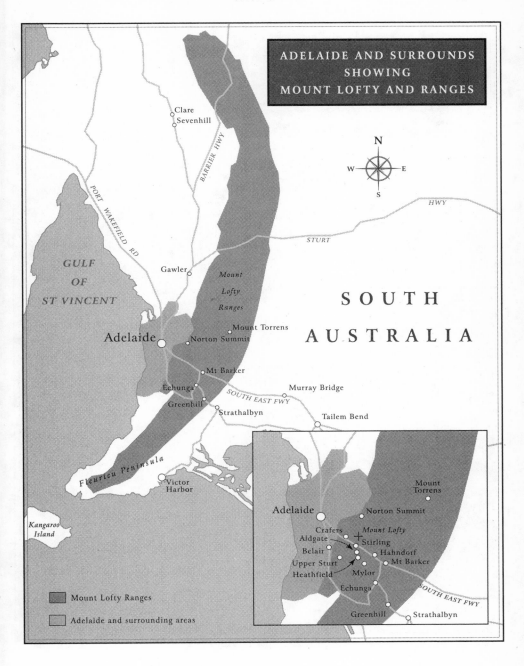

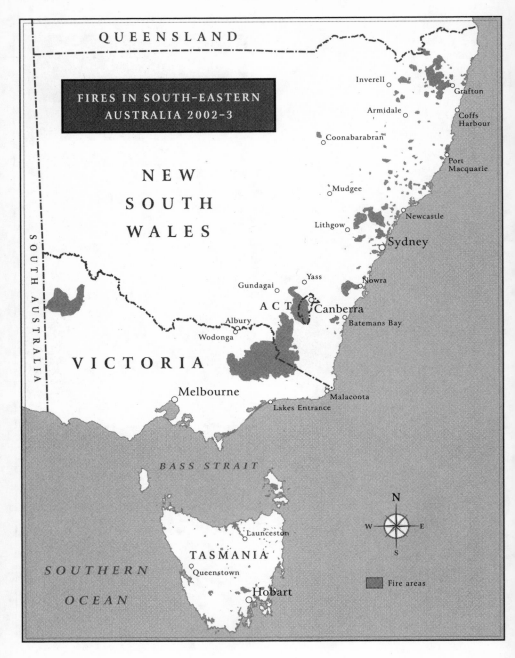

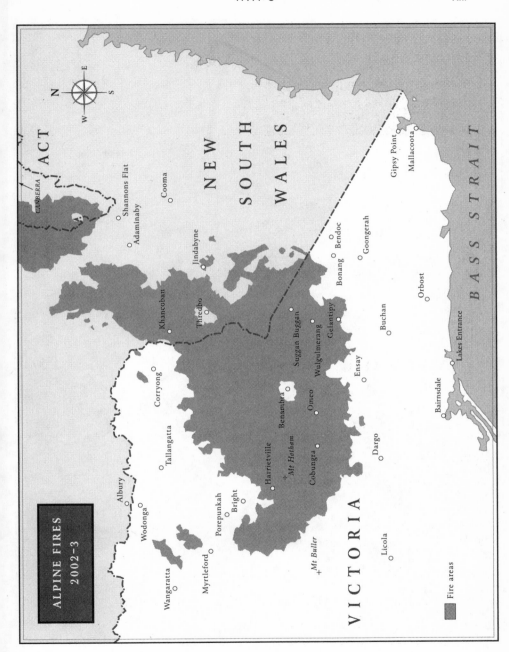

BURN

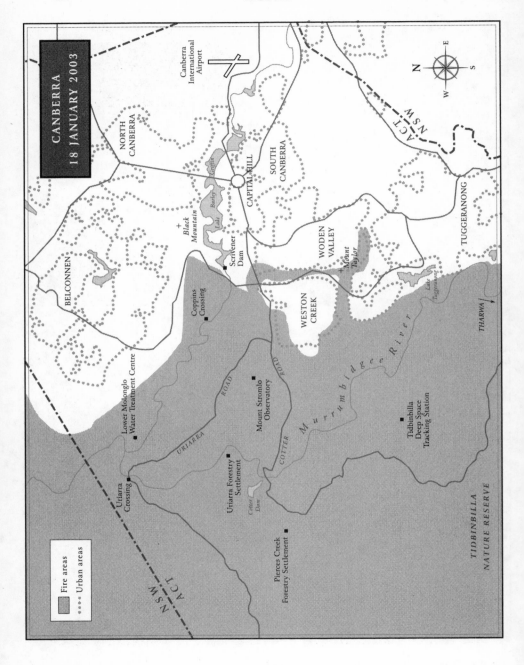

MAPS

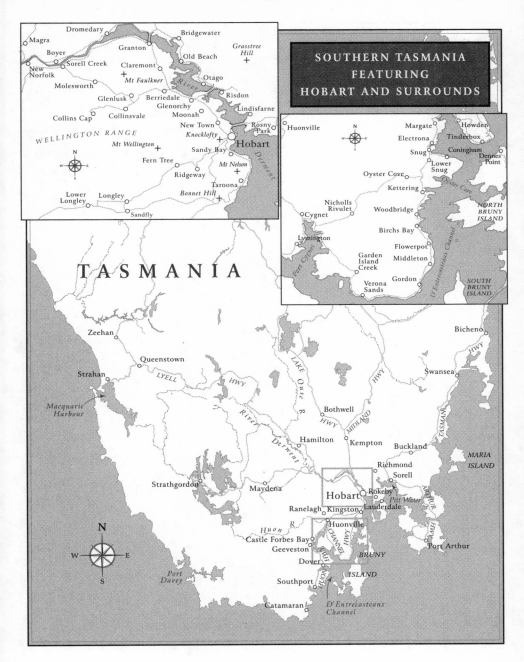

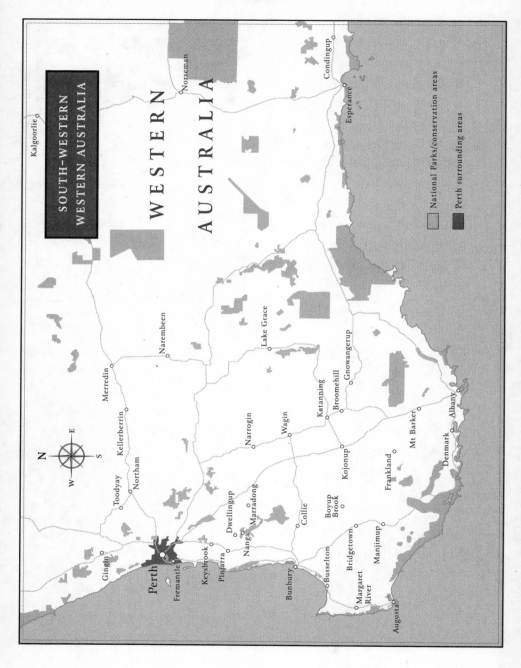

INTRODUCTION

For most of my life bushfires remained at the edge of consciousness. Like most Australians I knew they were a fact of life but I never got any closer to them than images on a television screen. Then, in November 2002, I bought a bush block in the Snowy country of New South Wales. Two and a half months later it was completely burnt out and reduced to cinders in probably the biggest bushfire since European settlement in 1788. Just two weeks before I had walked through parts of the Canberra suburbs that had been destroyed in the fire-storm of 17 January 2003. I suddenly saw the impact of bushfires on both suburbia and the bush. What is interesting is that the flora and fauna on the block have recovered wonderfully, whereas many houses in Canberra are still to be rebuilt and the area where there was a radiata pine forest still looks like a wasteland. In other words, fire is part of the natural life-cycle of Australia, and our plants and animals, although tested by it, adapt and recover. But those of us who have come since 1788 and the exotica we imported with us have much more difficulty recovering and are still far from being 'natives'.

We have yet to accept that fire is part of the very fabric of our continent. Australia would be completely different without it. In my view, fire should be seen as a positive force; it does not come to destroy but to sort things out, to probe our strength, and ultimately to renew.

When I started research for this book I very quickly learned that fire is one of those things that everybody has an opinion about; some have particularly strong opinions and wanted to know exactly where I stood on issues like fuel reduction burning. So from early on I decided to go my own way. Part of my professional training was as a historian and it was inevitable that I would approach bushfires historically. However, the book is not an exhaustive or comprehensive directory to all fires; many more recent fires have been omitted as there is usually accessible information available on them. So some will be disappointed that 'their' particular fire has been omitted. Unfortunately the length of a book is a real constraint on an author. But all major fires have been included.

The book is also about people, so their individual stories became important. Many are tragic and, like fire, dramatic. I have structured the book around what I think are the two archetypical fires since European settlement: 'Black Friday', 1939, and the 2002–3 alpine fires in Victoria and southern New South Wales. The core of my argument is that the regularity and intensity of fire has enormously increased since 1788 and that until recently Australians were almost 'pyromaniacal' in their approach to the environment. People lit fires everywhere. The great change that has occurred recently is that most, except malicious arsonists ('firebugs'), have adopted much better fire discipline. The next step will be to integrate an ecological approach to fire into our national consciousness. I hope this book will be a small step in achieving that.

The idea for *Burn* was suggested to me by Matthew Kelly, who realised there was a gaping hole in our historical literature in this area. I am very much in his debt for the idea and periodic discussions about it. At first I was not enthusiastic about bushfires, but eventually decided to take on the task after encountering some of the dramatic stories that make up the fabric of this history. Actually, there is a fine history of fire in Australia by the historian of world fire, Stephen Pyne. Pyne is an American and his *Burning Bush* (1991) is full of the kind of insight into Australian culture and mores that comes from an outsider's perspective. His new book *The Still-Burning Bush* (2006) came too late for consideration in *Burn*. My approach is different because my sources are different. I have used newspaper sources extensively to build up the narrative and have supplemented these with a range of other printed and oral material when available. I am very much in the debt of the Newspaper Room of the National Library of Australia in Canberra and, in particular, the knowledgeable and unfailingly courteous librarians of this fine institution. I also acknowledge the tremendous work newspapers do in preserving important elements of the nation's history. I have acknowledged newspaper sources in the text.

Many people helped me, but I am particularly in the debt of two: Professor Brendan Mackey of the School of Resources, Environment and Society at the Australian National University (ANU), and Australia's premier fire scientist, Dr A. Malcolm Gill, now a Visiting Fellow in the same school at the ANU. Both have been very supportive of the project and have helped me with many scientific questions. Also generous with their help were Dr Michael Roderick and Professor Graham Farquhar, both of the Research School of Biological Sciences at the ANU. Another who helped enormously by regularly discussing the issues and often pointing me in the right direction was fellow author and environmental historian, Bill Lines. Others who assisted with specific issues were Dr Carolyn Rasmussen on A.E. Kelso; Professor Hugh Stretton, who spent a day talking to me about his father, Leonard Stretton; Deirdre Hawkins, historian in Kinglake; Angela Nichols of the Channel Historical and Folk Museum; Marjorie Baulman, who advised me on 1930s fashion; Dr Bill Gammage; Dr Jeff Brownrigg and Dr John S. Benson. Detective Sergeant Paul Barclay and Detective Senior Constable Nicole Mulready of the New South Wales Police Service, who assisted the New South Wales Coroner, were particularly helpful to me in establishing the details of the story of the 2003 fires. Caren Omachen, librarian in the Victorian Department of Primary Industries, facilitated my work enormously by getting the Minutes of Evidence of the 1939 Stretton Royal Commission copied onto CD. This will make this precious record more widely available for research. Thanks also to Brian Walters SC, Patricia Leon of Omeo, Frank Purcell and especially Dr John Coulter, who helped with information about fire-fighting aircraft.

Those who have been through bushfires can speak from experience and I was fortunate to be able to talk to some of them. Basil Barnard of Warburton survived Black Friday, Kerry Wellsmore of Paupong survived the 2003 fires, Eda and Tony McGloughlin and Des Fooks survived the Canberra fires, and Nick Goldie fought them around Michelago. His wife, Jenny Goldie, provided me with an extract from her father's diary recording Ash Wednesday 1983 at Lorne. Tritia Evans, producer for ABC Local Radio in Canberra went to a lot of trouble to provide me with a text describing her unique experience producing the live radio coverage on the day of the Canberra fires. Louise Darmody of Sound Memories kindly provided me with a CD recalling the stories of the 2003 fires in the Snowy Mountains. All mistakes and inaccuracies in the book are entirely my own work!

Finally a word of thanks to Mary Cunnane, my agent; Rebecca Kaiser, managing editor at Allen & Unwin; Karen Ward, who edited *Burn*; Angela Handley, who guided it through the production process; cartographer Ian Faulkner, who designed the maps; and to all my friends, especially Marilyn Hatton, who participated in endless discussions and arguments about bushfires.

Canberra
January 2006

PART 1

A NATION OF FIRE-LIGHTERS

1

BLACK FRIDAY, 13 JANUARY 1939

i

No one in Noojee knew exactly where the fire came from originally. The rumour was—and, as the police later established, it proved to be partly correct—that it had begun almost a fortnight before, somewhere near Kinglake East, and that it had been lit by a farmer who thought he could do some burning off in the worst fire weather ever experienced in the European history of south-eastern Australia.

For the one hundred or so people still left in Noojee on 'Black Friday', 13 January 1939, all that was irrelevant now.[1] What mattered was that the tall mountain ash forests that surrounded the town were fiercely ablaze. Everywhere people looked, to the south-east, east, north, and north-west, there was a massive fire burning. Even though it was just after midday, the sky was so black that people needed lights and torches to be able to see. Flames leapt over 35 metres into the air above the tree-line and the wind direction changed constantly as the fire created its own wind patterns. The noise was deafening and smoke blinded both people and animals and choked them as they tried to breathe. The heat was so intense it blistered any exposed skin. There was only one narrow escape corridor, which fortunately paralleled the railway line and dirt road, both of which ran due south out of the valley in which Noojee nestled, and up over steep hills towards Neerim Junction, 8 kilometres away, in cleared and open dairy country.

Early that Friday afternoon, for the second time in its short history Noojee was totally destroyed by fire. It all happened very suddenly. Between 11.00 am and midday tremendous fires, whipped up by an almost hurricane-force wind, descended on the town from the north, west and east. Just before one o'clock the local policeman requisitioned several cars to take as many of the women and children, most of whom were vomiting from heat and nervous excitement, through the still unburned forest to the south out to safety at Neerim Junction.

Besides the policeman, several other people remained on duty. Among them was the assistant postmistress, Mrs Gladys Sanderson. She continued to work in the wooden post office, keeping contact with the wider world by phone and telegraph until just after 2.00 pm when the building itself caught fire. Most phone and telegraph lines had already burned down, but those connecting the town with Warragul were still open. Mrs Sanderson was the custodian of the only fireproof safe in Noojee and people in panic had mobbed her, begging her to put their valuables in it. Her final telegram to her superior, the Warragul postmaster, testified to her calm and stoic self-possession: 'I am about to close down now as the flames are licking the building. I have locked valuables in the safe and am going to the river. If the worst comes to the worst, you will find the keys of the office and safe strapped to my wrist.' She then ran the short 20-metre distance to the Latrobe River, which flowed besides Noojee's main street, and joined other remaining residents of the town sheltering in the relatively shallow water.

As the fire surrounded the town there was still a train standing at the platform of Noojee station. By 1.30 pm it had built up sufficient steam to depart, although it was not due to leave until 3.10. It consisted of a dirty grey-black steam engine, six dark-red and battered four-wheeled open freight wagons, four of them filled with logs and other goods, and a six-wheeled, freshly painted bright-red guard's van. The guard, Arthur Armstead, was becoming increasingly concerned and afraid. The 21 women and children who were left behind after the requisitioned cars departed went straight to the station, begging Armstead to get them out of Noojee by train as quickly as possible. The guard got them to climb aboard the two empty freight wagons.

The situation was getting worse by the minute with increasing heat, smoke and darkness. Armstead decided to call the stationmaster at Warragul to ask what he should do and in turn the Warragul stationmaster phoned Victorian Railways headquarters at Spencer Street in Melbourne for instructions.

Just before 2.00 pm the reply came from Spencer Street. 'Get the train out of Noojee immediately and take it to Warragul', Armstead was told, to save the rolling stock. He responded immediately. Reassuring as best he could the women and children sheltering in the last two wagons, Armstead waved his green flag and blew his guard's whistle and when the train began to move, jumped toward the running board and open swing-door of the guard's van. The train slowly gathered speed, pulling out of the station.

It was a nightmare journey. The single track ran up a long, gradual incline out of Noojee through thick forest which was by then beginning to catch fire. John Woolstonecraft, the Noojee postmaster, later complained that the Victorian Railways had been repeatedly asked by the Progress Association to cut the bush back from the track, but the railways had cited lack of funds and nothing was done.[2] The driver pushed the 120-ton N-class locomotive uphill as hard and as fast as he could, but both he and Armstead knew they had to get to 'bridge number seven', a massive wooden structure standing on nineteen trestles over a deep, narrow gully just under two kilometres out of Noojee, before the fire did—or at least before the bridge's structure was weakened so badly that it could not carry the weight of the locomotive and the cargo-laden freight trucks.[3] They also knew that the train had to cross six other wooden bridges on the way to Neerim Junction.

By now the ground itself seemed to be alight and the wooden sleepers were catching fire. Burning branches and leaves were falling all around the train or being blown into the freight trucks. 'Several times the blaze raged furiously around the trucks, only a few yards behind us', Armstead was to recall later. 'We were afraid the train would catch fire.'[4] The red guard's van in which he was riding was a wooden construction built on an iron frame attached to two-wheel bogies, and it easily could have caught fire.

When they got to the 'cobweb ladder', as the 102-metre-long and 21-metre-high number seven bridge was called, the driver stopped the train. He and Armstead ran ahead to try to check the state of the structure by leaning over the edge and peering through the dark and choking smoke to see what was happening down at the base of the wooden piles. It seemed safe enough, although there was burning vegetation down at the bottom of the bridge. Gingerly, the driver eased the train across.

They made it, but less than a kilometre up the track they stopped again to check another smaller bridge. It was a terrifying process and they were not safe until they finally broke out into the cleared, open country just to the north of Neerim Junction. Several of the women and children had suffered burns to their faces and almost all were suffering from smoke inhalation and some from smoke blindness. Less than fifteen minutes after their crossing the trestle bridge caught fire and was completely destroyed. One of the classic photographs of the Black Friday fires is of the still-smouldering remnants of the bridge and pylons, with the contorted steel track and burning sleepers hanging across the narrow valley, suspended 20 metres in the air.

The train and its passengers might have escaped, but the town didn't. Once the flames got a grip, Noojee was destroyed in just twenty minutes. The station was burned to the ground and the four freight trucks left standing on a siding were completely consumed. Only the Noojee Hotel, the butcher's shop and a house remained. This was even worse than the disastrous 'Black Sunday' forest fires of 14 February 1926 when most of the town had been destroyed, and 31 people were burned to death in nearby Warburton.

In 1939 there were about 60 people left in town when the fire arrived and most of these, mainly women, children and exhausted firefighters, took final refuge in the Latrobe River. They were there for four hours with the fire burning on the banks on both sides. The river was not particularly deep. It would have afforded some protection from the flames, but not from the heat and the smoke. There they stood huddled together near a water-wheel, covered with wet blankets. The *Age* reported the following day that the group included 'a mother with her nine months old baby, and Mrs Padgett, who is 92 years of age, who had been discovered in a state of collapse in her home. Sewing machines, wireless sets, bedding and other miscellaneous household goods, which had been hurriedly collected from their homes by the fleeing townsfolk, were thrown into the river, and three motor cycles and a motor car were driven into the water for safety.' The *Herald* reported that Mrs Padgett and her daughter were in a car to be driven to Warragul but 'before they could be driven away . . . fallen electrical wires blocked the exit. [So the driver] . . . drove the car with the passengers straight into the river.' After the fire, three Noojee men were taken to Warragul Hospital with severe burns and were listed as 'critical', but all survived. Mrs Sanderson emerged unscathed from the river and was back on duty as soon as phone and telegraph lines were restored. She was awarded the Order of the British Empire for devotion to duty.

After the fires died down later that night another heroine emerged to rival Mrs Sanderson. Mrs Chamberlain was a survivor of the 1926 fires and was the licensee of the Noojee Hotel. She turned it into a refugee centre. The hotel was inundated with people suffering from burnt hands, scorched faces, partial or total smoke blindness, and others vomiting from nervous excitement. The *Argus* wrote glowingly about her generosity:

> Suffering from blisters on the face and with one eye bandaged [she] has been on her feet for sixty hours giving assistance to all who have requested it. She has provided scores of refugees and volunteers with excellent meals, and has resolutely refused to accept payment for one of them. Nursing babies so that their mothers could rest and comforting sufferers from smoke blindness, she has won the admiration of the district. (16/1/39)

While no one in Noojee was killed, a terrible tragedy occurred 24 kilometres away near the village of Tanjil Bren where Alfred 'Ben' Saxton and his brothers, Eric and John Godfrey Saxton, ran a successful timber mill in the midst of mountain ash country. The fire descended on the Saxton mill at about the same time as it hit Noojee. At the mill site there were three dugouts, covered trenches cut into the earth that were usually from 4 to 7 metres long and between 2 and 4 metres wide, and deep enough for people to stand in, covered by about half a metre of earth dumped on top of a cast-iron or occasionally cement roof built over a timber frame. Dugouts could also be cut into the sides of hills or embankments and were protected from the flames and radiant heat by waterlogged blankets or hessian strips at the entrances or occasionally by steel doors, although any exposed timber was still in danger of catching fire in the intense heat. Containers of water were stored in dugouts to cool those sheltering inside and to keep the protective fabric wet. Sometimes containers of oxygen were left to assist breathing.

Just after 2.00 pm on Black Friday it was almost pitch dark and of the 39 people at Saxton's mill, 30 crammed into an unusually big dugout (15 metres long, 7.25 metres wide and 1.8 metres high) and six into a smaller one (7.6 metres long, 3 metres wide and 1.8 metres high). Over near the house, Ben, his wife and nineteen-year-old Mick Gorey, who had gone over to help Mrs Saxton clear valuables out of the house, clambered into another tiny dugout.[5] The big dugout was well stocked with food and water. By any standards it was safe. The same could not be said for the other two.

The main fire front hit the mill just before three o'clock. 'Impelled by a roaring north-west wind flames leaped across the [mill] clearing. Big lumps of wood flaming against a sky as dark as night ignited everything they touched', an *Age* reporter wrote on 16 January 1939. 'Four mill horses and a pony, which were free, dashed around the clearing screaming with pain, but after enduring the heat for half an hour they went mad and galloped off into the timber where their charred remains were found later.' On the mill site itself some 350,000 super-feet of timber burst into flame. This generated extraordinary heat—hot enough to melt iron. Meanwhile in the big dugout wet blankets were held over the entrance, but often they fell in charred pieces even before the water on them was dry. The heat was so intense that men could only hold the blankets for two minutes before they had to be replaced, and those who ventured outside with soaked hessian bags to protect the timber supports of the dugout could only stand it for a minute until, reduced to semiconsciousness, they were forced to retreat once more. Two men broke down completely under the terrible strain and several were blinded by the smoke. The wind and the noise were so bad, the *Age* reported, that 'it seemed like the concussion of great trees falling all around'. For more than three and a half hours the men were stuck in the two dugouts farthest away from the house. At one stage the smaller dugout caught fire, but a tank collapsed on top of it, pouring water over the burning supports. The six men there scooped up the mud from the ground to fill in the cracks and keep out the fire, eventually taking refuge at one end, burying their faces in the wet dirt.

But over near the house, which quickly caught fire in the surrounding inferno, the Saxtons and Gorey were already dead. Ben Saxton had been hit with a falling beam near the dugout's entrance when the supports caught fire and his neck was broken. When his wife and Mick Gorey went to his rescue they were overcome by smoke and quickly suffocated. When the men from the big dugout were finally able to run across the mill site to the tiny Saxton dugout, they found the three occupants dead. Red-hot pots and pans were lying at the entrance, and the dugout was still far too hot to enter.

Just over three kilometres away, out on William Rowley's farm, another tragedy had struck a whole family. Fifty-six-year-old Rowley was overconfident. He had survived the 1926 bushfires, and the area surrounding his farm had already been lightly burned in a fire in December which had dried out the undergrowth, but left the leaves dead on the trees. John Saxton described it as 'a tinder box and the whole thing would go like gunpowder'.[6] Although the Saxton brothers had strongly advised Rowley to dig a dugout, or retreat to the mill as soon as a fire threatened, Rowley refused. He was sure that he and the family were safe, despite the fact that they were surrounded by forest and undergrowth as dry as tinder.

Some time after five o'clock some of the mill hands pushed through the still-burning forest to the Rowley homestead. They found everyone dead. The family had obviously tried to shelter in the house, but when it caught fire Mrs Rowley had made a run for it with her six-month-old baby, Agnes. Her body, burnt beyond recognition, was found lying against a tree with the infant still in her arms. Her husband's body was at her feet.

The remains of four-year-old John Rowley and two-and-a-half-year-old Benjamin Rowley were found in the ruins of the house. Not far away the body of 30-year-old Frank Poynton, a paling-splitter from Saxton's mill, was found in the burnt-out forest where he had died alone.

Almost 65 years later Noojee is still rather isolated, nowadays with less than a hundred residents, even though it is only 107 kilometres by road due east of the centre of Melbourne. It is one of those places that you have to make up your mind to visit. Established as a timber town in 1919, it was burnt out first in the bushfires of 1926 and then in 1939. It lies well off the beaten track in a shallow valley surrounded by tall eucalypt-clad mountains. The predominant tree in these forests is mountain ash, the tallest hardwood trees in the world, highly valued for their fine timber, with the appropriate botanical name of *Eucalyptus regnans*—reigning eucalypt. Most of the highest and oldest of these kings of the forest, some of them more than 500 years old, had already been felled by 1939. There is much debate about their exact height, although there is little doubt that a number of the very oldest of them grew to over 100 metres. The explorer and botanist Baron Sir Ferdinand von Mueller reported one as high as 480 feet (146.3 metres) and a fallen tree in the Dandenong Ranges measured just over 400 feet (121.9 metres).[7] Nowadays mature trees average 60–80 metres in height. These tall, straight trees are held in place by a massive root system and a broad, bark-covered lower trunk. From this base the white or light green-grey barked spine of the tree tapers upward like a thin marble column to a very high but comparatively small and sparse leafy crown. From about a third of the way up the trunk, ribbons of bark peel off and hang like discarded skin.

Most of the forest that we see today across the whole mountain area north and east of Melbourne is the product of natural regeneration and trees planted after 1939. There are occasional older trees that survived the fires, but these are rare.

ii

What happened at Noojee and Tanjil Bren were just incidents in a much broader calamity, the worst that the state of Victoria and the whole of south-eastern Australia had ever experienced since European settlement. How did these fearsome wildfires come about?

The most terrible bushfires in the European history of Australia did not come out of the blue. The massive conflagrations of January 1939 had their origins several years previously. Rainfall for 1938 was drastically below normal; in fact, in many areas of south-east Australia the lowest on record. Southern Victoria experienced its driest ever July to December period. In northern Victoria and south-central New South Wales the water shortage was described as 'acute'. The wheat yield was exceptionally low and there was a shortage of stock feed. In vivid contrast it was pouring rain in northern Queensland in January 1939, with Innisfail and Babinda experiencing 48 hours of tropical downpour.

Linked to the drought and water shortages were above-average temperatures. The *Age* of 4 January 1939 reported that 'Temperatures were, on the whole, considerably above normal, the autumn and spring months being exceptionally warm'—about two to four degrees above average. The heat intensified as 1939 began. The whole border area between New South Wales and Victoria was sweltering. Yarrawonga experienced four successive days of over-century heat, but this was easily beaten by Wagga Wagga with more than twenty days over 100°F (37.78°C). On Sunday 8 January the temperature reached 112°F (44.44°C) in Melbourne and 116°F (46.6°C) in the small north-western Victorian town of Sea Lake where butchers were reporting that pigs were dying in the heat and that 'yabbies in shallow dams have crawled onto the banks to die, as the water was too hot' (*Age*, 9/1/39). Nearing Melbourne that same day on a flight from Sydney, Pilot Edgar Dorwood, flying a two-engined Ansett Airways Lockheed Electra 10-A, descending to land at Essendon Airport through severe turbulence caused by the heat, reported that the outside air temperature at 1000 feet was 110 degrees (43.33°C), although he said it was reasonably cool in the cabin. On the following Tuesday the Melbourne temperature reached the highest ever recorded to that point—112.5°F (44.72°C). As a result of the intense heat three people collapsed and died in Bourke Street, Melbourne, and four people died in Mildura where the temperature peaked at 117°F (47.22°C). Adelaide also reached an all-time record of 116.9 degrees (47.16°C). In the Riverina, Jerilderie's water supply, Billabong Creek, dried up completely and the council had to remove three tons of dead fish from the dry bed.

Throughout the whole of southern Australia the humidity was also very low, plummeting to below 10 per cent. A high pressure system that established itself out in the middle of the Tasman Sea from the first week of January allowed very hot air from the desert in the interior of the continent to flow south, bringing with it extremely high temperatures. This system remained predominant until about 14 January. On the morning of Black Friday wind velocity increased rapidly and, driven by fierce gales, the bushfires that were already burning came together in an ocean of fire.

The first major conflagration of the 1938–39 fire season began on Saturday 17 December 1938 in the Toombullup forest south-east of Benalla between Mansfield and Whitfield in north-central Victoria. It had been deliberately lit by prospectors. This fire eventually joined up with another in the King River Valley, probably deliberately lit by graziers.[8] By Thursday 5 January, a week before Black Friday, there were fires burning in south-western Victoria in the Otway Ranges between Apollo Bay and Cape Otway, at Woodend where an arsonist (or 'incendiarist' as they were called in 1939) had lit the fire, at Mount Macedon north-west of Melbourne, and at Penshurst in the Western District. The whole landscape was as dry as tinder and everyone, especially campers, had been warned to take great care. All over the state the sky was obscured by smoke haze from bushfires, many of them burning hundred of kilometres away. Despite the warning signs, people still lit fires. It seemed to be almost a cultural tradition, as vividly described by Paul Christensen, a director of the Delatite Saw Milling Company, to the royal commission sitting in the courthouse in Mansfield on 16 February 1939. After a succession of nine graziers extolling the virtues of 'practical bushmen' who 'burn the country to make it safe', Christensen told the commission:

> The whole of the Australian race have a weakness for burning. I do not want these remarks to be taken as an insult, but that weakness is inherited in the whole race. After all, it is only

100 years since the first white man came here [presumably he means this part of Victoria] . . . Timber became his enemy. The children and grandchildren of these men have grown up with minds opposed to timber. Fire is the easiest way to get rid of it . . . It was said by one of the graziers that he did not think that anyone would burn at the wrong time of the year. I would like to have that opinion. I would like to think the best of everyone; there are exceptions, however. There are men who are selfish and irresponsible, and they light fires that should never be lit.[9]

By the beginning of the week that was to culminate in Black Friday there were fires everywhere. People camping down at the beach at Dromana at the southern end of Port Phillip Bay had to take refuge in the sea, some driving their cars, trailers and caravans into the water to escape destruction, when a bushfire jumped Point Nepean Road and swept through the tinder-dry undergrowth and tea-tree on the foreshore. The *Age* reported on 9 January 1939 that many had to 'return to their city homes clad only in their bathers', having lost all their belongings in the fire. Driven by a gale-force wind the fire then moved inland towards Arthurs Seat, destroying 43 houses, most of them holiday homes. The whole town of Dromana would have been destroyed except for a change in wind direction. Fires, supposed to have been started by a household incinerator, also broke out further up the bay towards Melbourne at Frankston. Again, a large number of houses were destroyed.

Things got even worse. There were fires in the Dandenong Ranges at Montrose, Gembrook (where eleven firefighters were forced to retreat and shelter in a creek), and Cockatoo, at Erica in the north-west Gippsland mountain ash country, south-east of Powelltown and Noojee, and at Moe, Yallourn (where 28 cars were trapped on the Princes Highway and three were destroyed), Toora (where the smoke became so thick that they needed lights at midday), as well as Heyfield and Bairnsdale in Gippsland. The Toombullup fire was causing 'great anxiety' with thousands of hectares already burned out, and up in the Victorian high country there was a serious fire at Harrietville on the Bogong High Plains near Bright. Victoria's highest mountain, Mount

Bogong was ablaze, resembling a huge torch at night. There was a fire at Faithfuls Creek near Euroa, another on the Bonang road north of Orbost and south of Delegate, New South Wales and another around Woodend and Mount Macedon. The whole of Victoria was covered in a smoke haze and aircraft coming in to land at Essendon Airport were having trouble with visibility. Alfred Vernon Galbraith, Chairman of the Forests Commission (established in 1918 to provide consistent management of Victoria's state forests and particularly to supervise sawmillers operating in these forests), attributed the fires directly to 'human causes'. He warned that a fine of £200 or imprisonment for two years or both could be imposed on anyone lighting a fire in the open except in prescribed places and according to regulations.

The fires claimed their first human victims about midday on Sunday 8 January. Two Forests Commission officers, Charles Isaac Demby, 54, and James Hartley Barling, 31, were burned to death after they went to check the progress of a fire in the Toolangi State Forest. The fire had probably been deliberately lit near Kinglake some time before New Year. It had blown up and spread on New Year's Day and local Forests Commission employees and volunteers had controlled it until 8 January. Around midday the wind picked up and Demby and Barling had gone to check if the fire had jumped a break, but became separated from the main party. The other eight men had to retreat and were lucky to survive themselves.

Barling and Demby's bodies were not discovered until the next morning on a disused track. Barling's watch had stopped at 1.20 pm. He had three children, the youngest six months old. Demby was an ex-soldier, an experienced bushman and had been a Forests Commission employee for 30 years. The leader of the search party, Senior Constable Slatter, was reported in the *Age* of 10 January as saying: 'I have never seen a fire like it. The bush has been burned as clear as a floor, but limbs of trees are still burning everywhere, and there was considerable danger from falling trees.' This fire continued in a south-easterly direction across the hills through Toolangi towards the water catchment area of Maroondah Reservoir, threatening Healesville from the west and the north, and then moved north up the Maroondah Highway across The Blacks Spur, destroying the tiny villages of Fernshaw and Narbethong before moving east into isolated mountain country and south-east across the Acheron Way towards Warburton. The threat to Healesville was a continuing one. Guesthouses and homes were destroyed, and a group of schoolboys at a nearby camp had to take refuge in Chum Creek to escape the engulfing flames.

The Healesville fire provided one of the few amusing incidents to come out of this dramatic week. The *Age* of 12 Janury 1939 reported that the Ward family who lived in East Healesville had a pet cow named 'Biddie'. She was characterised as 'a friendly soul who likes companionship, especially in moments of stress'. As the fire approached the Ward house, Biddie spontaneously took refuge on the verandah, but as the heat increased she moved right inside the house and sought sanctuary in the bath while the small township was ringed by fire. Rushing in and out of the room to soak towels in the bath as he fought the fire and tried to save his house, Mr Ward constantly had to circumnavigate Biddie, who only emerged again, with not even a singed tail, once the house was saved by the efforts of the Wards.

The day Demby and Barling died was the beginning of a rapid descent into a week of fiery hell in Victoria. But information was slow in getting through to Melbourne. There were serious fires all over the state and communications had virtually collapsed. People in the bush were on their own and had to survive as best they could. By Tuesday 10 January it was clear to both the authorities in Melbourne and the newspapers that Victoria was facing a massive disaster. Almost all of the forests of the Great Dividing Range were on fire, as well as many other areas in the state.

There was a massive outbreak around the town of Erica, north-east of Moe, and at Powelltown between Yarra Junction and Noojee. At Erica a 26-kilometre front threatened eight bush timber mills and about 85 people, including women and children, were trapped. Many others had already taken refuge in the small town itself. The *Age* of 11 January reported that 'the roaring fires have ceased to be confined to a specific front and are practically ringing the town. There was thick smoke, intense heat and dust. Combined with this was thunder which gave the false promise of relief and the sky was so dark that the officers at the Erica post office were forced to light lamps.'

Even more dire was the situation at Powelltown. Just before 8.00 pm police at Yarra Junction received an SOS for assistance from Powelltown and then communications collapsed. By midnight women and children from Powelltown began arriving by foot at Yarra Junction as fires threatened on the north and south sides of the town and at one stage burnt right up to the school. It was saved by a change in wind direction. The Ada numbers 1 and 2 mills in the mountains to the north-east of Powelltown were destroyed as was the settlement of Nayook West. At the Ada no. 2 mill, twelve men, a woman and two children took refuge in the dugout where the smoke and heat were so intense that they almost lost consciousness. There were ten men and two women at the no. 1 mill dugout. Thirty people took shelter in 'The Bump' tramway tunnel on the Powelltown to Nayook West line under the mountain between the Little Yarra Valley and the Latrobe River valley.[10]

Just 10 kilometres east of Yarra Junction is Warburton, which was ringed by a raging fire that swept down on all sides from the surrounding mountains. About twenty homes were destroyed and many people took refuge in the Yarra River, but the town was saved due to the frantic efforts of firefighters and residents with only the most minimal equipment such as beaters and tree branches. The only 'high-tech' equipment they had was a motor pump which had been brought in from Lilydale. The situation was so dangerous that a refugee train remained at the station all afternoon with steam up.[11] Mills in the mountains around Warburton were destroyed. The countryside around Erica was still on fire with the small town menaced on all sides by a fire driven by an 80 km/h gale. After a week of defending the town it was not until Sunday 15 January that authorities considered it safe. Closer in to Melbourne almost 100 homes were destroyed at Warrandyte.

Despite the headlines in the newspapers about the fires, normal life continued. The countryside might have been bone dry, with people in Corowa—just across the border in New South Wales—paying two and six for a furphy cart of water, but this was early January and Christmas holiday time. Down on the beach at Inverloch on the Bass Strait coast, children were having 'a fairy-tale touch of adventure' riding camels, and attractive young women were basking in the sun in 'revealing' one-piece swimsuits, or taking it easy in more modest dress at Mornington beach, the *Age* of 4 and 5 January reported. Back in Melbourne, despite the searing heat, at Hoyts' De Luxe Theatre the Viennese-born Hedy Lamarr was, according to the *Age*, 'seductive and beautiful with a soft musical voice' in the film *Algiers* playing opposite the philandering Frenchman, Charles Boyer, and the State Theatre in Flinders Street provided an Australian flavour with Cinesound's 'hilarious comedy' *Dad and Dave Come to Town*. Several pantomimes were available for the 'kiddies', and even the Tivoli had suspended its usual risqué vaudeville shows over the Christmas holidays to stage the pantomime, *Adriana Casolotti, the Voice of Snow White*. For those ladies looking for more participative entertainment, there were daily matinees at Luna Park near the beach at St Kilda where the manager, George Curwen, announced that 'all women, irrespective of age, are eligible to enter the beautiful hair competition'.

The Australian mania for sports also continued despite the heat. All four Melbourne newspapers devoted several pages every day to horseracing, and Don Bradman's achievements with the cricket bat were widely reported. For those seeking more intellectual pursuits, the 72-year-old English writer, H.G. Wells, was in town causing both controversy and amusement, being described by the *Herald* on 5 January as 'a man of simmering humour and ready wit who declines to be serious about himself or his works, or to speak with feeling about anything but the muddle-headedness of mankind with which he has been grappling all his life'. Wells had already upset Prime Minister Joseph Lyons by referring to Adolf Hitler as a 'certified lunatic' and to Benito Mussolini as 'a fantastic, vain renegade from the Socialistic movement'. Lyons felt that all national leaders, including fascist dictators, should be treated with respect.

Most Australians in the big cities knew about the fires, which were reported in the newspapers in detail, but it was the intense summer heat, water shortages and smoke haze that really impacted on them. Certainly, as all the Melbourne newspapers of the time show, city folk were generous in supporting the many appeals that were set up to assist bushfire victims, but most of the fires were well out of sight in rural areas.

Those most affected were timber workers and their families, people who worked and lived in the forests, especially in the mountain ash forests to the north and east of Melbourne, and in the Otway Ranges to the south-west of Port Phillip Bay. These mountain ash forests extended eastwards right along the Great Dividing Range to the far end of Gippsland. The mills were in the forests for economic reasons; owners were convinced that it was cheaper to mill the timber on the spot and then transport it to Melbourne. Unlike today, when most people have retreated from the bush to live in country towns, at the time of the 1939 fires the countryside and the forests were full of single men living away from their families. Some men had brought their wives and families with them into the logging communities. As well as the loggers, there were farmers on blocks cut out of the virgin forest, gold prospectors, eccentrics and hermits. Their numbers had greatly increased throughout the 1930s as a result of the Depression and widespread unemployment, when many people took to the bush to eke out a hand-to-mouth subsistence existence outside the mainstream economy.

When the 1939 fires came thousands of people, including many families, found themselves right on the front line. These were strong, resourceful people, not easily panicked. The men typically were dressed in either a working shirt or singlet, and trousers held up by either a thick belt or braces, and almost every one of them had a battered, broad-brimmed felt hat. Shorts were unknown. The women who lived at mill sites always looked as though they were ready for Sunday church, in calf-length cotton or silk, long-waisted dresses, with an occasional frilly petticoat showing, stockings and 'practical' shoes with medium heels and a strap across the instep. All wore hats.

The most crippling problem in the 1939 bushfires was communication. Many mills were isolated and could only be reached by rough tracks or by narrow-gauge tramways that snaked their way through the forest. This was the era before bulldozers. Telephones were not common and, like the telegraph, were completely dependent on overhead lines that could be easily burnt out.

iii

Unable to retreat, people needed a place of safety in the forest. Dugouts were felt to be the answer. While they saved lives, having to take refuge from a raging bushfire in a primitive, quickly constructed, covered hole in the ground would have been a claustrophobic and frightening prospect for anyone, even for the bravest and most bushfire experienced. In that horrific second week of January 1939, hundreds of men, women and children, some accompanied by their pet animals and birds, had to make the choice between trying to outrun the fires, or retreating into primitive and claustrophobic dugouts hoping to survive the maelstrom in what many felt could easily become a suffocating underground deathtrap.

Clarence Patterson, a sawyer at Ruoak Timbers no. 6 mill in the Rubicon forest, told the royal commission inquiring into the fires of 1939 that at his mill 25 people, including his wife and two daughters who ran a boarding house for single men, crowded into a dugout measuring 'about twelve feet long and seven feet wide at the most' (3.65 by 2.1 metres). The shaft was not driven directly into the side of the hill but slanted. 'It was all timber on top and about six inches of dirt at the deepest part thrown over the top', there was no water laid on, although 'we had previously filled tubs and kerosene tins and placed them in the dugout'. The structure caught fire three times and they had to use some of the water to extinguish the fires. 'It was catching alight at the corners where there was no dirt on the wood.' They had a wet blanket at the entrance and no ventilation at the back. Asked what it was like, Patterson replied, 'It was fairly warm, I can tell you that . . . It was also smoky . . . We went in at half past nine [at night] and at half past three we came out.'[12]

Not every timber mill in the bush had an adequate dugout, even though this had been mandated by the Victorian Forests Commission in 1932. And while the dugouts saved lives, occasionally they were indeed deathtraps, especially if the area around them had not been cleared of flammable material. One of the dangers was that they were often sited near large piles of cut or uncut timber which would generate immense heat when burning. Some mill owners stupidly or carelessly stored drums of petrol and oil just outside a dugout. But the main problem was that many people found that retreating into a dugout seemed like doing nothing when action was needed, like surrendering when one needed to be decisive.

Late in the afternoon of Tuesday 10 January 1939, decisions about making a run for it or retreating to a dugout became more than theoretical questions at Feiglin's numbers 1 and 2 mills in the isolated

Acheron Valley between Marysville and Warburton, about 100 kilometres east of Melbourne. These two mills were in a recently opened up area for logging, and their owner was not typical of the second or third-generation Anglo-Celtic Australians who dominated the timber industry in the late 1930s. Moses Feiglin was a Russian Jew who had come as a refugee to Australia via China in 1912.[13] After successfully growing fruit near Shepparton, he moved into the fruit crate-making business, and with his son Judah got into the logging industry near Warburton in 1927. They were Orthodox Jews and popular with their workers because of the extra religious holidays they celebrated.[14] Moses opened two mills in the bush off the then newly built Acheron Way, a tourist road constructed by 'susso' workers (men employed by the government during the Depression on subsistence wages) that ran from just north of Narbethong on the Maroondah Highway, 33 kilometres across the mountains to Warburton. It was, and still is, a rather narrow road surrounded by forest, just wide enough for two cars to pass.

The no. 2 mill was less than a kilometre back in the bush on a track off the Acheron Way, very close to the headwaters of the Acheron River. The fire that originated in Kinglake had been burning for a couple of days to the north, west and east of Feiglin's isolated mills. Basil Barnard was working with his mate Snowy Vennell at the top of the lowering-down gear at the summit of the Poley Range, loading logs onto the wheeled bogies that were mechanically winched down the mountainside on steeply inclined rail tracks to the mill.[15] Bas recalled that at about 8.30 in the morning up on the mountain 'we started being showered with burned leaves and twigs, and we heard a roar in the distance out over Healesville way. There was a roar, like a high wind in the distance, and Snowy Vennell said "There's a bloody big fire over there", and anyhow we just worked on', ignoring it.[16]

Thirty-five-year-old Kenneth Kerslake and his brother-in-law, 23-year-old Frank Edwards, were paling-splitters. They were a bit more concerned about the fires because around 10.00 am they rode one of the bogies up to the summit, telling Snowy and Bas that they wanted to bury their valuable tools in case the fire came through. The four of them were eventually joined by Percy and Stan Isaacs, Jim Dowling and Jim Woodall, who was a tractor driver.

Meanwhile Ken's wife, Eileen Kerslake, had become so concerned about the fires that she and their six-year-old daughter Ruth left their rented house at Somers Park, a short distance further along the Acheron Way towards Warburton, to walk along the road to Feiglin's no. 2 mill through the hot and smoky forest. When she got to the mill she rang Ken on the communications phone to ask him to come down because there was fire everywhere. But Ken and Frank had already left, jumping on a log and riding it down to the mill. Reunited, the family decided to drive to Narbethong which was 15 kilometres away and safe, having been burnt out on the previous day. They couldn't get Edwards' small Vauxhall started because the heat was causing the fuel to vaporise, so they left in the Kerslakes' old Packard. But for some reason when they got to Narbethong, they decided to come back for the Vauxhall.

Eileen Kerslake had been right about the fires, because by late Tuesday afternoon the situation in the area had become acute. There was fire everywhere. All along the Maroondah Highway north from Healesville, from the top of The Blacks Spur about 10 kilometres to the west of Feiglin's mills, around the hamlets of Narbethong and Buxton, and on to the town of Alexandra, 50 kilometres to the north, the country was a blazing inferno. Local people were left to work out what was happening around them, and then do their best to fend for themselves. Most assumed that responsibility for the direction of firefighting lay with the Forests Commission but, as the subsequent royal commission discovered, its responsibility outside the area of public forests was legally unclear and it was hopelessly under-resourced.

At Feiglin's mills, because of the smoke and the mountainous nature of the country and the fact that there was no clear fire-front, no one could ascertain exactly which way the blaze was heading or where it had been. By now communications had broken down completely and roads were very dangerous with fallen and burning trees. So when a fast-moving fire-front swept down from the north-west over Mount Strickland towards Feiglin's mills at around 6.30 pm, decisions had to be made very quickly. By then the men on the mountain had come down and most decided either to make a run for it in available cars, or retreat to the dugout in a cleared area away from the machinery.

By now the Kerslakes and Frank Edwards had returned from

Narbethong to try to start the Vauxhall. The facts are not entirely clear, but they apparently got the car going and decided to make a run for it along the Acheron Way back to Narbethong.[17] It is hard to know why they made this decision.[18] If they had stayed at the mill they probably would have been safe. Perhaps they thought that Narbethong would be more secure because the fire had already been through. But the fire was still burning fiercely on the top of The Blacks Spur just above the village. So by heading back that way they were actually going into the fire rather than running ahead of it.

At about the same time Allan Spencer, a mill worker who owned a big Buick, left the no. 2 mill with two of his mates, also with the intention of trying to get through to Narbethong. The three cars would have been close to each other on the Acheron Way, with Spencer's Buick leading the way, followed by the Kerslakes' Packard, and Frank Edwards' Vauxhall open tourer bringing up the rear. However, they would have been unable to see each other even with their headlights on because by now it was increasingly dark as the fires descended. Smoke blocked out the light of the sun, and a premature blackness seemed to descend—not pure darkness, like night-time, but a kind of malicious dinginess.

As the cars drove along the road all around them the towering mountain ash forest was ablaze. Bas Barnard described the approaching fire as a kind of 'twister . . . spiralling smoke all gathered into a big whirlwind'.[19] The massive trees were exploding into flame, or being lifted out by the roots and thrown onto the ground by the sheer power of the wind. Smaller trees simply melted in the intense heat. In the extreme temperatures of such wildfire conditions iron and steel melt and twist, and even the soil itself can burn until its nutrients are destroyed. The energy and heat released has been compared to the intensity of an atomic explosion, but bushfires are different in that the energy is dispersed over a much wider area for a longer duration.

The three cars were completely surrounded by fire. Flames were reported to be reaching up to 60 metres above the top of the tree-line. There were constant showers of sparks, flying embers and burning leaves. The noise was deafening and terrible: they would have had to shout to be heard. The smoke was both blinding and choking, and the heat seared and blistered the skin. The drought meant that the normally wet forest undergrowth of ferns, wattle, blackwood and shrubs was bone-dry and provided a massive fuel load to sustain the fire. If this situation was terrifying for the adults, it must have been doubly so for the six-year-old Ruth. Yet there were reports from all over the fire-ravaged areas that even very young children did not panic.

As the three cars drove along the Acheron Way they passed Manuel's Quarry where a small party of Australian, Maltese and Greek men had been working. As the fire-storm approached, these men crossed the road and took refuge in the headwaters of the Acheron River which at that point was not very deep. When three of the Greeks—Chris Solaaris and the brothers Peter and Antonio Igoshus—heard the cars they bolted out of the river and ran begging the Kerslakes and Edwards to stop and take them into Narbethong. With fire all around and knowing there was no time for discussion, Kerslake had little choice, the three men squeezing into the back seat of the Packard. With five adults and a child now in the car, Kerslake tried to pick up speed.

Even today, the Acheron Way is narrow and twisting, roofed by a canopy of mountain ash. That January evening it must have been horrendously frightening driving in the searing heat, artificial darkness and high winds with walls of trees on fire on both sides and above.

It was probably the wind rather than the fire that eventually trapped them. In fire-storms massive wind velocities can be generated ahead of the main blaze, and big trees, some of them not even on fire, can simply be ripped out of the ground by the force of the wind-blast. Given that the forest around them was already aflame it is likely that the mountain ash that fell in front of the Kerslakes' car had probably been already weakened internally by burning. Spencer and his two mates in the Buick later reported that they had negotiated this same section of the blazing roadway at 60 miles an hour just a couple of minutes ahead of the Packard. There was no way to get the car around the fallen tree. They would now have to try to make a run for it on foot.

They did not know it, but behind them Frank Edwards was already dead. He was less than a kilometre back, but apparently blinded by smoke, he lost control of the Vauxhall which was later found half-off the road. The car must have immediately caught fire for his charred body was found lying beside the wreck among the remains of saucepans, a kettle and other household goods.

The Kerslakes and the three Greeks now faced a horrifying prospect. The only means of escape was to run along the narrow road surrounded by towering fires on both sides. The heat would have been intense, the smoke dense, and breathing very difficult. They would have been subjected to an incessant shower of burning debris and in constant danger from falling branches and burning trees. According to the report in the *Age*:

> The Kerslakes and Solaaris left the car, and for half a mile struggled along the road, which now formed a narrow ribbon of safety between two walls of fire. Mrs Kerslake was the first to collapse. Her husband, running some yards ahead with his little daughter, tried to struggle back to her, but was overcome on the way. Solaaris had outstripped the Kerslake family, achieved a further forty yards and then succumbed (12/1/39).

The body of 35-year-old Peter Igoshus was later found some distance away in the bush above the road. Seemingly he had tried to escape by running straight through the fire. His clothing was not burnt and he probably died from suffocation. His brother's body was never found. Chris Solaaris, who had been caught in the fire and whose body was badly charred, had a watch which fixed the time of the calamity: it had stopped at exactly 7.30 pm. The catastrophe was compounded by the fact that 65 metres further on from where they died there was a sizeable area that had already been burned in which the seven might have found some shelter. When the bodies of Eileen and Ruth were found the next morning they were not burned and their clothes were only charred. The child had a scorched but colourful ribbon in her hair, and she was grasping some money in one hand and her dad's tobacco tin in the other. 'Jackals', as Bas Barnard later called them, were caught by some timber workers attempting to strip the burnt-out wreck of the Kerslakes' car before the police arrived.

It took almost 30 hours for a police rescue party to reach Feiglin's mills. Those who stayed behind at the no. 1 mill spent a terrifying night in the dugout, regularly throwing water over each other, and holding soaked blankets over the entrance to keep out the fire-storm that was raging around them. The heat was intense but the bottom of the dugout was like a mud-filled swamp. Everything at the mill was destroyed. A special reporter from the *Age* who accompanied the police rescue party wrote on 13 January that at the no. 1 mill 'Piles of thousands of planks which had been fifteen feet high were reduced to mere four inch heaps of ashes—the rest had vanished! Heavy tractors and machinery were just twisted masses of metal, while the tram route consisted of the iron rails in grotesque curves lying on the ashen shapes of sleepers.'

Eleven kilometres further on after a struggle through an almost impenetrable mass of smouldering trees, still-burning undergrowth, and the putrefying bodies of animals and birds, the search party reached Feiglin's second mill to find George Unger and Jim Huie. The other five men who had shared the dugout with them had already begun the trek out to Warburton on foot. George and Jim had an extraordinary story to tell.

There were seven men left at the no. 2 mill when the fire-storm descended about 11.00 pm on Tuesday night. They all took refuge in the dugout where there was a good supply of water. 'We hung a wet blanket at the entrance and we kept it wet', Unger told the *Age*:

> Apart from about an hour when a large wooden hut close to the dugout was burning fiercely, we managed along pretty well. But that hour was terrible. We had a good flue through the roof for ventilation, but that did not stop the place from becoming terribly hot. Still, we could breathe and we were safe. We didn't save anything but ourselves. A cat and a dog that we had cleared out when the fire came, and we didn't know what had happened to them. Our two horses we let loose before the fire got to its worst to take their chance, which did not seem to be worth much, but by some miracle for which no one can account, they got through and were found later without so much as a hair on their tails singed. The only company we had in the dugout was provided by hundreds of blowflies. Some of the boys were swearing at them being there, but I suppose like us they knew where they were safest and we could hardly blame them. When

the smoke in the dugout got really bad we got towels—it was [Bill] Bromley's idea—and soaked them in water and put them over our heads. In that way we managed to breathe fairly well. The wind was blowing a gale all the time. (*Age*, 13/1/39)

iv

The fires that extended across northern, central and eastern Victoria from Sunday to Wednesday (8–11 January 1939) that took the lives of the Kerslakes, Frank Edwards and the three Greeks, were really only an overture. The worst was yet to come, and the death toll was to more than triple, reaching 71 in Victoria alone. New South Wales, the Australian Capital Territory, Tasmania and South Australia were also to suffer from extensive fires.

In Victoria, Tuesday 10 January was the pivotal day. The stationary low pressure system over the Indian Ocean and the anti-cyclone off the Queensland coast were both intensifying and the ridge of pressure between them poured hot air out of central Australia, creating strong, intensely dry winds right across the south-east and west. The almost gale-force winds out of central Australia stirred up the smouldering fires. The Toombullup forest fire had now burnt out 200,000 hectares and was still raging. There were continuing fires at Erica, Noojee, in the Rubicon forest south-east of Alexandra, at Mount Macedon, Ballarat, Yea, Pakenham East, Yallourn, Omeo, on the Bogong High Plains and around Bright, in the Kiewa Valley where there was a fire on a 40-kilometre front, in the Gembrook–Cockatoo district, at Fern Tree Gully, North Croydon, and Healesville, in the Otway forest, and at Mornington, Daylesford, Korobeit (near Ballan) and at Cobden in the Western District. In the Dandenong Ranges there had been fires since 4 January. Olinda, Sassafras, Kalorama, Mount Dandenong and Belgrave were threatened at various stages. Many homes were lost and the whole of a state forest was wiped out.

While the wind and the weather were the immediate causes of the fires, the fact is that most had been lit either deliberately, carelessly or maliciously. Alfred Vernon Galbraith, Chairman of the Forests Commission, was quoted in the *Argus* of 10 January, stating unequivocally that 'Every bush fire which occurred in Victoria during the week-end—with the possible exception of one—was caused by human agency . . . It takes only one careless person to set a forest ablaze. I wish I could drum that into the heads of all the people of Victoria.'

At almost the same time as the Kerslakes were facing death on the Acheron Way, another terrible tragedy was being played out just 40 kilometres due north in the Rubicon forest, an important area for quality timber, east of the villages of Buxton and Taggerty and south of Alexandra and Eildon. Fires had been burning in the forest from early January. No one knows how they originated. It may have been embers blown from other fires in the area, or the fires may have been caused by lightning, or lit by locals. The police view was summarised by Senior Detective Charles North for the Melbourne City Coroner, Mr A.C. Tingate. North believed that the fire was caused by lightning, although he says that the police 'traced the fire back through Wood's Point and Matlock to Kinglake [where the police found] evidence to support [the view] that the same fire had come right through from Kinglake to the Rubicon and continued on to Matlock and Wood's Point'.[20]

If the origin of the fire is uncertain, what actually happened in the fire-storm is even more confusing. We do know that twelve men died, none of them actually at the seven destroyed mill sites, but close to them. Senior Constable Olaf Rawson told both the Coroner and the subsequent royal commission into the 1939 fires: 'Four men were found on the tram track to the [Ruoak Timbers] no. 3 mill, and they would be a quarter of a mile from the mill site on the track down to the Tin Hut. At [Clark and Pierce's] no. 2 mill, the [eight] deceased [men] would be approximately one and three quarter miles from the mill site.' Asked if they could have got to dugouts at the mills Rawson replied, 'With regard to no. 2 mill they would not be within reach of the dugout. At no. 3 mill I understand they were leaving the mill to make for the safety of the Tin Hut.'[21] The Tin Hut was the tramway terminal in a cleared area where three lines from scattered mills met about ten kilometres from the town of Thornton.

The evidence gathered by the Coroner shows that no one in the Rubicon area expected the conflagration to explode as quickly and intensely as it did. Most men had just returned from holidays. Some were not even back on the payroll and had just gone to the mills to collect belongings they had left behind. One witness, Reginald Yeo, a mill hand, told the Coroner: 'In my ten years' experience, in common with most people about the district, I have not known fires of this intensity before . . . Nobody anticipated a fire that simply lit up the whole of the air as well as the trees . . . I never dreamt that a fire of such intensity and velocity could have gone through that area . . . It was almost like a vortex, a hurricane'.[22]

That evening there was a change in wind direction. It swung around from the south-west and as timber worker Alfred Biddle told the Coroner 'blew a gale about between 60 and 70 miles an hour'.[23] Around the same time as the Kerslakes were setting out on their last journey, the foreman of the no. 3 mill, Fred Mitchell, was instructed to 'Get everyone out'. Twenty-three men from the mill, led by Mitchell, began running down the tramline toward the Tin Hut area, a distance of about two and a half kilometres. There was a quite effective dugout at the no. 3 mill, but the mill workers had already filled it with their possessions so there was little room left inside. A frustrated Alfred Cherry who designed and built this dugout, told the royal commission: 'I do not think any man could get in. It was full of furniture.'[24]

But about a third of the way to the Tin Hut, first one, then three other men seemed to turn back. Biddle told the Coroner:

> We started to run. We ran for our lives; we had to. We had to run through the fire. We met it about half way from the mill. The fire had certainly crossed the track. I know the four men that lost their lives . . . My son-in-law passed these four men. That was before we had run through the fire. That is the last time I saw them . . . When I last saw them alive it appeared to me that they had gone back. They must have turned back when they saw they had to go into the fire. I cannot understand why they did. [Alfred] Neeson had a dog with him when he was running. I saw the dog. I heard he had gone back for his dog.[25]

The *Age* of 13 January reported one of the men saying: 'Neeson stopped and said he was returning for his dog. Despite the fact that the flames were almost at their heels, [Baden] Johnston and [John] West [a Forests Commission foreman] fell back to dissuade him. The flames were only ten yards behind when [Peter] Murdoch dropped behind in the race down the timber [tram] line. It was thought he went to succour the three others.'[26]

Unlike their colleagues at the no. 3 mill, the men over at the lowering winch in the bush close to the no. 2 mill did not have an escape route. Lewis Ware had been working at the winch all day. He told Coroner Tingate, 'A few minutes after leaving the winch . . . I heard this roaring of the wind. We looked across to our left . . . The fire was going up the side, hundreds of feet the flames were. That is what was making the noise . . . It was burning towards the winch where the men had been left behind.'[27] They could not see the fire coming down on them and Ware contemplated running back to warn them. But the speed of the fire beat him. The eight men at the lowering winch who stayed to back-burn a break around the machine were burned to death.

It was not until early the next morning that news of the tragedies filtered out. At 3.00 am a badly burned mill worker, Walter Angus, staggered into Thornton and raised the alarm. A search party soon found the bodies of Neeson, West, Johnston and Murdoch on the tram track. But it was not until later on Wednesday afternoon that the men at no. 2 winch were found. Reginald Baynes was part of the search party:

> They were all charred up so that you could not recognise them, except one or two. There were three of them in the dugout, what you call was a dugout, three of them in a group about twenty yards further up on the side of the hill, and another ten yards further up was the seventh man. As to this dugout, as far as I could see it was just an opening in the ground . . . Three men were lying dead in that.[28]

Many people, including wives and children of workers, or those who provided shelter and domestic work, were also caught out at the mills scattered throughout the Rubicon forest. Certainly dugouts at the mills saved lives, but the experience usually was a terrible one. W.S. Noble, who reported on the fires for the Melbourne *Herald*, interviewed Mrs

Foster Potter, who survived a dugout and a four hour walk through the smouldering Rubicon forest.

> [Mrs Potter's] voice was still little above a whisper as a result of the smoke she had breathed into her parched throat. Here is what she said: 'When the fire came roaring down on Tuesday afternoon we rushed to the dugout. The last woman, Mrs Mason, got there black with smoke and covered with glowing cinders. We could hear the fire roaring over the top of the dugout, and several times the timbers holding the roof caught fire. As the hours dragged on, the dugout filled with smoke. There were four women, five children and eleven men. One of the children was my baby, Ronald, only eight weeks old, and there was another baby of eight months. We held the children close to the ground so that they could get as much air as possible. We kept Ronald's face plastered with plenty of mud from the bottom of the tunnel, to keep it from blistering. A cockatoo and a parrot which we had taken in with us died within twenty minutes. As the smoke became denser we could not see one another, but called out in the darkness to find out how the others were standing it. Gradually the replies became fainter. I could hear Mr Bert Murphy reciting the Lord's Prayer, and then his voice drifted off. Several of the men, who were at the back of the tunnel, where the smoke was densest were out to it when at last the fire began to ease and we could drag them out. None of us could have lasted another half-hour. We tried to sleep in the bush that night and then came down on the tram-track. My husband was blinded by smoke, but I led him as he carried the baby. It took us four hours to walk seven miles.'[29]

Frank Sims, a former manager for Clark and Pierce, told the royal commission that the dugout measured 3.65 by 3.65 metres. When asked if this was inadequate he replied, 'I do not know. It all depends on how long you had to stop in there'.[30] What eventually emerged from a procession of witnesses at the royal commission hearings into the 1939 fires was a picture of build-ups of dry, unburnt heads (timber scrap) right next to mill sites, of inadequate dugouts, of desultory application of rules, of confusion and discontinuity on everybody's part. Safety was not a big issue and mill owners kept pretending that a big fire would never happen.[31] Locals tried to shift the blame to the Forests Commission.

v

On the morning of Black Friday itself the Melbourne papers were almost optimistic. The headline in the *Age* was 'Fires under control'. The *Argus* declared that the towns of Healesville and Erica were safe and had begun an appeal for bushfire victims. The usual post-fires 'blame game' had also started: 'Murderers blamed. Minister Bitter' splashed a headline. Albert Lind, Minister for Lands and Forests, was quoted as saying, 'Those who lighted [*sic*] the fires . . . are murderers, consciously or unconsciously, and we shall do everything possible to run them to earth . . . A man who would burn off country now is a lunatic.' But remembering that he represented an East Gippsland electorate, he quickly added: 'I do not think that graziers would be so utterly foolish'. A similar sentiment was voiced by the New South Wales Commissioner for Forests, E.H.F. Swain, who said that the fires in New South Wales were the result of criminal negligence and irresponsibility. He particularly blamed small farmers and campers.

But the temperature was already climbing toward 114.1°F (45.6°C) in Melbourne, and a 65 km/h, intensely hot, northerly wind was blowing. The *Argus* of 14 January reported that, 'A dense pall of smoke carried from the bushfire areas hovered over Melbourne all day and with scorching winds made conditions almost unbearable. Occasionally the sun, dull red in appearance, shone through the smoke. Because of the sullen half-light offices and shops had to turn on lights.'

Terrible things were happening that day in the bush. One of the most tragic was the death of the four Robinson children near Barongarook, a mixed farming and forest area north of the Otway Range and south of Colac, 145 kilometres south-west of Melbourne. John and Mary Robinson lived in a corrugated iron house abutting the forest. John and his eldest son Jack were woodcutters. 'There was a fire burning well back for about a week, you could see the smoke going up', Jack reported. 'We never took no notice of it . . . But boy, on the day of the fires when the smoke started, she really went up . . . It was like being in hell, the fire was that hot. Blood red, the sun was blood red.'[32]

In a letter written after the fire Mary Robinson described what happened to her family:

> The 13th of January was different—hot north winds and oppressive heat were felt early in the day . . . My husband and son comes home and my husband says 'It looks bad all around'. We had no wireless to tell us about the fire or which direction it was coming from . . . About 11 the smoke and the heat were worse but we still did not believe there was any danger. We were all together about 10 minutes to 12 when the fire struck the house.

In the confusion of escaping from the house the family separated into two groups. Four of the children made a run for the nearby road, perhaps confused by Mary Robinson shouting, 'Run for your lives'. The parents and the other four children ran into the front garden where they found a bare patch of ground about three metres wide.

> We lay for hours below the terrible heat of the burning house and hot iron, red hot, missing us by inches. I held the baby down to the earth . . . and then she went unconscious. My shoes were burned off my feet and I never felt them. My husband battled to keep the other boys down in case they panicked, and his clothes were on fire time and again and he still kept telling us under the noise of the fire that we would get out of it. But what of our four lambs, where were they? I said among the fire [if our four children were dead], I never would believe in God again or have faith in him, but believe it or not, I knew Christ lived and I could not deny him. The roos and rabbits and birds all came around us to die . . . They must have chocked with the heat but we lived at a terrible cost.
> We lay on the ground about three hours and all of us had the terrible thought—what of the kiddies?'[33]

In fact the 'lambs'—Teresa, 13, Mary, 12, Vera, 10 and Paul, 8—were already dead. Apparently terror-stricken, they had started running down the road where they became trapped by the fire. After John Robinson found them he covered the unburnt bodies (they had died of suffocation) with an old coat, and the remaining family walked four miles out to the main road 'through burning sand and falling trees . . . I had no shoes on and the heat of the ground was torture. It took us hours to get out . . . It was like a nightmare,' Mary wrote.[34]

Mary Robinson's strong faith was not destroyed, but her grief was profound.

> The terrible sorrow; I wished I could die and that I had some awful disease, that my time would be short. My husband kept falling backward, but God must have taken pity on us or else we would have lost our reason, and day by day we lived. And the people around us were very good and kind, they brought us clothes, food and money and poor pensioners offered us their pension. We believe God's way is best, and that we will be united with our Loved Ones in a happier sphere.

In his grief John Robinson used to sit alone by the well. One day 'he noticed a very large tiger snake lying beside him. It never attempted to fight but just crawled into a hole beside the well. [My husband] said afterwards it must have known he was in sorrow, as he felt no fear of it.'

The police later collected convincing evidence to show that the fire which trapped the Robinson family was not the fire that had been burning some distance away for a week, but one that had been deliberately lit that morning by a local woodcutter. Detective Francis Raper told the royal commission that the Geelong police suspected that Thomas Neal, who lived about four kilometres from the Robinson home, set the bush on fire on Friday 13 January and simply trusted that the fire would not get out of control.[35] When interviewed by the royal commission Neal was not accused because the case could not be proved, but it was clear that he was the leading suspect.[36] He showed no remorse.

The newspapers and the royal commission uncovered an epidemic of fire-lighting in the Otway district where the Robinson family lived. At the royal commission on 20 February, forestry officer Neal Oldham, who had 'just come in from fighting bushfires today to attend and give evidence', said that there were numerous fires across the Otway district from October to December 1938.[37] He said that the people who caused fires were fishermen burning along the Gellibrand River to get access to fishing spots, settlers burning off, and graziers lighting fires for green pick (the fresh grass that springs up after a fire). 'They burn at every opportunity they get . . . usually a day with a hot north wind and low humidity'.[38] Oldham succeeded in getting one conviction when he caught a grazier in the act of lighting the fire on 2 January—the same man had already lit another one in the State Forest in which he had no grazing licence. Oldham said there were 32 fires in his district of the northern Otways in January 1939 alone. Thirteen were lit by graziers for green pick, eleven by settlers and farmers burning off, three by fishermen, one by an arsonist, and four were of unknown origin.

A similar pattern was noted by another forestry officer, Henry Irvine. 'Twelve of the fires were put down to grazing interests. Two were put down to farmers burning back in fear of their own properties. One was put down to deliberate lighting to cause loss to Hayden Brothers' trams and bridges. [There had been a contentious strike at this timber mill in 1938.] The cause of two other fires were unknown.'[39] Another witness, Arthur Kellner, an engineer from Colac, confirmed the forest officers' evidence. He said people tended to put a match to undergrowth even in proclaimed periods as long as it would burn. 'Some of them seem to wait for a good north wind to come along. They think it will make a good clean burn and leave it clean for a few years to come'.[40]

The police actively pursued and charged fire-lighters wherever possible. First Constable Malcolm Mildren of Apollo Bay told the royal commission that he had charged Clarence Leslie Marriner, a farmer, who admitted to lighting a fire with a 14-kilometre front on 2 January on Cape Otway to get green pick. He was also involved in investigating a deliberately lit fire in a timber mill sawdust heap on Black Friday, as well as investigating four other fires that were lit in the latter part of January *after* the conflagration of Black Friday.[41]

vi

Even though they are only about 170 kilometres east of Melbourne, nowadays Matlock and Woods Point, like Noojee, are isolated places you have to make up your mind to visit. On Black Friday both were burnt out. Sixteen men died at Matlock, only seven of Woods Point's 150 houses were left standing that Friday night, and a local woman was killed. The worst disaster, in terms of loss of life, on Black Friday occurred at James Fitzpatrick's mill in the Matlock forest. Fitzpatrick's, and the neighbouring Yelland's mill, had only been operating for two and a half years. Both were totally unprepared for the massive conflagration they faced. Neither mill had a dugout or a telephone. Although it was a licence condition for mills to have dugouts, a legal loophole allowed mill owners to get around this. In addition, the Forests Commission did not expect widespread and intense bushfires in the mountainous Matlock–Woods Point area due to its alpine vegetation.[42] Only one person died at Yelland's, but at Fitzpatrick's fifteen men died including the owner, James Fitzpatrick, 66, of Box Hill, as well as his two sons, George, 32, and Cecil, 27. Only one man survived.

So what happened? 'Borne across the mountain tops on a tornado-like wind', as described by the *Age* of 16 January, the fire destroyed all five mills in the area. It came generally from the west, from the Warburton direction, burning on a wide front. Yelland's and Fitzpatrick's were about one and a half kilometres apart on either side of the Warburton–Matlock–Wood's Point road. As in so many other places there was a false sense of security at the mills and people hung around doing desultory clearing despite the fact that there were fires everywhere. Work at Yelland's stopped at midday when the boilers had to be shut down because of a shortage of water. The men working in the bush returned to the mill just before 3.00 pm. There was no one in charge and no decisions were taken to get people out.

There were 27 people at Yelland's and as the fire approached in mid-afternoon almost complete darkness descended. William Francis, a timber worker, told the Coroner: 'Between 3 and 3.15 the fires appeared round the mill. Sparks were carried by the wind which was gaining in velocity all the time.' The fire constantly changed directions.

'Sometimes it would be southerly, then northerly again and so forth. This high velocity of wind reached about 90 mph [145 km/h] and was more or less a whirlwind coming from any direction . . . as it chopped around . . . We soon realised that to fight the fire was apparently impossible'.[43]

By 4.15 pm everyone at Yelland's had retreated to the main brick-built house because it was too hot to remain in the open. After the group had endured about an hour of breathing suffocating smoke inside the house, 24 nearby drums of fuel oil and petrol caught fire and created such an intense heat that the roof of the house caught fire and began to collapse. Everyone had to make a run for it. The cook, Mabel Maynard, an older woman from the Melbourne suburb of Fitzroy, refused to leave the house and died when the burning ceiling collapsed. The 26 others, including four women and three children—a twenty-month-old baby girl and a four-year-old boy who were both uninjured, and a three-year-old girl who suffered acute burns on her legs—took refuge in an open, burnt-out area in the middle of the mill. 'The fire had swept over it previously, half an hour or an hour before. It was just possible to exist there with the wind blowing. It perhaps helped to purify the air . . . It was impossible to see any distance because of dense smoke', Francis told the Coroner.[44] With extraordinary courage and self-discipline they remained in the open for an hour with the fire raging around them.

> As night approached the women and children were made as comfortable as was possible in a small shelter improvised from galvanised iron and odd pieces of timber, into which had been placed, before the fire occurred, the camp's medical supplies and a few blankets and personal belongings. Here they remained for the night while great logs smouldered around them and burning trees crashed dangerously nearby. (*Age*, 16/1/39)

They had no food or water. The *Argus* of 16 January reported one of the male survivors saying that 'one of the reasons for so complete an escape was the wonderful behaviour of the women, who kept their heads and bent all their energies on preserving the lives of their children and helping the men'.

Two of the men from Yelland's went over to Fitzpatrick's to see what had happened there. They were shocked to find only one person alive: George Sellers, a mill-hand who was semi-conscious wrapped in a wet blanket. Sellers later told the Coroner that under Fitzpatrick's direction they had been working all day making a fire-break on the western side of the mill. They knocked off around 3.00 pm:

> It was getting very dark, almost as dark as night . . . The horses had been put in the stables. We got things out of the huts. They were stacked on the eastern side of the mill . . . I noticed the flames coming at a quarter past three. They came up like a racehorse; a tremendous gale was blowing at the time. No one appeared to take charge of the men. It was a case of every man for himself. Apart from the brick boiler supports there was no place to seek shelter . . . If they had not put bedding on the eastern side, the cleared side, they would have had a chance there. It got alight and you could not get out there'.[45]

'When the mill caught alight it became unbearable', he told the *Argus* of 16 January, 'and most of the others rushed behind the saw-dust heaps further back. I was by a drum of water in which some blankets were soaking and I wrapped these around myself. It became too hot to remain near the blazing mill, so I grabbed fresh wet blankets and ran into the burnt out clearing through which the fire had passed and rolled myself on the ground.' With the wet blanket wrapped closely around him and holding a corner to his mouth to prevent smoke inhalation, Sellers lay in the open for nearly two hours as the flames swept overhead. It was the most terrifying experience of his life and as he told the *Age* on 16 January he 'would not part with this blanket for anything'.

The others had no chance. The bodies of ten of the men were found huddled behind the sawdust heap. One young lad, Michael Rogers, nineteen, from Deloraine, Tasmania, climbed a ladder and dived into a 15-foot water tank where he became trapped and boiled to death. Four other bodies were found around the edges of the mill. Two mill workers had made a run for it to a cleared paddock just over two kilometres away. One of them, Bill Ellingsworth, 60, got to within 300 metres of it before collapsing. The other man died in the bush. Ellingworth's son Henry, 25, died back at the mill. The family's tragedy was completed when another son, Maurice, who had just married, lost everything when his home burnt down at East Warburton. A prospector, James Rusden, 45, who lived in a tent, died alone in the bush. His body was found four days after the fire.[46]

Out on the rough, unsealed road between Matlock and Warburton a former soldier, Bill Dafter, a mill hand at Richard's mill, was driving his old 1924-model car hoping to escape the fire. He became trapped after the fire surrounded him and the fire penetrated the car's petrol tank. Rushing away from the vehicle he lay in the drain by the side of the road. 'All of the debris and embers were getting blown on him [but] . . . he wouldn't give up . . . he was so mentally tough and physically tough. At one stage . . . if he could have got up he would have ended his life because of the pain from the burns'.[47] He lay there in the intense heat for several hours. The next day he was found wandering along the road, delirious. He had a long recuperation in Warburton Hospital and claimed that it was his military training that saved him.

It is a short 6 kilometres from the Matlock district to the town of Woods Point, an elongated settlement in a very narrow valley. Fires had been burning in the surrounding mountains for a week and around the district since December. With minimal resources the local Forests Commission officer, Hubert 'Roly' Parke, struggled to get the fires under control, a couple of which were re-lit. On New Year's Day a witness 'had seen a car leave the vicinity where the fire was thought to start. Apparently a tourist had been there.' It was a billy fire not properly put out.[48] By early afternoon on Black Friday all of the fires had merged and the temperature in town reached 120°F (48.89°C). Fires were burning down both sides of the valley and a gang of volunteers from the Morning Star mine were desperately trying to shift a magazine full of explosives into a tunnel; the task had to be abandoned with 150 cases still in the magazine. Men, women and children took refuge in the tunnels, shafts and mining inlets in the side of the hills. One group, mainly women, took refuge in the local swimming pool.

Charles McKay, a Morning Star compressor driver, told the *Age* that eventually:

[I]t was a case of every man for himself. Many reached the tunnels but many had time just to jump into the Goulburn River under the bridge. I ran for an inset and helped along two mates who had collapsed . . . We were safe but the heat was unbearable. I saw the battery catch alight. Large lumps of red charcoal and large sheets of flame from burning gasses floated over the open ground. Then above the terrific roaring of the flames came a tremendous 'Bang'! Up went the magazine about 4.00 pm . . . We thought we had borne as much heat as human endurance could stand, but it was a Melbourne cool change compared to the blast from the magazine. A hole of 25 feet wide by 20 feet deep was left where the magazine had stood. After consuming the hospital, the post office, the hotel and most of the houses, the fire travelled over the forest to the south after four hours of hell. (16/1/39)

Miraculously, only one person died: Miss Ellen O'Keefe, 69. Apparently delirious, she ran up the main street right into the conflagration in full view of the horrified people in the swimming pool. They could do nothing to help her. Her clothes caught fire and a fence fell on her. Her charred body was later recovered.

Survivors in Woods Point were astonished on Sunday afternoon when an 80-year-old bushman and prospector, Tom Adam, ambled into town commenting, 'It's rather hot!' The *Sun* of 16 January reported that Adam was at Jericho due south of Woods Point when the fire came through. Everyone assumed he was dead. 'I just made for a clearing', he said, 'and waited for the fire to go through, and then made a bit of a detour to get here'. The 'detour' was a 65-kilometre trek through rough, fire-devastated country. He had no food. After a stint in Mansfield Hospital, he said he was 'going back to a bit of a reef' he knew about. Equally phlegmatic in attitude was Neil Ross, a trained teacher, who worked in his parents' hotel in Woods Point. Commenting on the bodies, many burnt into grotesque shapes, that were brought into town from Fitzpatrick's mill on an open truck and parked in the main street, he said that officials had trouble getting them into coffins. Limbs had to be sawn off and many of the locals witnessed this. Ross later bluntly told the ABC: 'If that had happened nowadays, there would have been all sorts of advisers and counselling and so forth; and obviously there were no psychiatric diseases that we could suffer because they hadn't invented them yet!'[49]

The tiny mining villages of Ten Mile and A1 Settlement, north of Woods Point, were burnt out. Aberfeldy, to the south, was also destroyed. Locals fought to save the village and only at the last moment retired to a mine tunnel for safety. Three men were badly injured; one, William Bolton, later died. Every mill between Warburton and Woods Point was destroyed.

The fires continued to burn eastwards across the mountains. One hundred and sixty kilometres away, after a fiercely hot day (the temperature reached 115°F (46.11°C)), the fire came through the town of Omeo in north-east Gippsland at 8.30 in the evening of Black Friday. Just before the hospital burnt to the ground two men and three women patients, including a woman who gave birth to a baby girl an hour and a half later, were removed from the hospital to a partially completed concrete building. The Golden Age Hotel, about 50 homes and eleven shops were also destroyed, and at one stage 5000 gallons of petrol in 44-gallon drums exploded, creating heat so intense that the fire was unstoppable. At Cobungra Station on the Omeo–Mount Hotham road the homestead and outbuildings, three other houses, and the post office were destroyed. Several people suffered severe burns and firefighters and people from the station had to take refuge in the Victoria River. A station hand, Ernest Richards, 30, was burned to death along with his horse when he set out looking for his wife and baby not knowing that they had already been taken to Omeo for safety. Thousands of head of cattle were also lost.

Massive fires swept across the Bogong High Plains with flames 35 metres above the mountain tops. The village of Harrietville was threatened and residents were amazed that it was not destroyed. When the fire approached their home in the nearby Buckland Valley, James Lowry, 41, a prospector, and his wife and fifteen-year-old nephew took refuge in a nearby dugout he had constructed. During a lull in the fire Lowry ran back to the house to see what he could save. Panicking, his nephew followed. Both were cut off from the dugout when the fire suddenly intensified. They made a dash for the Ovens River, but James

Lowry tripped and fell into blackberries. He was quickly overwhelmed by the fire and his body was terribly burned. The boy reached the river bank, but collapsed there and died. The whole of the Buckland Valley was burnt out around Myrtleford, Bright and Harrietville. The post office in the village of Freeburgh and the Mount Saint Bernard Chalet at Mount Hotham were destroyed with the owners taking refuge in the Dargo River. Around the town of Corryong on the Upper Murray a fire which had been confined to the mountains for a week seriously threatened the town several times on Black Friday night and Saturday morning.

By Monday 16 January the full impact of what had happened was dawning on the whole nation. Thirty-six people had died in Victoria on Black Friday alone, bringing the total number of deaths in the state to 64. It was to climb to 71. Stories poured in of destruction and death. William Doig, 70, was trapped in a blazing shed at Woodend and died in Kyneton Hospital. A pensioner, Frederick Topping, 72, was found dead in his burnt-out home in Warrandyte. Charles Catternach, 58, was found lying near his hut with his dog and burnt cart at Moyston West near Ararat. His horse was running loose nearby. William Cyril Bolton, 75, died from burns and shock on his way to West Gippsland Hospital after being caught in a fire near Aberfeldy, south of Woods Point. And near Hall's Gap in the Grampians the Habel family from Nhill were suddenly surrounded by fire when they tried to repair a blown tyre on their trailer. Freda Habel and the three children were badly burned. The eldest child, Eric, thirteen, died later in Stawell Hospital. The other son, Rex, and his mother died soon afterward. At Lake Mundi just west of Casterton, a farmer's wife, Mrs McGinty, was home with her two children when a fire swept across some of the farm outbuildings. Wrapping her baby son in a wet blanket and holding her four-year-old son by the hand, she left the house and ran in front of the fire in the direction of a neighbour's home. Her terrified little boy broke away and ran into a clump of ferns. His mother tried to save him but was badly burnt. The four-year-old died later in hospital. The infant was uninjured. The tragedy was all the greater as the McGinty house was not burned.

Around the state on the Sunday after Black Friday no doubt many priests and ministers in their sermons reflected on the fires. In Melbourne at Scots Church the Rev. J. Golder Burns told his congregation that tragedies like the bushfires were permitted so that the grace of human sympathy would not dry up. And at Collins Street Baptist Church the Rev. Reginald Kirby took a more apologetic approach and argued that the sufferings of the bush-fires should stimulate a re-examination of the grounds of human confidence. He argued that the only foundation for this was belief in God which came through putting all one's trust in Christ.[50]

The economic losses in Victoria were enormous. Between 1 and 16 January 1.3 million hectares of forest and farmland were burnt in 479 fires. The loss of building infrastructure is staggering. Probably more than 800 houses, schools, churches, post offices, chalets, hotels, and the Omeo Hospital were destroyed.[51]

vii

Victoria was not the only state on fire. South-eastern South Australia, southern New South Wales, the Australian Capital Territory and Tasmania all confronted the same fire conditions as did Victoria in January 1939.[52]

Between November 1938 and mid-January 1939 the ACT experienced the driest period since 1918. Water was short. The Molonglo River had virtually dried up. Some scattered thunderstorms brought insignificant rain. Yet despite the heat, tourist numbers in Canberra were up and 'exceptional demands were made on refreshment and eating houses', according to the *Canberra Times* of 9 January 1939. The small city (population 10,000) was hosting over 1100 guests from the UK, the United States, and Commonwealth countries attending the jubilee ANZAAS (Australian and New Zealand Association for the Advancement of Science) conference. At 108.5°F (42.5°C) the delegates received a 'warm welcome' to the national capital. The *Canberra Times* reported:

> Under this broiling heat scientists and their wives gathered in the parliamentary gardens and made a gallant effort to take an interest in a garden party arranged in their honour by the Commonwealth Government. Most of the interest of a scientific character centred on a courageous prophecy by Mr Inigo Jones,

the famous Queensland weather forecaster, that by [the next day] it would be cool and raining in Canberra. (12/1/39)

For once Jones was wrong. The heatwave and dry conditions continued. On Friday it was 107.4°F (41.9°C). Three people were treated for sun-stroke, and a 45-year-old man died from heart failure in 109°F (42.78°C) heat in nearby Yass.

More ominously there were fires in the Brindabella mountains, especially around Mount Franklin. These fires were an overture to a terrible weekend. Early Saturday morning reports were coming in that the fires were coming down from the mountains driven by fierce winds of up to 70 kilometres per hour, with spotting occurring up to 9 kilometres ahead of the main blaze. The fires were on three fronts. One came down from Mount Franklin, a second near Mount Coree, and the third from the north near Horseshoe Bend.[53] For a while the fires were held at the Murrumbidgee River, but by mid-morning they had crossed in several places. The worst affected areas were around Weetangera, Hall and Uriarra Station. 'The fires were so scattered and capricious that the prayer for a change of wind could not be general. While a change might relieve the situation in some places it would cause real and near danger in others', reported the *Canberra Times* on 16 January. Major public assets were threatened including the Mount Stromlo Observatory; Government House, where staff were warned to be ready to evacuate; and even Parliament House, which was showered with wind-borne burning debris. During Saturday afternoon the fire at the Murrumbidgee was on a 70-kilometre front.

Given Canberra and nearby Queanbeyan's small population it was difficult to get a sufficient number of firefighters. Many at the ANZAAS conference abandoned their deliberations in order to help. 'The spectacle of learned professors . . . fighting side-by-side with Canberra citizens . . . was one of the features of the day . . . Sir Douglas Mawson of Adelaide, Mr J.R. Darling, Headmaster of Geelong Grammar School, Professor D.B. Copeland, Professor Wood, and Professor George Brown of Melbourne, and others, fought the Weetangera fire throughout Saturday afternoon and late into the night', noted the *Canberra Times* on 16 January. But nothing could save the area's extensive pine forests which were largely wiped out; they were valued at £300,000. Eventually Inigo Jones' predictions came true. On Sunday afternoon a cool change moved through the ACT with rain and fog over the mountains. The fires died down. By Monday two other fires, one near Tidbinbilla, and a much larger fire between Braidwood and Queanbeyan, were also under control. Around 55,000 hectares of timbered and grazing land had been burnt.

viii

Two of Australia's first fire experts, R.H. Luke and A.G. McArthur, consider the 1938–39 fire season one of the worst ever experienced in New South Wales.[54] Throughout December 1938 there were large-scale fires around Dubbo. On 11 December thirteen houses were lost in the Sydney suburb of Lugarno on a peninsula in the Georges River. But the real killer was the heat, especially in western New South Wales. In the week from 7–14 January, 116 people died of heat-stroke. For weeks the temperature in towns like Collarenebri, Cobar, Wentworth, Bourke, Hillston and Wilcannia did not drop below 100 degrees. Cobar had a run of 40 days above 100, and Ivanhoe had 27 days straight when the temperature was over 104 (40°C). On 9 January it reached 122°F (50°C).

Water shortages became acute. Ice had to be flown on an Ansett Airlines Lockheed Electra from Melbourne to Narrandera, where seven people had died, to preserve essential medical supplies. To make matters worse, there were dust storms and thick smoke, the latter of which was even interfering with aircraft navigation.

By Sunday 14 January the Sydney temperature reached 113.6°F (45.28°C). The city was a furnace, and the sand became so hot on the beaches that people could not lie on it. The heat was made worse by the humidity and the fact that the temperature remained high at night and, as the *Herald* pointed out, locals insisted on living as though they were in another climate:

> Tweed or serge suits and hot felt [hats] were worn by as many perspiring men as in winter; just as many chops and steaks were

ordered in stuffy restaurants; even more whisky and spirits were consumed in crowded bars, and record weekday crowds baked on the beaches in the burning sun. To the Australian all this is commonplace, but recent arrivals from Europe are astounded. They think that, as the climate cannot be altered, men should adapt themselves to it . . . They suggested that Sydney men do not take their summer intelligently. (10/1/39)

From mid-week onwards fires were burning in the Sydney suburbs, especially around Parramatta and Lidcombe. On Wednesday 11 January the highest unofficial temperature recorded in the state was 122°F (50°C) at Tebbutt's Observatory at Windsor. The next day saw serious fires around Katoomba in the central Blue Mountains and at Valley Heights. Tremendous efforts were made to bring these fires under control, but on Thursday Sydney was surrounded by fires and these tested firefighters to the limit.

Black Friday was a scorcher. Across New South Wales, twenty people died from heat-stroke. There were fires in Mosman, Willoughby, Manly, Frenchs Forest and along the north shore railway line. The whole area around the northern beaches from Narrabeen, along the Barrenjoey Peninsula and around Pittwater was alight. To the south there were fires at Sylvania and in the Port Hacking area, where many homes were burnt, as well as in the west and north-west of the city around Mount Druitt, Windsor, Castle Hill and Pennant Hills. There were even fires in the inner suburbs: scrub was burning at Bankstown, and Chatswood, Petersham, Earlwood, Croydon, Peakhurst and Sefton Park all reported fires.

In the Blue Mountains fires extended west to Mount Victoria and across the plains towards Bathurst. There were outbreaks around Lithgow and Oberon, and in the Bathurst–Orange area almost 1000 men were fighting fires. To the south of Bathurst, the small town of Trunkey was almost burnt out. One local resident, William Smith, 74, died from burns after he tried to rescue his dog, which had wandered into the bush. Closer to Sydney, Percival Davis, 33, and a young boy, Snowy Medcalfe, died after being trapped asleep in a hut near Rouse Hill; Harry Martin, 53, was overcome by smoke trying to release horses near Castle Hill; and

John Roach, 70, an invalid pensioner, collapsed crawling up from a creek to his camp at Yalwal, west of Nowra. There were fires in the Southern Highlands, around Goulburn, in the Snowy Mountains, on the South Coast from Wollongong to Nowra and Ulladulla, and at Nelligen near Batemans Bay where two churches and eight houses were destroyed. Further south a badly-burnt firefighter died in Bega Hospital after fires surrounded the town. At one stage on Saturday all telephone lines between Sydney and the South Coast were down.

At the Warragamba Dam construction camp six men risked their lives fighting a fire for three hours to prevent a gelignite magazine exploding. The nearby village of Wallacia was threatened when fires crossed the Nepean River. At Seaham, north of Raymond Terrace, the Presbyterian church, state school, and houses were destroyed, while in the Yass district over 60,000 hectares were burnt. 'Horses were seen galloping through the flames with their tails in flames. In the burnt country the skeletons of many kangaroos, opossums, and other bush animals were found', noted the *Sydney Morning Herald* on 17 January.

Some of the worst of the fires were in the Southern Highlands. At one stage Picton was surrounded and there was a danger that the town would be destroyed. At nearby Thirlmere the Queen Victoria Hospital was threatened by burning scrub and grass and 50 women TB patients had to be evacuated. A fire that originated near Berrima moved swiftly into the Mittagong–Bowral area and 50 homes were destroyed. At one stage it seemed as though Mittagong might be burnt out. The tiny village of Penrose, on the main north–south railway line south-west of Bundanoon was almost completely destroyed. The blaze started when an unattended billy fire got out of control. Cyril Hayward, 39, was later sentenced to three months' jail for having left a fire in the open. He was most repentant and did everything he could to fight the fire, but it got away and quickly spread to the Hume Highway, and along the main Sydney–Albury railway line. The line was only saved when an engine pulled trucks with water tanks along the track spraying the wooden sleepers. Things became exciting for passengers on the Saturday night train from Sydney to Albury when it ran through a blazing stretch of country between Mittagong and Goulburn. The *Spirit of Progress*, which continued the journey from Albury to Melbourne, ran for most of the journey through dense smoke and arrived an hour late at Spencer Street Station.

In New South Wales the fires directly killed six people, but 113 people, mainly the elderly and babies, died from the heat and an estimated £300,000 worth of damage was caused.[55]

ix

Adelaide was also sweltering in the week that culminated on Black Friday, with temperatures soaring to an all-time record of 117.7°F (47.61°C). On Tuesday, a blisteringly hot day with north-westerly winds that reached 60 km/h, fire surrounded Adelaide on three fronts and because of the wind the path of the flames was completely unpredictable. The fires centred around Meadows, Macclesfield and Strathalbyn to the south-east of Adelaide, and around Mount Torrens, due east of the city. There was another fire between these two extending eastwards in forest country. The town of Mount Torrens was evacuated, and Clarendon, east of Morphett Vale, was only saved by hundreds of volunteer firefighters. There were ashes and cinders in the main street.

On Black Friday itself the fires were closer to Adelaide in the Mount Lofty–Stirling district, with the town of Crafers encircled by flames. A number of houses were destroyed. There was extensive damage at Echunga near Mount Barker. A group of firefighters had a terrifying experience, reported the *Advertiser* on 14 January, 'when they were trapped by a fire on the Adelaide side of Meadows. They had just sufficient time to soak their clothes and the ground with water and to cover themselves with wet sacks when the fire was upon them. For three hours they were forced to lie face down on the ground while the flames swept over them.'

Relief came with rain and a cool change over the weekend. On Monday 14 January the *Advertiser* editorialised with the usual encomiums on the generosity of the human spirit and heroic firefighters. However, there was nothing in the editorial about the 'moral lesson' to be learnt from the people who deliberately lit these fires. At least the Melbourne *Age* of 13 January described the person who started fires for their own interests as 'the man with no conscience'.

In Tasmania the 1939 fires came later than on the mainland. There had been minor outbreaks in northern and south-eastern Tasmania in the first part of January. But between 22 and 25 January the situation in the north-west had become extreme with over 5000 hectares of forest and farmland burnt out. Fires broke out all over the state between 28 January and 1 February, as well as on King Island. On 7–8 February the north-west experienced its worst fires ever, and much valuable forest was destroyed. Even the Prime Minister was affected. During the January heatwave the rather dour Joseph Lyons was playing one of his occasional morning rounds of golf near his home in Devonport. Looking up from the green he noticed clouds of smoke coming down the hill from near his house. A nearby radiata pine plantation was on fire and his home was seriously threatened. Lyons, his son Kevin, his private secretary, Franz Schneider, and his press secretary, Claude Dawson, abandoned their golf, rushed to the house and began fighting the fire. They were joined by the Prime Minister's wife, Dame Enid Lyons. Then Miss Lenham, the official typist, and four of the Lyons daughters, Sheila, Kathleen, Enid and Moira, joined in as the flames reached to within 18 metres of the house. About an hour after the outbreak the volunteer fire brigade from Devonport arrived and saved the day, and when all had settled down and the fire was quelled, Dame Enid and the Prime Minister personally served refreshments to the firefighters.

x

By mid-February the worst fire season since European settlement had finished. These six weeks saw an unmitigated national disaster in south-eastern Australia. Eighty-five people died, almost 1.6 million hectares of forest and farming land had been burnt out, and close to 2000 houses, shops and other structures had been destroyed. Many domestic animals and millions of native animals and birds had been killed. Governments and thinking people began to realise that something had to be done about bushfires. Astonishingly, for 150 years settlers on this most fire-prone of continents simply pretended that the periodic massive conflagrations were really isolated aberrations that were not to be taken seriously, and had nothing to do with their agricultural activities and fire practices. Certainly fire has been an integral and constitutive element of the ecology of Australia for thousands of millennia. But European settlement changed the whole balance of nature, and massively increased the intensity and frequency of fire.

In the environmental history of Australia, Black Friday was to be the day of judgement for the 'colonialist' vision of Australia. It was the moment when the attempt to reconstruct the ecology of Britannia in the Great South Land was shown up for what it was: an unrealisable and destructive enterprise. It was not just weather that created the 1939 fire-storms. These extreme fire conditions had been largely created by the economic, social and ecological attitudes that predominated in Australian culture prior to 1939. For the whole of the European history of the continent, settlers had refused to accept the pervasive presence and influence of fire throughout the environment. As the historian of fire, Stephen Pyne, says so accurately in his book *Burning Bush*: 'Ultimately a fire conscience would only arrive after the emigrants had established a commitment to Australia, after they attached themselves permanently to the island continent'.[56]

In the process of attempting to realise this totally inappropriate vision whole species were driven to extinction, sometimes even before their existence was scientifically recorded. By 1939 Australia already had the worst record of faunal extinction in the world. Irrigation and land-clearing on a vast scale were already leading to salinity and the degradation of a country that was always only marginally fertile. It was these cultural attitudes that the Black Friday fires were to challenge and confront. On 13 January 1939 it was just a fortnight short of 151 years since the First Fleet had arrived. During that second week of January the colonial mentality was confronted with a fire that was to become the benchmark by which all other fires were to be measured. The natural world was taking revenge for 150 years of ecological abuse by settlers.

Not that everything changed after the 1939 fires. The Second World War quickly became a priority, followed by a period of massive post-war development, and it took decades for some of the lessons of these fires to be re-learned. But the searing experience of Black Friday was never forgotten, so much so that when the fires of 2003 were compared with those of 1939 many Australians knew exactly what was being talked about. For there is a real sense in which the 1939 fires were a historical turning point. People were so shocked by them that they realised something had to be done about the way most Australians lived in and related to the natural environment.

In Victoria, praise for the bravery shown by the utterly ill-equipped and disorganised firefighters was followed by an orgy of blame. But the Victorian Government acted quickly and forestalled the blame-game by appointing an outspoken, intelligent and literate judge, Leonard Stretton, as royal commissioner to investigate the fires. He completed his work in four months and his report is perhaps the most penetrating, hard-headed, practical and elegant of any royal commission report produced in Australia. Stretton got to the heart of the matter and debated the issue that underlies the whole Australian dilemma: how are we to live in a way that fits in with the most fire-prone place on earth? Three volumes of evidence were taken by the royal commission, and throughout the hearings Stretton consistently asked penetrating questions about people's fundamental approach to fire. In teasing out the underlying cultural presuppositions, it was he who first said that Australia was a nation of pyromaniacs, a people who lit fires everywhere without any sense of the consequences.

It was the Stretton Royal Commission that helped consolidate the sense of a need for change.

OVER A CENTURY OF FIRES, 1788-1938

i

The Kinglake Hotel is a medium-sized, single-storey timber structure with a gabled roof and a wide verandah. It had much to recommend it in the 1930s. At that time Kinglake was 68 kilometres by road from Melbourne via Whittlesea, and the district was a mixed forest and farming area. The Kinglake National Park was established in 1928 and it protected a lovely spur of the Great Dividing Range that was popular for outdoor activities, bushwalking and picnics. According to the hotel's advertisement it had all modern conveniences: 'First class cuisine. Hot and cold water. A tennis court. Electric light throughout, and a moderate tariff'. In 1939 the building was just twelve years old, having been rebuilt in 1927 after it had been burnt to the ground in the 1926 bushfires. Just the place to get away from it all.[1]

But on the dry, hot Wednesday morning of 8 February 1939 its large dining room with a big fireplace and wide windows was handed over to an unusual function. It became the temporary seat of the Royal

Commission into the Causes and Origins and Other Matters Arising Out of the Bush Fires in Victoria During January 1939. The royal commission was presided over by a tall, well-built, athletic, almost bald and rather severe-looking 45-year-old judge: Leonard Edward Bishop Stretton. His serious visage and horn-rimmed glasses masked a witty, intelligent, compassionate man with a lot of sympathy for the underdog and a strong sense of social justice. As he strode onto a makeshift bench in the dining room at 10.00 am, dressed in an ordinary dark suit, tie and vest, the crowd of forestry officers, police, landholders, prospectors, farmers, reporters and sundry spectators shuffled to their feet. In an unusual twist for a legal hearing you could smell lunch being prepared in the kitchen directly behind Stretton's chair.

This was the fifth day of evidence given to the royal commission. It was an important day because Stretton and Counsel-Assisting, Gregory Gowans, KC, were trying to ascertain the origins of the massive wildfire of early January that had spread north, then east-south-east from Kinglake through the mountain ash forests for 130 kilometres as far as Erica, Matlock and Woods Point. Had it been deliberately lit on or just before New Year's Day, or did a tree stump that had smouldered since an October 1938 burn-off until severe north-westerly winds and dry conditions caused it to re-ignite on 1 January 1939 set the blaze on its way?

It was important to get to the bottom of this because this fire had led directly to the deaths of two forest officers as well as the Kerslake family, Frank Edwards, and the Greeks on the Acheron Way, and indirectly to the deaths of many other people, including those at Saxton's and Fitzpatrick's mills. This was also the fire that had burnt out Noojee and caused incalculable damage to the Victorian mountain ash forests. Evidence before the Coroner had suggested that a particular individual was responsible for lighting this fire and a barrister representing the Kerslake family said bluntly that such a person would be guilty of manslaughter. The *Age* of 4 February ran the headline 'Manslaughter Charge May Follow'.

Four days later at the Kinglake royal commission hearing, Stretton was at pains to reassure the locals. He began by telling witnesses that he wished 'to make it clear that nobody need be nervous about giving evidence before this Commission. This is not a detective inquiry. We are merely here for the purpose of arriving at the broad cause of these fires, and if any witness feels he might incriminate himself . . . he need not give evidence. We are not here to trap any man in any way.'[2]

Leonard Stretton was an ideal choice for royal commissioner. He had come from a poor background and had an unstable childhood with a father addicted to gambling but had made his way to the University of Melbourne where he studied law. A friend of Robert Gordon Menzies, Stretton and Menzies fell in love with the same woman, Norah Crawford, a clergyman's daughter from Brighton. Both proposed and Norah chose Stretton in preference to the prime minister-to-be. Called to the bar in 1929, he was appointed an acting County Court judge in April 1937. Leonard developed an informed but broad contempt for politicians in general, although his son Hugh Stretton reports that 'he had considerable respect for some of his Labor mates. His sympathies tended to be on the workers' side in industrial disputes.'[3] It was this compassion for ordinary people and tolerant understanding of their difficulties that made him an excellent choice for this royal commission.

ii

As royal commissioner, Stretton adopted an unusual approach. He did not sit in Melbourne and wait for witnesses to come to him. Comparing his commission to 'the greatest show on earth', Barnum's travelling circus and menagerie, Stretton journeyed across Victoria in extremely hot conditions through burnt-out country to see the results of the fires for himself and to meet those affected.[4] Accompanying him were Gregory Gowans, KC, P.A Carbines, the commission secretary and a bevy of barristers and representatives of bodies with a particular axe to grind before the commission. Stretton did not like driving, so he was accompanied by Senior Constable J.E. Hutchinson as chauffeur and orderly. The commission began and finished in Melbourne where the judge heard the expert witnesses.

Stretton held 34 days of hearings in Melbourne, Healesville, Kinglake, Marysville, Alexandra, Mansfield, Colac, Forrest (in the

Otway Ranges between Colac and Apollo Bay), Lorne, Willow Grove (north of Moe in Gippsland), Noojee, Maffra, Belgrave, Woods Point, and Omeo. The commission met in courthouses where available, shire or local halls, hotels, and in Woods Point in a muddy-floored marquee that also comprised the temporary schoolhouse. The 200 willing and unwilling witnesses were a mixture of intelligent and sensible men, together with the misinformed, opinionated, stupid, and avaricious. Some were recalled. The only woman to appear was Elizabeth Ashmore of Healesville whose husband Henry Ashmore also gave evidence. As Stretton himself observed in his final report: 'Much of the evidence was coloured by self-interest. Much of it was quite false. Little of it was wholly truthful.'[5]

The commission also visited burnt-out areas near Warrandyte, Ballarat, Silvan and Dromana. Throughout proceedings Stretton kept everyone focused on the primary issues at hand: what had caused the terrible fires of 1938–39, and what measures needed to be taken to prevent them in the future? But that did not mean that he did not allow debate on a wide range of issues about bushfires.

Two attitudes toward the causes of fires and how to deal with them repeatedly emerged in evidence to the commission. The first held that it was best if the bush, that is the forests, be cleared and eliminated completely by burning in order to open the land and remove the danger of fire altogether. As Robert Code, a senior inspector of the Forests Commission said: 'We have inherited a point of view from our ancestors. They had to clear the land and they regarded every tree as an enemy. Unfortunately, we still have many people who think that way'.[6] This view reflects the settler-grazier mentality which by 1939 had become an unshakeable conviction among most people in the bush. If forests and woodlands were to be preserved at all, they must be burnt as often as possible to make sure that the forest floor was 'clean' of any form of undergrowth. William Irvine, a farmer, road contractor and councillor of the Shire of Eltham, is typical. He complained to Stretton that:

> Growth should not be allowed for ten years as it has been in the past in both Forests Commission and Board of Works

[water catchment] country . . . many areas within a few miles of Healesville is [*sic*] covered with scrub you could not walk through. I have tried to get down fishing—with a permit—but I could not get down to the river . . . You cannot walk across the bracken and other scrub that is 15 to 20 feet high. I have lived here for practically fifty years . . . Any practical bushman will tell you the same thing . . . The accumulation of such growth is the worst possible thing for fires.[7]

He said that landholders were so worried about the penalties for burning off in the prohibited season that they simply set fire to the bush and let it go, disappearing as quickly as possible before the police or Forests Commission officers arrived.

The law since 1926 was that fires could not be lit anywhere in Victoria between 1 December and 15 March without Forests Commission permission. Technically, the penalties were two years' imprisonment and a £200 fine. However, these were never imposed. The evidence before the commission was that hardly anyone was ever charged. As the police pointed out, the difficulty was getting evidence, so unless someone was caught red-handed or pleaded guilty there were unlikely to be convictions. There was also widespread confusion as to the dates of the proscribed period and penalties incurred. So rural fire-lighting continued unabated despite the law. Some people, like forests inspector Robert Code understood the consequences of continual burning: 'A forest policy has to be a long-sighted policy, and if you are going to burn, burn, burn you exhaust the humus from the soil and the seed supply until eventually the forest floor becomes barren'.[8] Erosion follows.

Into this 'unquenchable restlessness', as fire historian Stephen Pyne calls Australian pyromania, came the professional foresters. They represented the second point of view put to Stretton. They wanted to bring fire under disciplined control, to use it as a tool. They realised that Victoria was the most fire-prone place on earth and that controls were required. The foresters were driven by an ethical and technological imperative to conserve forests as natural and exploitable assets. They 'found laissez-faire pandemonium and created system'.[9] But even among them there was disagreement.

By the time of the royal commission the pivotal question had become: should Victorians use controlled fire to protect themselves from wildfire? In other words, was the only way to stop conflagrations like Black Friday a process of regular, controlled burning to create a stable situation? Or was there another, more long-term preventative mechanism which allowed the bush to recover something of the natural stasis that it had before the arrival of Europeans?

The argument was that if the bush was burned too regularly, these very fires would encourage the growth of bracken, weeds and undergrowth that provided fuel for the next wildfire. Reginald Needham, a forestry officer from Gippsland, put it succinctly for the commission: 'Fire creates undergrowth, and undergrowth increases fire'.[10] But if the bush was allowed to return slowly to its natural state and was rigorously protected from any fire the forest canopy would close over and prevent light penetrating to the forest floor and gradually eliminate the most inflammable undergrowth. The ground cover would return to its natural, more open state. This argument was put by the most distinguished forester in the country, the English-born Charles Edward Lane-Poole, Inspector-General of Forests for the Commonwealth. He argued that this was especially true of the mountain ash forests of central Victoria. Stretton tried very hard to get a time frame for this process, but no one was willing to hazard a guess.[11]

Opposed to scientific forestry were those who held the first view: settlers, farmers, graziers and others who considered themselves born and bred in the bush, the men who presumed they had inherited a 'practical' understanding of nature. These made up the majority of witnesses heard by Stretton. For them the Forests Commission was the enemy, a corps of youthful professional theoreticians from 'the city' who tried to impose their view on 'experienced bushmen'. George Purvis, storekeeper, grazier and resident of Moe in Gippsland, typified the rural witnesses. In fact, he was not a bushman at all, but a businessman who owned sixteen stores across Gippsland from Trafalgar to Bairnsdale, but that did not stop him assuring the commission that 'country men, born and bred' understand firefighting. 'They know that when they come out to fight a fire, they must bring matches in their pocket' and fight fire with fire.[12] 'I know the big majority of forest officers personally and

I think that they comprise as fine a body of men as we have in our public service', he told Stretton.

> [I]t is only that they are so keen that they hate to see even a little gum tree destroyed, and that is where we think differently from them. I believe that a lot of the fires around this area on the bad day in January were due to the fact that so many farmers had failed to burn off and protect their own properties. They did not burn off because they were afraid of the Forests Department coming down on them. Everybody used to burn off many years ago. We would meet a few of our neighbours and say 'What about a fire?' We would get together, burn off and protect each other . . . There was no danger and no trouble was caused. Nowadays if we want a fire we nick out in the dark, light it and let it go.[13]

Another witness, Alfred Webb, a Gippsland dairy farmer, told Stretton that he had had experience of bushfires since 1905 and that, legal or illegal, people had to burn 'for the protection of the public . . . If it were not for local people and "criminals" as they are called at present—I should call them the reverse—there would not be a living being in that country, because every individual would have been burnt alive'. Webb argued that none of the forest officers were real 'bushmen'. 'To send a man direct from college up here means he will abide by the laws and regulation . . . He would know what to do if he lived amongst the people here long enough.'[14] Underlying much of the evidence is an assumed but hardly ever articulated presupposition that the bush is the enemy. People were fascinated by, even addicted to, burning and it was a ritual of rural manhood to go out and struggle with the elements, particularly the most unpredictable and dangerous element of all: wildfire.

The question of the use of fire to prevent fire still confronts us today, although we debate through the rhetoric of modern ecology. But there was someone present throughout the commission hearings who represented what we would now call an 'environmental' position: this was Alexander Edward Kelso, the engineer in charge of Melbourne's water supply, who represented the Melbourne and Metropolitan Board of Works (MMBW) throughout the hearings. He consistently argued that if the forests were protected from fire and they could re-attain their original pre-European integrity they would be naturally fire-resistant. A number of other witnesses also represented this point of view, including the well-known ophthalmic surgeon and man of influence, Sir James William Barrett; Alfred Douglas Hardy of the Victorian Branch of the Australian Forests League; and Lane-Poole, the Commonwealth Inspector-General of Forests.[15]

Ultimately, Stretton was to disagree with this viewpoint. While quite tolerant of a procession of self-interested and opinionated rural witnesses, he seemed bad-tempered and impatient with 'environmental' advocates. There was abundant evidence before the commission that graziers were the worst culprits for fire-lighting. They constantly abused their leases and lit fires to get green pick that were often let get out of control.

iii

The behaviour of graziers and farmers was certainly the specific issue at the Kinglake hearing. The central question before the commission that day was the origin of the New Year's Day fire. It was clear that the fire began in the no. 3 Mountain Creek area, a stream which rises east of Kinglake on the northern slope of the Kinglake plateau and described in a 1930s tourist brochure as, 'Quite a fine little stream in its lower courses, where the fishing is good before it joins the Yea River . . . [with] numerous fine fern gullies'.[16] According to the evidence of John Blackmore, a foreman of the Forests Commission in Toolangi, he spotted the fire early in the afternoon of New Year's Day, and eight Forests Commission staff and 30 to 40 volunteers managed to keep the fire under control until Sunday 8 January. But they could not get it out. According to Blackmore the problem was, 'The wind was changing all the time'.[17] About midday on that Sunday a severe northerly gale arose and the fire suddenly took on a new intensity.

No amount of pre-season burning would have stopped this fire. 'It was everywhere in no time . . . If a break had been a mile wide, it would have gone over it', Blackmore said.[18]

When the inquest into the deaths of the Forests Commission officers and those killed on the Acheron Way was held in Healesville on 3 February, allegations were made that the fire had been deliberately lit and this was when references to manslaughter charges arose.[19] However, the rhetoric was toned down when the royal commission arrived in Kinglake. The commission heard that Senior Detective Charles North, accompanied by an experienced fire investigator, Detective Craig, had found the source of the fire on the bank of no. 3 Mountain Creek. Here the bush was 'exceptionally dense', there had been no campers or prospectors in the vicinity, and the police could not see how burning the bush at this time would 'benefit anyone, or make anyone's property more safe'.[20] The fire began on the edge of an unoccupied property known locally as 'Andrews'. The nearest occupied property was that of John McMahon and his son James who were both interviewed by detectives. North told the commission:

> There was a very old gum tree with a circumference base of probably 18 to 20 feet. It . . . had burnt right up the centre. There was an accumulation of ashes about one foot deep around the base, and the tree had obviously been burning for a considerable time. It was not burning when we were there; but I believe that the tree may have caught alight and the old fire [that is one legally lit by James McMahon in October 1938] continued burning. I believe that sparks from that started the new fire.

Questioned, he admitted that this was 'merely theory'.[21] North said that he considered the possibility that the fire had been deliberately lit on 1 January but could find no evidence.

At the Kinglake hearing Leo Gamble, a farmer, told the commission that there 'had been numerous signs of fire in [the Mountain Creek] direction, Toolangi way' on New Year's Day.[22] There were fires everywhere and people had become indifferent to them. In evidence John McMahon admitted that a fire had arisen 'somewhere in the vicinity' of his allotment 'within a few days of New Year's Day'. He said he saw smoke on the evening of 30 December but could not 'say exactly where [the fire] was'. It was not 'a raging bush fire . . . It would be New

Year's Day before it got any volume to it'.[23] His son, James McMahon, confirmed his father's evidence that he burnt some heads (off-cuts from logs) in early September on the no. 3 creek on his property but then admitted: 'I did some burning-off in October' to safeguard against bushfires. When Stretton asked him: 'If there is a fire burning out in the uninhabited area, is anybody interested to go and see how it is getting on, and try to put it out, or do they wait until the weather is starting to get hot?' McMahon replied: 'I have not heard of anyone going out'. Stretton asked: 'What is the principle on which you act? If it is going to miss your place, it is all right?' 'It is a case of waiting until it comes', McMahon said.[24]

No one was charged over this fire. But there was a sudden rash of charges brought against other individuals in early February. The only grounds for prosecution was burning within the proscribed period. On 4 February Clarence Marriner of Apollo Bay was jailed for three months and fined £25 for lighting a fire in the Otway Ranges that was still burning at the time of his conviction and that had caused £20,000 worth of damage. On the same day in the Ferntree Gully local court twenty men and two women defendants were charged. Most received fines of between £25 and £30 plus costs, and one man, John Davie of Cockatoo, was sentenced to one month in prison.[25] Charges were also brought against others in Sale and Traralgon. But most fire-lighters were never charged. It would have involved prosecution of many of Victoria's rural population.

What emerged at the commission was that Australia was a nation of fire-lighters who constantly, deliberately and often carelessly lit fires everywhere and in all seasons, including the height of summer. 'The most disastrous forest calamity the State of Victoria [had] known' was caused by fires 'lit by the hand of man', Stretton reported.[26] Alexander Kelso told the commission of one particular horrifying mid-summer experience:

> On one occasion I was travelling along [the Alpine] country with an employee of a grazier. We were still some distance away from the hut to which we were making and he took to throwing matches into the forest . . . I talked to him and discussed the

matter with him. He said 'Well, there is one good reason, that the boss to which we are going will know by the fire we are coming'. He was actually carrying out a practice he had regularly carried out.[27]

iv

The tradition of constant burning went back to the origins of European settlement and what Stretton was dealing with was a history and folk practice rooted in this process. The evolution of this burning regime is the key to understanding Australian attitudes to fire. Male identity particularly was forged in fighting wildfire, but it was a peculiarly symbiotic procedure: wildfire was fought and conquered by deliberate burning. And the landscape was finally conquered only by clearing, which itself involved fire.

There is no doubt that the settlers enormously increased the frequency and spread of fire on the continent, perhaps five- or six-fold. Country that might have burnt in a natural rhythm every 30–50 years before 1788 was being burnt every three years. Certainly Aboriginal people burnt some country on a regular basis but this was focused and low-intensity burning. What becomes clear from a study of fire in European Australia before 1939 is that the settlers introduced a massive dislocation into the established ecological patterns of the landscape. While they would have claimed that they were using controlled fire to fight wildfire, their burning patterns were encouraging the very vegetation that fed increasingly massive conflagrations. It is a tribute to the toughness and staying power of the landscape that it has to some extent survived the European onslaught.

The first few years of settlement in New South Wales were relatively wet, but by 1792 the white settlers of Sydney Town were beginning to experience the more usual pattern of drought, dry, hot summers, and the ever-present threat of fire, mostly generated by themselves, or used as a weapon by the Aborigines. There had been heavy rain early in 1792, but by Wednesday 5 December the temperature reached 114°F (45.56°C) in the sun, and the Deputy Judge-Advocate, Captain David Collins, reported:

a day most excessively sultry. The wind blew strong from the northward of west; the country, to add to the intense heat of the atmosphere, was everywhere on fire . . . the grass on the back of the hill on the west side of [Sydney] cove, having either been caught or set alight by the natives, the flames, aided by the wind . . . spread and raged with incredible fury. One house was burned down . . . The whole face of the hill was on fire threatening every thatched hut with destruction.

Fires were burning at Parramatta and Toongabbie, where wheat crops were destroyed. In a colony scarcely able to support itself in food, this was a serious loss. In the evening the heat abated and some rain fell.

In January 1797 the area around Toongabbie was again alight. 'The fire broke out about eight o'clock in the evening; the wind was high, the night extremely dark, and the flames had mounted to the very tops of the lofty woods which surrounded a field called the ninety acres, in which were several stacks of wheat', Collins wrote.[28] As a result of the efforts of a jail gang only 800 bushels were lost. Having witnessed the Toongabbie fire personally, Governor John Hunter told the Colonial Office in London that it moved so quickly that, 'Trains of gunpowder could scarcely have been more rapid in communicating destruction, such was the dry'd and combustible state of every kind of vegetation, whether grass or tree'. He also reported that 'the noise it made thro' lofty, blazing woods was truly terrible'.[29]

With fire-threats to food supplies, Hunter issued a Government and General Order directing that all farmers enclose their wheat stacks with a paling or wattle hedge and plough fire-breaks and dig a ditch around their fields and dwelling-places. Describing them as 'pests and villains', Hunter said that escaped convicts who had joined Aboriginal people were lighting fires indiscriminately across the landscape to cause trouble.[30]

'Many parts of the country' were on fire again in early summer in September 1798. Drought and high temperatures in December that year led to more fires in early January 1799. Collins reported that 'the country was now in flames; the wind northerly and parching; and some showers of rain which fell . . . were of no advantage being immediately taken up again by the excessive heat of the sun'. As a result of both fire and drought there was a massive loss of pasture for cattle, and water reserves of the colony were so depleted that much of it became brackish and scarcely drinkable.[31]

By the time of Governor Lachlan Macquarie (January 1810 to December 1821) settlement had spread along the road to Parramatta and beyond to the Hawkesbury River. While 11,000 hectares on the Cumberland Plain had been cleared for agriculture, there were still large areas of untouched forest and unoccupied wilderness in the Sydney basin. By now government and settlers were beginning to realise that fire was endemic in south-eastern Australia, but they were unaware of how much worse their activities had made the situation. The regularity of Government Orders about precautions against fires throughout the first two decades of the nineteenth century indicate that bushfires were a continuing problem. They were particularly intense in 1826. The *Sydney Gazette* of 29 November said that:

> The heat and the hot wind of Saturday last excelled all that we ever experienced in the colony . . . In some parts of the town . . . [the temperature] reached 100 degrees [37.78°C], and in others 104 degrees [40°C]. To traverse the streets was truly dreadful; the dust rose in thick columns; and the north-west wind from which quarter our hot winds invariably proceed, was assisted in its heat from the surrounding country being all on fire—so that those persons obliged to travel felt themselves encircled by lambent flames. Sydney was more like the mouth of Vesuvius than anything else. Sunday, however, brought a change of wind, since which the weather has been somewhat more endurable.

The bushfires that summer seem to have been particularly bad on the inner-north shore of Sydney Harbour. James Milson's house at Neutral Bay 'was consumed'. In a pattern that was to become common in many subsequent bushfires, his wife was home with their five children, and she 'was much burnt' in the effort to rescue them. The *Sydney Gazette* commented that:

never before was anything equal to the devastation which the fire has effected; the grass is destroyed for miles upon miles, and the trees are continually falling, to the danger of the traveller beneath . . . Mr Wollstonecraft's lovely retreat at the North Shore has shared the ruin which many poor people have experienced, the fences and huts of the men being destroyed. It is feared that cattle in the neighbourhood have been burnt to death as many head are missing. We have just heard that a poor man was burnt to death on Saturday at Kissing Point in endeavouring to save his little all. (29/11/1826)

In early December 1826, fires around the Sydney basin were attributed by the *Gazette* to human carelessness and particularly to the lower orders' addiction to pipe-smoking. Labourers and convicts often kept a fire burning in the open close to workplaces so they could easily light their pipes during 'smoko'. The *Gazette* went on to report fires in the Kurrajong, Richmond, Nepean and Hawkesbury areas in which homes, outhouses, a sty full of pigs, and various other buildings were destroyed, and the body of an old man named McNamara was found dead in a creek 'very much scorched'. The same fire destroyed most of the property of the Coroner, Thomas Hobby, in what is now the south Penrith area.

A decade later that acute observer, Charles Darwin, crossed the Blue Mountains on 19 January 1836. He reported that, 'In the whole country I scarcely saw a place, without the marks of fire; whether these may be more or less recent, whether the stumps are more or less black, is the greatest change, which breaks the universal monotony that wearies the eyes of a traveller'. Three days later he was in the Bowenfels–Wallerawang district not far from Lithgow, and he saw 'at evening great fires'. They were still 'raging' the next morning: 'We passed through large tracts of country in flames; volumes of Smoke sweeping across the road'.[32]

In late November 1842, the countryside was bone dry after a long drought and was just waiting for a deliberately lit fire to get out of control. It happened on a very hot day with high winds when a gardener 'imprudently commenced burning-off some scrub'. The fire spread rapidly and major damage was only avoided when the wind subsided in the evening. But the *Sydney Morning Herald* of 23 November warned that, 'The whole of the surrounding country, owing to the prevailing drought and intense heat, is in a very inflammable state, and at such a time fires of this description are exceedingly dangerous, nothing but rain being able to suppress them'. Things had got worse by December. Early that month there were extensive fires in the Illawarra, at Castle Hill, Baulkham Hills, Seven Hills and Dural. Other fires were reported to have been burning for several weeks to the north and south of Port Stephens. To the north of Windsor fires were burning along the Hawkesbury River from Pitt Town to Colo. 'The weather has been immoderately hot, and the dry, parching north-west winds have put everything in the condition of tinder. The thermometer has frequently been above one hundred in the shade. It is to be feared in some cases these fires have been caused by the wilful or careless use of tobacco and fire-sticks' reported the *Sydney Morning Herald* on 9 December. The fires persisted for weeks and lasted well in to the first quarter of 1843.

In Van Diemen's Land the same pattern was emerging. By February 1834 the Black War against Tasmanian Aborigines was over and the use of fire as a weapon against white settlement had ceased. But fire, mainly caused by ignorance and carelessness, was endemic. This is vividly illustrated by the account of the Austrian Baron Charles von Hügel, who visited Hobart in late January and early February 1834. A botanist, diplomat and scholar, von Hügel visited Australia as part of a world trip to assuage his sorrows after a broken engagement. He kept a detailed journal.[33] He had already witnessed bushfires in Western Australia, but it was around Hobart that he was to experience them directly. While collecting botanical specimens, von Hügel saw a lot of burnt-out country resulting from 'the driest [year] yet experienced by the colony'. The fires were caused by convict carelessness: their 'supreme and only pleasure is tobacco', he wrote. He found the heat suffocating and from a high point on the north-eastern shore of the Derwent he experienced a 'gale' which fanned the bushfires on the western, Mount Wellington side of the river. 'The sun could not penetrate the smoke from the fires . . . It was like a volcanic eruption.' Two days later on the slopes of Mount Wellington von Hügel actually found himself in a bushfire.

He instinctively made the right decision: he decided to go straight through the flames. Blindfolding the horse he led it through on foot. They were both saved, 'but my beard, hair and clothes were singed and the horse's feet, mane and tail were shaved quite smooth . . . The scene on the other side of the fire was terrifying. The very tallest trees were blazing from root to crown. I watched while a whole tree was enveloped in flames . . . the fire ran up the trunk as if it had been covered in sulphur. The whole countryside was charcoal and ashes.' As he travelled south from Mount Wellington towards the Huon Valley he found the whole area burnt out.[34] The next day it was pouring rain, such was the irony of the Australian climate.

v

A massive expansion of the colony of New South Wales followed the first crossing of the Blue Mountains in June 1813. From the mid-1820s to the 1840s more and more country was opened up for settlement. This meant that the frontier moved constantly and settlers, well beyond the control of the government, found themselves in regular conflict with Aborigines. The history of this constant, low-level, vicious warfare is now well documented.[35] What has not been fully recognised is the impact of white settlement on the landscape.[36]

This seizure of Aboriginal land took the form of squatting.[37] Squatters were men who decamped up-country with herds of sheep on unoccupied crown land not already taken by others. The *Sydney Gazette* reported on 17 October 1829 that 34,505 square miles (9 million hectares) were already open for squatting. This frontier constantly expanded, especially after the settlement of the Port Phillip District (now Victoria) after 1835. Squatters were usually men who had just arrived in the colony with some capital, adventurers who were willing to leave the towns behind and live in the bush far from civilisation. The squatters, like the selectors who later carved small farms out of the big squatter runs in the decades after 1860, constantly used fire to clear the bush in an attempt to impose European animals, plants and agriculture upon a resistant and unsuitable land. This is where the myths of endemic rural burning began.

But for the settlers wildfire was something else altogether. It was a destructively frightening force, uncontrollable and unpredictable. In European consciousness it came to represent the inscrutable nature of Australia, its alien otherness. It was a terrible threat because wildfires were overwhelming and settlers were bluntly reminded that they lived in a fragile and vulnerable relationship with the most fire-prone place on earth. They struggled against the otherness of the bush, the 'scrub' as they derisively called it. They aimed to create Britannia in the Great South Land, to turn Australia, like Europe, into a vast monocultural, agricultural-industrial site, with wild landscape completely eliminated.

The clearing, feral vegetation and burning practices brought by Europeans had only one result in Australia—disastrous wildfires. The squatters' ignorance of the environment led to a massive disruption of the flora and fauna, as well as the life cycles of the Aboriginal tribes they supplanted. The low intensity and scattered Aboriginal patterns of burning, to which the landscape was well adapted, were disrupted or ceased altogether as squatting and settlement pushed the Indigenous people off their traditional lands. Long-term, natural patterns were abandoned and Australian biota were changed forever. Native grasses were quickly eaten out, hard-hoofed exotics like sheep, cattle and horses compacted the soil, destroying the native vegetation long-adapted to soft-footed marsupials, and foreign plants and vegetation, much of it highly fire-prone, invaded the landscape. The settlers often used fire as a tool to clear their 'runs' but they did not understand the dynamics of burning in such an incendiary environment. With the introduction of non-native vegetation and the disturbance of natural ecological rhythms, the scene was set for monumental wildfires. Our fire narrative prior to 1939 is a long history of deliberately lit fires getting out of control and, when weather conditions were right, coalescing to form massive wildfires.

The first of these great fire outbreaks occurred on 'Black Thursday', 6 February 1851. While the fire was almost totally confined to the newly established colony of Victoria, it set a pattern for the great conflagrations of Australian history and became part of folklore until Black Friday replaced it in the national consciousness. The infant city of Melbourne had only been settled in 1835, following the exploration of south-eastern

Australia by the New South Wales Surveyor-General, Thomas Mitchell, who described the fertile land he discovered as 'Australia Felix'. The population of the Port Phillip District had increased to 77,000 by 1851. But the growth of the human population was nothing compared with the increase in the number of sheep. In 1843 there were 1.6 million. This had grown to 5.1 million in 1848, and to over six million in 1851.[38] Victoria had become a separate colony from New South Wales six months before Black Thursday.

Black Thursday was preceded by a drought.[39] Settlers found that food and water for stock became increasingly scarce. Waterholes and creeks dried up and even though occasional thunderstorms brought heavy rain to particular regions, the precipitation was quickly absorbed by the dry ground. The Melbourne *Argus* of 21 January 1851 reported that one day the previous week in Kilmore,

> from sunrise to sunset, the wind was strong at North . . . in consequence of the long drought, accompanied by dense clouds of dust, so impervious to rays of light that at midday the sun was darkened . . . so that instead of walking with head erect and your eyes acting as a guide to your movements, your chin reclined upon your breast, and with eyes closed, you were compelled to feel your way from one end of the township to the other.

Reports were also coming in that the drought extended right up into the Edwards River country, present-day south-western New South Wales around Moulamein. Throughout January 1851 there were very high temperatures and bushfire outbreaks occurred in isolated, unsettled country, a few of them started by dry lightning strikes, but most lit by settlers opportunistically trying to clear land or encourage green pick, which the cattle and sheep loved. Some fires were lit by itinerants like bullock-drivers and swagmen who made campfires and left them burning.

At this time there were only a few scattered towns besides Melbourne, the small capital of 23,000 people. There was Kilmore 60 kilometres due north, with Seymour just 36 kilometres further up the track to Sydney. Past Seymour there were small settlements, usually with a police station and a few stores supplying the surrounding squatters, and a shanty that served as an 'inn' and supplied liquid refreshments and accommodation to people on the road. Kyneton was 84 kilometres north-west of Melbourne and Geelong, on Corio Bay, was 75 kilometres to the south-west. The town and surrounding area had 8000 inhabitants in late 1849. Smaller centres were Belfast (Port Fairy) and Portland, almost 350 kilometres west-south-west of the capital on the south coast, and to the far south-east was Port Albert on the south-Gippsland coast. Otherwise, settlement was very thin with just a few areas thickly settled.

In mid-January reports came in from Kilmore of deliberately lit fires burning in unsettled country to the north-west, and also from the Goulburn River district where outbreaks were already 'overwhelming in their destruction'.[40] These had been burning for a month. Other contemporaries blamed the Black Thursday fire explicitly on a couple of careless timber splitters (or in another version of the story, bullock-drivers) in the Plenty Ranges north of Melbourne who left a campfire burning.[41] Black Thursday in Melbourne dawned scorchingly hot. Various figures were quoted, but between 11.00 am and midday the temperature probably reached between 117°F and 119°F (47.22°C to 48.33°C). The draftsman and artist, William Strutt (1825–1915), recalled in his *Australian Journal* that:

> ... the wind increased with stifling heat, that it must be felt to be realised ... At the breakfast table the butter in the butter dish melted into oil, and the bread when just cut turned to rusk. The meat on the table became nearly black, as if burnt before the fire, a few minutes after being cut ... Cold water you could not get, and the dust raised in clouds by the fierce wind was sand which penetrated everywhere ... The flies swarmed everywhere and settled on the black coats of the pedestrians in the streets in hundreds ... But the fire settled them all on that day ... The sun looked red all day, almost as blood ... with immense volumes of smoke.[42]

Many who experienced Black Thursday reported that the dust and the hot wind from the north-north-west obscured the sky and brought about a kind of midday darkness. Even in northern Tasmania and far eastern Gippsland the sky was black and everything was enveloped in smoke with falling ash and burning debris. It appeared that the whole of the colony of Victoria was on fire. James Fenton reported that people living near Devonport, Tasmania experienced strange natural phenomena in the afternoon of Black Thursday:

> Early in the afternoon clouds came rolling over the heavens, obscuring the light of the sun in a most ominous and mysterious manner . . . There was a lurid glare in the sky [and] . . . There began to fall thickly over the ground in every direction charred fragments of vegetation resembling fern leaves which had been burnt to ashes but retained the distinct form of the leaf . . . By four o'clock the whole face of Nature was enveloped in utter darkness . . . It never occurred, I think, to anyone that a bushfire could possibly be the cause.

People felt the end of the world was coming. One of Fenton's neighbours said: 'I never prayed before, but that brought me to my knees, and I never left off till daybreak!'[43]

Information about what occurred in the bush that day only gradually got through to Melbourne from up-country because what limited communications there were had broken down completely. What can be established is that between one-third and half of the colony was involved in the conflagration. In a pattern that was to become familiar on days of extreme fire danger, from early in the morning a dry, dusty, almost gale-force hot wind from central Australia stirred up and united the scattered fires that were already burning across the colony. Other fires were ignited when the north-west wind carried burning embers, causing spot fires to break out.

With no organisation and fire-fighting equipment and minimal means of communication, local people were quickly overwhelmed. One squatter, H.J. Kerr, said that 'So frightful was the celerity with which the fire travelled that the swiftest horse could barely gallop away from it . . . On the great plains many flocks of sheep fell a prey to the devouring element, and drays that had been left standing, some perhaps with their helpless teams still yoked, were totally consumed, and only the iron work left remaining.'44

William Howitt wrote a vivid description of the fires in a series of articles in *Cassell's Illustrated Family Paper.*

> Soon the people had to flee from the remorseless enemy in all directions . . . over an extent of many hundreds of square miles. The women and children fled from their blazing huts; the shepherds left their flocks to perish unable to drive them to any conceivable place of refuge. Cattle in vast herds were seen careering madly before the fires, which not only leaped [*sic*] from tree to tree, but travelled at once with its velocity and deadliness. Troops of horses wild from the bush, with flying tails and manes, and neighing wildly, galloped across the ground with the fury of despair. Flocks of kangaroos and of smaller animals, leaped desperately along, to escape the conflagration, and hosts of birds swept blindly on, many falling suffocated headlong into the flames, and the rest raising the most lamentable cries. Horsemen, seeing the raging sea of fire advancing with whirlwind speed from almost every quarter, galloped madly and for scores of miles, till their horses fell under them. Drovers conducting mobs of cattle and horses . . . were compelled to leave them to shift for themselves, and fled away at the highest speed of their horses for their own lives. The destruction not only of farms, crops, shepherds' huts, cattle, horses, and sheep, was immense, but the destruction of the wild creatures of the woods, which were roasted alive in their holes and haunts, was something fearful to contemplate. People . . . rushed into waterholes and creeks—happy were they who had any near them—and sunk themselves to the very mouths in them, were yet in some instances so scorched and broiled as to perish from the efforts.[45]

While many believed that the whole colony was alight, all of the surviving evidence indicates that the fires were concentrated in several broad, interconnected areas. If we take Port Phillip Bay as a pivotal point, Howitt estimated that the fire extended 300 miles (480 km) from east to west, and about 150 miles (240 km) from north to south. Around Melbourne a series of fires extended in a wide arc from Western Port Bay and south-west Gippsland in the east, to Corio Bay, Geelong, and the Bellarine Peninsula in the west. This fire extended as far north as the Goulburn River valley. The area between what is now Geelong and Ballarat was badly burned. Massive fires also occurred in the Western District around Portland and Port Fairy. Other fires were reported in the Wimmera and Mallee, but details about them are scarce. Ash from the fires even fell on ships at sea in Bass Strait and the sails of one caught fire.

We know most about the fires in the arc immediately around Melbourne. The Dandenong district, extending south and east toward Western Port Bay, was burnt and this conflagration extended north-east through the areas that are now the southern and eastern suburbs of Melbourne to the mixed rainforest and wet sclerophyll woodland of the Dandenong Ranges, which were almost completely burnt out. While no one was killed in this area, many had dramatic escapes and several people were very badly burned. An *Argus* correspondent reported from this area on 10 February 1851 that a schoolhouse with nineteen children and a woman teacher was 'saved at the risk of life', and that he saw 'pigs and dogs running loose . . . [who were] burned to death, birds dropping off the trees before the fire in all directions' and that 'opossums, kangaroos, and all sorts of beasts can be had today ready roasted all over the bush. Fully one half of the timber in this neighbourhood had been burned or blown down, and all the grass has been burnt.'

In the Plenty River district the McLelland family ran sheep on the Upper Diamond Creek between present-day Hurstbridge and Eltham. About noon on Black Thursday the fire swept down on their farm and the only possible escape for Richard and Brigid McLelland and their five children was in the nearby creek. Driven by a strong north wind, the fire moved quickly towards the McLelland's wattle-and-daub house. The family were too slow in leaving. The bodies of Bridget McLelland and her five children, John, James, Joseph, Mary-Anne and William, were found not far from the house. Richard, as he ran through burning trees trying to rescue his son John, was overcome by smoke and exhaustion and was badly burned on his arms and legs. The boy died in his arms. Leaving the body, McLelland eventually reached Diamond Creek and he remained there in a traumatised state until he was discovered in the afternoon by the family's shepherd, Alexander Miller. The two men remained all night on the bank of the creek and reported the tragedy to authorities the next day. More than 100 other families in the district lost their homes and farms.

The arc of fire around Melbourne extended across to the 1000-metre-high Mount Macedon, which was burnt out. The *Argus* (10/2/1851) reported that:

> A bullock-driver named Bill fetching a load of timber from the mountain got enclosed by the fire, he unyoked his bullocks to give them a chance to escape, [and] seeing all hope cut off for himself, he held hold of the tail of one, and giving a shout, which along with the instinct of the animal, cleared him of all danger. Cattle are found in every direction dead and dying, many of them with their entrails protruding.

Travellers were particularly unprotected and vulnerable. The *Argus*'s Kilmore correspondent reported that, 'A company of Thespians consisting of Mr and Mrs Evans, Mr Elrington and Mr Moss *en route* for Sydney with a cart filled with the necessary paraphernalia for their vocation, which they intended following at the various towns upon their journey, were surprised by the flames on the Big Hill, and the whole of their wardrobe, etc., was destroyed. The only articles snatched from the burning being a cornopean [portable organ] and a violin' (11/2/1851). In another incident a carrier from Yass named Edward Dodswell:

> was brought into the [Melbourne] Hospital . . . [His burns were] . . . of a very serious nature, extending the whole length of his back and over the greater part of his chest. He was brought in from the ranges near the Plenty River, where it appears he was encamped with his wife and a bushman in his company. The whole team of bullocks and the dray were burnt. The other man was saved by running until he gained some ground that had been [already] burnt. Dodswell and his wife laid themselves down in a creek, which had but about six inches in depth of water, but

fortunately it was running fast, and by turning themselves over in it for about six hours, they escaped with difficulty from being burnt to death. (*Argus*, 17/2/1851)

Another series of intense fires was reported around the settlement at Geelong. Down on the Bellarine Peninsula, reported the *Geelong Advertiser* on 8 February 1851, two firebugs, probably swagmen with a grudge against a local squatter, deliberately lit a fire and when challenged 'scattered the burning embers about in all directions . . . [and] soon set the whole place in a blaze . . . with the rapidity of lightning the fire ran along and in its progress burning hurdles, huts, house, and everything valuable to the ground'. Five deaths were reported from these Geelong district fires.

Altogether twelve people died as a result of Black Thursday. There may have been more who died alone in the bush, unknown to anyone. Given the size and intensity of the fires, this is a remarkably small number of fatalities. While the physical injuries eventually healed, it is clear that the psychological scars affected people for the rest of their lives. The fact that Black Thursday became so much part of national consciousness indicates that many people were mentally scarred by what they had experienced. A large number also faced financial ruin. There was no insurance. Many squatters with smaller holdings were forced to abandon their runs and sell off their remaining goods and animals.

Perhaps the most vivid picture we have of Black Thursday is a painting by William Strutt. He sketched a number of studies in preparation for his large, dramatic work commemorating the fire.[46] Strutt's description of the painting reads: 'The terrified settlers and squatters hastily made their retreat, leaving everything. The sick, put into drays, were hurried off, it was now a stampede for life . . . Kangaroos and other small animals, immense flocks of birds of all kinds mingled in mid-air, amidst the flying sparks, and the stifling smoke heading for the south. A number dropped dead from terror and exhaustion'.[47] The painting vividly conveys the settlers' and their animals' complete alienation from the Australian landscape. It is as though they were fleeing from a hostile natural world that was trying to chase them right off the land altogether.[48]

It is difficult to assess exactly the extent of the full damage caused by Black Thursday. A figure of 1 million or more sheep destroyed is usually quoted. There was also a considerable loss of cattle, although the number was probably only in the tens of thousands. In addition, there was widespread destruction of homes, buildings, equipment, outhouses, sheds, crops, haystacks and animal feed.

Black Thursday was the result of the concatenation of circumstances. The first was a very strong north-north-west wind that swung around to the south-west and drove a whole lot of disconnected fires together. Second, the country was a bone-dry wilderness of forest and scrub with high summer temperatures. But the real villains were those who lit the original fires. Sensible people at the time were certainly conscious and highly critical of the fire-lighting habits of their fellow colonists. One frustrated Mount Macedon correspondent wrote to the *Argus* (12/2/1851) saying that the fires were deliberately lit by settlers burning off to get green pick. He was willing to name names to the authorities, although there is no record that they did anything about his claims. Essentially, he said, the problem was selfish burning at the wrong time. 'Every man who has more grass on his run than he has stock to eat it . . . [puts] a fire stick in, merely because he may wish for something green for a lambing flock'. A similar story was told by the *Argus* Kilmore correspondent. He reported that everyone had agreed to burn off stubble on the Tuesday before Black Thursday but one stubborn settler 'fired his stubble on Wednesday morning, just before the commencement of the gale, and to that may be attributed the extensive loss which has been sustained'. The ingrained habit of pyromania was thoroughly established by 1851.

vi

While the 1851 Black Thursday fire was seared into the consciousness of Australia, particularly in Victoria, the practical lessons that could have been learned from the disaster were quickly forgotten, setting a pattern that has been followed after all major bushfires since, except perhaps in 1939. Six days after Black Thursday, gold was discovered near Bathurst, New South Wales. This was followed in July 1851 by discoveries in

Victoria, and hundreds of thousands of people rushed to the diggings from all over the world. Victoria's population increased from 76,162 in 1850 to 521,072 in 1860, a phenomenal seven-fold increase in one decade. However, Australia's gold rush was short-lived and soon many displaced diggers, who knew nothing about farming, were looking for work and land.

This period also saw the introduction of two of the most destructive feral animals, the rabbit and the fox. While rabbits destroyed native vegetation, foxes attacked native fauna. However, it was primarily human activities and broad-scale pastoral and agricultural practices that were most destructive to the environment. This assault on the landscape had catastrophic results. The frequency and intensity of bushfires increased with the spread of European settlement. The 'new chums' were ignorant of the rhythms of the landscape. They were here to escape the hardships of the old country and make their fortune. They were the products of a culture profoundly convinced of the infallibility of its technology, and were driven by the myth of material progress and British civilisation. These settlers were determined to bring the land into production as quickly as possible without any understanding of long-term consequences. For them fire was a primary tool in clearing the 'scrub' and civilising the landscape. 'Ultimately agricultural fire was less a means to live within Australia as to replace Australia', fire historian Stephen Pyne wrote.[49] Settlers became addicted to 'burning off' as a way of clearing land. Pyromania became endemic.

Until the next massive conflagration of 1898, bushfires were a regular part of summer life in the Australian colonies. This period saw many serious fires, almost all the result of the fire-lighting tendencies of the settlers. When a 'burn-off' either accidentally or deliberately 'got away' the settlers often tried to shift responsibility to nomadic bush workers or 'swagmen'. No doubt 'swaggies' were responsible for some fires, but the Melbourne *Argus* of 8 August 1865 was overstating it when it said that 'Some of the most calamitous bushfires owe their origin to these outlaws'. The counter-claim was that fires were often lit by squatters and selectors to claim insurance.

The end result was that vast areas of bush and agricultural land were destroyed by fire, allowing scrub, bracken and noxious weeds, themselves highly flammable, to take the place of native vegetation and grasses that were well adapted to fire. This lethal cocktail was further destabilised by soil compaction from hard-hooved animals preventing native vegetation from re-sprouting after fire. Pyne comments that 'Australian grasses were resilient to fire, but not to grazing. Australian soil could accept fire and drought but not compaction from hooves'.[50] Another consequence of massive vegetation destruction was severe erosion. This was understood at the time. As early as September 1840 the Polish-born scientist, Paul Edmund Strzelecki, correctly described for Governor Gipps the results of 'the innumerable flocks' and clearing of native vegetation. He said that the soil dried out and was 'abandoned [to] a most prejudicial practice . . . to the constant and wilful incendiarism, which, instead of producing the expected and former herbage and vigour of the soil, in fact only calcines its surface and eradicates even the principle of reproduction'.[51] Between fire and ringbarking of trees a third of all New South Wales' forests were destroyed before 1892, with similar proportions in other colonies.[52]

One of the ironies of the drama of great bushfires was that they created the myth of the heroic firefighter. In the writings of the nineteenth century 'mateship' school, much publicised by the *Bulletin*, wildfire was the enemy that broke down the 'deadly feud of class, and creed and race', so that selectors, squatters and bush workers eventually came together to confront the alien threat of nature. This is vividly illustrated by Henry Lawson's poem, 'The Fire at Ross's Farm', a Romeo and Juliet story of a squatter's son in love with a selector's daughter. As Christmas approached a bushfire burning in wild country threatened Ross's farm. At first the squatter rejects his son's pleadings to help, but at the last moment he races off with a dozen stockmen to help save the selector and his family from ruin.

> Down to the ground the stockmen jumped
> And bared each brawny arm,
> They tore green branches from the trees
> And fought for Ross's farm;
> And when before the gallant band
> The beaten flames gave way,

>   Two grimy hands in friendship joined—
>   And it was Christmas day.

Fighting fire became a kind of ritual in which Australian manhood was formed and the ethos of mateship developed in the struggle with nature.

It is clear from many nineteenth-century sources that bushfires had become an inescapable part of summer life in Australia. But the history of bushfires in the second half of the century is fragmentary because they were so common that they were not reported unless they were close to settled districts, widespread, or disastrous. We know there were conflagrations right across what is now central Canberra in 1858, 1861, 1865 and 1888.[53] The earlier fires are possibly the ones that we also know about from dendrochronological evidence (the study of tree age from tree rings) in the Brindabella Ranges.[54] A severe drought occurred in central and southern New South Wales from 1867 and in January 1870 there were vast fires in this part of the colony which were especially bad in the Riverina. From December 1874 to February 1875 the Sydney area experienced widespread and destructive fires as did southern Queensland.

But Victoria remained the epicentre of Australian fire, and from early January to mid-February 1877 there were conflagrations in various parts of the colony and in 1878–9 the South Gippsland rainforest caught fire. An early settler, Frank Dodd, says that, 'The summer was a dry one and the whole country got on fire . . . The smoke was dreadful, and though the moon was at full during this period we did not see it, and for days the sun was obscured.'[55] Typical of the memoirs and recollections of observant contemporaries from this period was Joseph Jenkins, whose *Diary of a Welsh Swagman* describes life in the goldfield area around Maldon, north-west of Melbourne, between 1869 and 1894.[56] His *Diary* reports a litany of fires throughout the period. Some of the worst were in February 1893 when the temperature was 110°F in the shade (43.33°C). He said that, 'Many die[d] of sunstroke. The atmosphere is laden with smoke from the bush. It is so dense that I cannot see the sun.'

The Tasmanian landscape has much in common with Victoria and the rainforests of the south-west were intimately related to those of the

Strzelecki Ranges of South Gippsland. From the 1850s an extensive network of tracks was established through south-western Tasmania and timber cutters, prospectors and explorers began to penetrate the area. Fire researcher Jon Marsden-Smedley says that 'there are numerous references regarding the use of fire in south-western Tasmania from this period. It appears that the normal method of moving through the region was to burn out a section of the country and then follow the resulting easy path . . . It appears that a large number of fires were lit in hot, dry (and probably windy) weather resulting in large, high-intensity fires in all vegetation types.'[57] Around 1850 or 1851 a deliberately lit fire burnt from the Pieman River to Maydena, a distance of about 180 kilometres covering an area of more than 400,000 hectares. Marsden-Smedley lists 42 known fires that occurred between 1837 and 1938–9 in the area, of which he classifies 22 as high intensity.[58] In other words there was a major fire about every four and a half years. Nineteenth-century bushwalker, solicitor and historian, James Backhouse Walker, observed the results around the junction of the Collingwood and Franklin rivers, just south of the present Lyell Highway. He says: 'To the tourist it is exasperating to see the exquisite native beauty of these forests desecrated and turned into grim blackness by fires which during this hot summer have swept over so many miles of bush. But doubtless the prospector views it with other eyes.'[59] In 1897–8 there were fires from the West Coast Range, to the Hartz Mountains, to the Southern Ranges, an area of about one million hectares, probably the largest conflagration in Tasmanian history. Six people were killed.

vii

If Black Thursday was Australia's first great fire, 'Red Tuesday', 1 February 1898, was the second. It came in the middle of the 1895–1903 drought, the worst Australia had ever experienced. From December 1897 to January 1898 the temperature was abnormally high. There were fires across south-eastern Australia, but Victoria was the worst-affected colony. Fires broke out in a wide arc from the west to the north-east of Melbourne including serious outbreaks in the mountain ash forests, at Euroa in the centre of the colony, in the Western District, and over the border in New South Wales. But the worst fires were in the rainforests of South Gippsland.

This conflagration was the central act in the destruction of one of the largest remaining cool temperate rainforests in the world. Of the original half a million or more hectares of rainforest, nowadays only about 2500 hectares survive in a couple of tiny national parks.

Surrounded by swamps on the coast and lowland forests of stringybark and peppermint, the central core of the rainforest covered the Strzelecki Ranges, a series of rolling hills about 600 metres high, and enveloped in moist rainforest, dominated by mountain ash and blue gum that towered over the understorey like cathedral spires.[60] These trees were often 50 metres or more in height; some of them are known to have reached 100 metres. The main canopy was filled with those relics of the ancient Gondwana forests that once covered much of Australia: myrtle beech and southern sassafras. Small trees like austral mulberry and banyalla and wet-gully shrubs such as mountain correa and Victorian Christmas bush made up the bulk of the lower vegetation. The forest was full of wonderful tree ferns, some 10–15 metres high. They grew in great profusion along the creeks and gullies. There were koalas, wombats, echidnas, wallabies, possums and lyrebirds. This was also the land of the giant Gippsland earthworm, which can measure up to more than a metre in length and is as thick as a man's thumb.

Certainly there were natural and probably occasional Aboriginal-generated fires around the edges of the rainforest and there were claims that the 1851 bushfires burnt much of South Gippsland. There is no record of this in the Melbourne newspapers of the time. While there may have been minor fire outbreaks on the western edge of the rainforest nearest Melbourne in 1851, there is no evidence that the fire penetrated the forest.

Settlement started to occur on the fringes of South Gippsland from the 1840s onwards.[61] The first crossing of the 'scrub'—as the rainforest was called—was by Count Paul Edmund Strzelecki and three companions in May 1840. But it was not until the mid to late 1870s that settlement began to penetrate this 'scrub country'. Most settlers were from middle-class Melbourne families, former British soldiers retired from India, or working men with little or no capital. Few had experience on the land. Their accounts make it clear that they saw the forest as a challenge and the struggle to 'tame' it in terms of warfare. The settlers were 'in the firing line of the stern battle waged against nature and adverse circumstances in The Great Forest of South Gippsland'.[62]

From about 1875 onwards selectors moved in from the south around Mirboo and began carving out small farms on lots of 130 hectares or less. They cleared land by 'scrub cutting', which involved chopping down the mountain ash or ringbarking them and clearing as many of the rest of the trees as possible. Fire was used to get rid of what was not cut down. According to one selector, W.H.C. Holmes, the burn in summer was an essential part of this process: 'After months of incessant toil with the axe with anything from 20 to 100 acres of scrub awaiting a favourable day for burning, how eagerly and anxiously did [the selector] weigh the chances of each hot day after the middle of January . . . although the sun poured down in the hot summer months with a fierce heat, still there were some years when there was not a sufficient number of consecutive dry days to ensure a good, clean burn.'[63] The burn was followed by 'picking up', when the debris left on the ground was physically heaped and burned. As clearing progressed much of the landscape came to resemble a scattered forest of tall, dead trees standing like giant matchsticks, mute reminders of what had been. The settlers almost 'hunted' the trees as they descended into steep, narrow gullies to destroy a forest giant for no apparent reason. It seemed 'as though the settlers actually despised the forests they were destroying' as one commentator Bernard Mace, has put it.[64]

By the summer of 1897–8 the great forest had already been carved up. Cleared areas intersected with thick bush all connected by muddy tracks. Well-established dairy farms and 'new chums' trying to carve out selections, were surrounded by dead and dying trees and drought-dried vegetation. December and January were dry and hot. On 16 and 30 December the temperature reached 107°F (41.67°C) and was never less than 90°F (32.22°C) throughout the month. On 11 January it reached 109°F (42.78°C), the highest temperature for sixteen years. Fires had been burning in wild country since early summer, but the serious conflagrations began in early January, the result of deliberately lit 'burn-offs'. Uncontrolled bushfires were reported right across Gippsland. On 14 January Neerim township was destroyed, 33 houses were burnt in

Thorpdale and twenty in Morwell. One settler reported that, 'People who have not lost everything regard themselves as lucky. On every hand there is nothing but black ruin, grim and hopeless . . . Except for the tall, gaunt tree-stems, charred from root to crown, the face of the country for miles around has been swept so clean that one would think that a blade of grass had never grown on it'.[65]

Describing how the fire began, a reporter from the *Age* wrote on 17 January 1898 that many of the big trees on earlier settlers' farms were still standing, but they were 'dry as tinder'. As soon as new arrivals started lighting fires, one of the big trees on a nearby selection would catch fire.

> That is enough. Up dart the flames along the barrel of a dry forest giant to a height of 50, 100 or even 150 feet, and in such heat . . . with a strong north gale away goes the fire from tree-top to tree-top in bounds of 60 to 120 feet at a time. Up aloft the flames roar like a furnace, and the rarified air, rushing in to feed the blast from below causes violent whirlwinds underneath, making a deafening roar, while a heavy pall of smoke blinds the terrified settlers, who, seeking to escape, are hemmed in by unexpected sheets of flame coming from all quarters at once, owing to the veering about of the furnace-fed conflagration above their heads.

A settler reported that 'the sky began to take on an effect so dreadful and threatening that it made one almost afraid. Its colour was of a strange shade of purple, tinged with blood . . . The air was full of dense smoke, and sparks were as thick as the flakes of a heavy snowstorm. Flames burned blue instead of red, and the great tongues of flame had no illuminating power . . . The forest had largely disappeared.'[66]

Some rain fell on 22 January, but then the temperature climbed again to 106°F (41.11°C) on Saturday 29 January. Fierce heat continued and bushfires broke out all over Gippsland, peaking on Red Tuesday, 1 February 1898. Dense smoke covered the entire area. On the afternoon of 2 February one witness said that, 'Warragul reports the town in imminent danger. South Warragul residents hemmed in with fire. Drouin is in a semi-circle of fire, and the whole country is ablaze.

Many homes are destroyed.'[67] A family with nine children had to flee their home in the forest. 'Placing the younger children on a horse, they led them through the bush, while the house, outbuildings and stacks that they had left were reduced to ashes.' The family took refuge with neighbours about half a mile away. Almost immediately this second house was threatened with fire and the two families now joined together and retreated to another neighbour a mile further on. 'The journey through the heat was terrible, the children crying with fright and the pain and suffocation caused by the blinding smoke.' Finding what they thought was a refuge in the forest, the group tried to rest. But before the children could be remounted the fire was on them again like a 'roaring tornado'. 'A frightful scene ensued, the women, fighting desperately for their families, stripped off their skirts to beat out the rushing fire and save their children from being burnt to death, and at length, after a desperate struggle, in which many of them received severe burns, they succeeded in getting the children mounted and once more on retreat.' The group then made for another farm, but before they could settle the fire was down on them again. 'Driven to distraction the four families, containing now 24 young children, mounted on horses, again set forth on a perilous journey' eventually reaching another farm. Here the lives of the now ten adults and two dozen children were only saved by a change of wind. And at a school in another part of the forest, twenty children sheltered all night in a hole excavated by an uprooted tree. They were rescued the next morning.[68]

The fires dragged on until 12 February when again rain brought some relief. Two days later the fires blew up again when the temperature reached 38°C with low humidity and winds up to 75 kilometres per hour. The following weeks saw more outbreaks, but the worst had already occurred. When the smoke eventually cleared, twelve people were dead, animal and stock losses were huge, and 260,000 hectares of cool temperate rainforest had been completely burnt out, never to recover. There was optimism for some, however: at last the central portion of the 'scrub' was gone. But the area never produced the longed-for agricultural bonanza. The Strzelecki Ranges are now primarily a marginal dairy and plantation forestry area, still known as the 'heartbreak hills'. Pyne's judgement is right: 'Red Tuesday was not so much an unprecedented aberration as it was a macabre, reckless parody of frontier land clearing, transience and violence. It crowded into days and weeks what had been occurring over decades.'[69]

viii

While Victoria remained the epicentre, the other colonies were just as fire-prone. South Australia is the driest place in Australia with two climate zones, an arid north covering about 80 per cent of the state, and a temperate southern zone which has much in common with Victoria and is also inflammable, especially in the Mount Lofty Ranges to the east of Adelaide. The earliest settlers in 1836 noticed Aboriginal-lit fires as they sailed up the Gulf of Saint Vincent. Fire was to become a regular occurrence in the hills above Adelaide. 'Scarcely a summer passed without the report of a blaze somewhere in the Hills, but often as not [the people of] Adelaide had only a vague idea of where the fire was raging.'[70] In February 1859 a huge fire raged through the district to the east of the capital and caused widespread damage to bush and settlements. There was another serious fire in the summer of 1876. There is not a lot of detailed information about fires in South Australia prior to the 1930s. However, the evidence is that 'serious fires could be expected in the state in six or seven years out of every ten'.[71]

The fire history of Western Australia has much in common with the east despite the vast desert that intervenes. Most fire occurs in the southern agricultural and forest zones which comprise a roughly triangular area reaching from Geraldton in the north inland along a line to Esperance on the south coast. This region was originally vegetated with open stands of eucalypts, acacias, mallee scrub and heath.[72] Most of this has now been cleared, resulting in widespread salinisation. The area with the highest rainfall is the forested region stretching south-westward along a ridge-line extending from just north of Perth to east of Albany. The predominant trees of these western forests are jarrah (*Eucalyptus marginata*) and, mainly in the south, karri (*Eucalyptus diversicolor*). Jarrah grows from north of Perth to Albany, and karri, which is a much taller tree growing up to 90 metres, is found predominantly in the wetter parts of the south-west.

OVER A CENTURY OF FIRES

There was Aboriginal burning prior to European settlement in 1829.[73] The first recorded white death from bushfires was on 28 November 1833 by Baron von Hügel, who reported that between Perth and Fremantle he came upon 'a burnt-out house and a grave nearby under a eucalyptus tree. One night recently a fire driven by a strong wind set the house alight. The only occupant was a girl of sixteen years. The owner was in Fremantle and the unfortunate girl was taken by surprise as she slept and did not manage to escape.' According to von Hügel this was not unusual. He said most farmers' houses 'were built very imprudently close to the forest or as they say here *the bushes* (the thickets), so that I was not surprised to hear endless stories of burnt-out houses'.[74]

While we do not know the details of specific fires in the period prior to 1914, we know there were many, and that most were deliberately lit and either got out of control or were let go. A bushfire ordinance passed in 1847 probably indicates that Aboriginal people were using fire as a form of defence against white invasion. However, 'land owners were exempt and were able to light fires on their own land whenever they pleased. Naturally, many of the settlers' fires escaped into the forest and burnt uncontrolled for many days.'[75] The forests were exploited from the 1840s onwards and 500,000 hectares, about half the total at European settlement, were logged in the nineteenth century. The industry was uncontrolled and 'heads'—timber scrap—were left to build up on the forest floor. The result was massive fires. It was not until the 1885 Bushfires Act that the colonial government attempted to impose authorised times for burning and these were further tightened in a 1902 act. The first twentieth-century fires we know about occurred around Gingin to the north of Perth, and in the Toodyay and Northam districts north-east of the capital in late February 1914. They were apparently quite disastrous.

ix

On 1 January 1901 the colonies federated to form the Commonwealth of Australia. With federation the fire situation became more complex. Bushfires were the responsibility of state governments but, with the exception of Western Australia, outbreaks were not confined by state boundaries. The situation was further complicated by the fact that after the First World War some of the prominent theorists of Australian forestry, Charles Edward Lane-Poole and Stephen Kessell, began to influence the way people thought about forests. Forests were no longer viewed as mere obstacles to broad-scale farming and grazing, but were seen as long-term commercial propositions in themselves. Timber was needed locally and for export. Forests therefore had to be protected from fire for their commercial and even aesthetic values. This was opposed by the grazing lobby and the scene was set for long-term, bitter clashes of interest, all exposed by Stretton in the 1939 royal commission.

In the four decades from 1901 to 1939, deliberately lit fire became endemic in Australia. In Victoria, Tasmania, New South Wales, South Australia, Western Australia and coastal Queensland it was usually bush or forest fire. But in outback New South Wales, South Australia, western Queensland, northern Western Australia and the Northern Territory fast-moving and unpredictable grassfires were the norm. Essentially the problem was that there was little or no legal control, or at most ineffective regulation over what people did. Lighting fires had simply become an ingrained habit and fighting wildfires helped form the essence of rural manhood. As noted earlier, the police and other authorities found it almost impossible to catch people in the act and prosecute. So there were literally thousands of fires, many of them local and quickly forgotten, but some of them destructively massive.

From 1901 to 1910 Victoria remained the pivot of fire in Australia. As Queen Victoria died on 22 January 1901, the whole state was engulfed in widespread, disastrous fires which spread north into the Riverina, reaching as far as Wagga Wagga in New South Wales. These fires were particularly severe in the Western District north of the Otway Ranges around Colac. In an eerie reminder of Black Thursday, the worst fires occurred on Thursday 7 February. *The Colac Herald* of 8 February 1901 reported:

> The morning turned out unbearably hot, and about 10 o'clock a strong wind commenced to blow, accompanied by clouds of dust. The thermometer rose to 103 [39.44°C] in the local post office . . . The dust storm gradually increased in violence

until about half-past four in the afternoon it was blowing with hurricane force . . . Huge volumes of smoke, rising principally in the Otway forest and other parts, clearly indicated that fires were raging which would inevitably bring disaster in their train.

At Birregurra, a small town just east of Colac, a fire was started by a spark from the engine of the morning train from Colac to Geelong. It spread rapidly, threatening Birregurra. By mid-afternoon it was only about 90 metres from the Colonial Bank with the line being held by volunteer firefighters. Eight men were badly burnt, and three of them subsequently died in Colac Hospital. But an even worse tragedy had occurred near the local reservoir. Three young boys, Francis McDonald, eleven, and the brothers John, twelve, and William McCullum, eleven, had attempted to run there on the advice of an adult. They were trapped and burnt beyond recognition.

Much worse was to come in 1904–5, and especially in 1906. Throughout January 1906 fires raged across Victoria. Temperatures were extraordinarily high on 6 January with Mildura reaching 124°F (51.11°C), Swan Hill 120°F (48.89°C), and the whole of the north-west was well over 100 degrees. Animals, fowls and horses left in the sun were dying in large numbers and people were affected by 'heat apoplexy'.

Evidence that fires were deliberately lit comes from scattered reports of 'incendiaries' being brought before the courts, including a drunken young farmer, William Fitzgibbon, who had previously suffered head injuries; he was charged with lighting fires in 27 different places near Strathfieldsaye. The court was lenient, fining him £3 with £1 costs. A large fire around Penshurst 'was believed to be the work of a young incendiary' who had lit the fire with phosphorus used for killing rabbits. By 17 January fires were being reported right across the state, and in one instance the whole congregation at Mass on a Sunday morning left the priest at the altar when they went out to fight a fire successfully at Pettifords Hill near Beechworth. There were also serious outbreaks across the border in the Riverina near Albury, where a fire that burnt along the main railway line was as fast-moving as the mail train that ran along beside it.

At Mount Gambier in South Australia serious fires had been

burning for over a week and 12,500 hectares of grassland were burnt. This period of heat and fire also led to increasing numbers of snake bites, some of them fatal. The need for water drove reptiles closer to human habitation. There was a report of a child bitten in bed by a tiger snake sheltering in the sheets.

On the morning of Tuesday 23 January 1906 it was South Gippsland's turn again. The 1898 fires had destroyed most of the western and central portion of the South Gippsland rainforest, but much of the eastern section around the towns of Foster, Toora and Welshpool north of Wilsons Promontory still survived because it had only been opened up for selection in the early 1900s. The fire began in the heavily timbered, mountain ash country of the Hoddle Ranges above Foster. The wind was blowing a gale and the fire spread around the forested hills forming an arc about 20 kilometres wide around the three towns. Flames were reported up to 35 metres above the tree line. The *Age* reported on 26 January 1906 that, 'The settlers who were in the midst of this inferno say that the effect of the fire and the noise which accompanied this terrible work of destruction were simply indescribable. The heavens above the conflagration were overcast with a sickening yellow hue. Within the fire zone the smoke made the air so black that it was impossible to see more than a few yards.' Panicking, crazed horses and cattle created pandemonium.

The most terrible incident occurred when the fire swept toward the small settlement of Mount Best in the hills above Toora. In a story headlined 'Six Children Roasted', the *Age* of 25 January described what happened:

> the fire swept over the mountain and down the side . . . Morning school had been dismissed a short time previously . . . Eye witnesses state that the fire suddenly rushed from the mountain top in one sheet of flame, and then the whole of the mount was ablaze . . . Three of the [Lonsdale] children were attending the school at Mount Best, but when the alarm of fire sounded, Mrs Lonsdale . . . thought her children would be safe in their own home, and sent a message to the teacher desiring their return.

The children reached home safely, but the fire soon threatened the house.

> The mother, having brought her children to what she deemed to be a place of safety [on a nearby road], took the eldest girl with her to an adjacent creek to obtain a supply of water . . . In the meantime, however, the flames came sweeping across the roadway, threatening the children on all sides. They first started to run one way, only to find a menacing wall of flame barring their progress. Then they started in the other direction only to be finally blocked. The distracted children seeing no escape ran hither and thither like terrified rabbits. The second eldest girl had the baby in her arms and she succumbed first to the intense heat, falling with the body under her. The little child was subsequently found alive and well, having been protected by her sister's body from the fire. The other children, too, fell victims to the intense heat and were found afterwards as they fell, two in one place and two in another. The boy, a lad of about eight years, was still alive when found, but he did not live long after the terrible ordeal. The father was coming rapidly to their aid, and was only a few hundred yards away . . . He came upon the scene half-blinded . . . Nestling between the feet of the children was a pet dog who had survived the terrible heat.

Only two of the eight children were alive. The eldest daughter, seventeen, had taken refuge in the creek, and the baby was sheltered by the body of her sister.

The local schoolteacher, Harry Vale, showed remarkable calmness and courage when he saved the other 28 children from the Mount Best school. After the schoolhouse had been burnt down and they were surrounded by fire, Vale, assisted by two other adults, kept the children covered under wet blankets in the open, moving them as the surrounding fire threatened them. This went on for almost four hours until Lonsdale, the father of the six children already burned to death, arrived and led them all to a place of safety nearby.

Six other people were killed in the area, including two other children and four men, one of whom was an invalid named Williams. He had to be abandoned by his rescuers who 'had to fight desperately to save their own lives'. There were some extraordinary escapes. The local shire president, William Clemson, managed to rescue his wife and three infant children, as well as another woman and two children, from his burning house. They took refuge in a galvanised iron tank which had been blown off its stand. Clemson and an Italian worker, Albino Clavarino, rushed backwards and forwards getting water to keep the blankets wet covering the women and children in the tank. When a nearby outbuilding caught fire Clavarino dashed out several times to extinguish the flames; it was so close to the tank that those inside would have been overwhelmed by the heat. Each time Clavarino managed to put the fire out and returned to the shelter of the tank with his clothes alight. The *Age* of 26 January reported that, 'The heat became so intense that the women became hysterical and had to be forcibly restrained from running straight into the flames. "Clavarino acted like a true Briton"', Clemson said without irony, '"and if it had not been for his heroism and self-sacrifice I don't think we would have been alive today."'

The fires also raced down the thickly forested hills toward Toora on that Tuesday morning. About 11.00 am the town was surrounded by fire. 'The heat was intense, and great suffocating clouds of smoke almost hid the town from view.' The town was eventually saved during the afternoon. The afternoon train from Yarram to Foster and Melbourne was also caught in the fire with passengers suffering from smoke inhalation. At various stages in the next week the townships of Foster and Welshpool were threatened. In the end it was established that 56 families were completely burnt out with the loss of everything, sometimes even including the clothes they were wearing.

But it was not just South Gippsland that was suffering. There were fires all across central Victoria which continued to burn until mid-February. The *Age* assumed that most of these fires were deliberately lit and, as in previous years, there were a few trials of 'incendiarists', but no systematic investigation. Once the tragedies were forgotten, life moved on.

The second decade of the twentieth century saw widespread drought from 1911 to 1916. The fire effects were first felt in southern Tasmania in early 1912. Mount Wellington experienced its worst conflagration for many years. There were also terrible fires on the west coast and on King Island in Bass Strait where, in a minuscule repeat of the destruction of the Gippsland scenario, the dense, cool temperate vegetation of the island was being cleared for dairy farming. In mid-May 1914, an unseasonal period for outbreaks, bad fires broke out in the north-west and central-north near Ulverstone and Devonport. Mount Wellington and the New Norfolk, Huon and Zeehan districts suffered extensive fires in October. In 1915 the fires were mainly in northern Tasmania which led the *Launceston Examiner* (20/2/15) to editorialise that:

> Hardly a summer passes without severe losses from bush fires . . . The principal cause is not so much carelessness as taking a deliberate risk which should not be taken. Settlers engaged in clearing, and anxious to secure a good burn, set fire to the scrub in the most dangerous month of the year, when everything around is dry as tinder. The law provides that burning-off shall not be started until after March, but it is more honoured in the breach than the observance, and we seldom hear of anyone being prosecuted for the offense . . . The chief offender is the settler himself. He wants the fire to sweep his holding but he cannot control it . . . A far more stringent administration of the law is required to check this pernicious fire lighting.

The lax attitude of the Tasmanian Government was revealed by the fact that the law forbidding fires between December and March was first passed in 1854 and remained without review for 80 years.

In Melbourne, 1913 seemed to begin with ideal weather conditions for summer holidays. Apart from a thunderstorm on Christmas Day it was mild and pleasant. But by the middle of January things were unbearably hot. The Melbourne temperature reached 104.5°F (40.27°C), a 'brickfielder' enveloped the city; a powerful hot northerly wind blowing a haze of dust and smoke from the interior.[76] Fires, strong winds and dust storms enveloped central Victoria as far east as Gippsland. Fires raged across central and southern Gippsland and in the forest north of Yarram just to the east of the 1906 fires. Some of these fires were started by young rabbit shooters smoking cigarettes.

The summer of 1914 brought disastrous fires around Albury and in the Wingham district west of Taree. In March 1915 the country west of Lithgow was alight. But these were just the beginning of a major fire season which hit New South Wales in 1915–16. There had been a severe drought throughout the winter and spring of 1915 which reached from the Queensland border to the Blue Mountains. By November 1915 huge forest fires and grassfires extended right along the eastern Great Dividing Range through Dorrigo, Grafton, Bellingen and Macksville, and later in the month to as far north as Murwillumbah. In January 1916 there were extensive grassfires around Holbrook in the Riverina and in the Howlong district on the Murray near Albury. These fires followed good rains in October which led to a prolific grass growth. These southern New South Wales rains contrasted with the north coast drought. In the 1919–20 season there was a significant fire to the south of Canberra reaching right across the Brindabella Ranges from the Goodradigbee River in the west to the Murrumbidgee near Tharwa in the east.

Little is known of fires in Queensland prior to 1916. Between July and September that year there were big fires in the semi-arid woodlands and grass country at Stamford in the Hughenden district, followed by a terrible grassfire started by lightning on New Year's Day 1917 on Warenda Station, about 90 kilometres north-east of Boulia where three men were killed. In mid-September 1918 an outbreak occurred in the central Queensland grasslands that lasted until late December. It covered a vast area between Charleville, Blackall, Barcaldine, Hughenden, and Longreach, and led to the deaths of five men at Saltern Creek north-west of Barcaldine in late October. More than 100,000 sheep were killed. At the same time there were disastrous and extensive bushfires all over southern Queensland.

Perhaps the First World War explains the fact that there was a lessening of bushfire activity from about 1915 to about 1919. Out of a population of less than 5 million, 421,809 men had joined up. Over 10 per cent of the population was in the forces, and of these 331,781 served overseas.[77] Since males were the principal fire-lighters, it means that there was a sharp drop in the number of people to light rural fires. Significantly, outbreaks seemed to increase again as these men were re-absorbed into the population. Throughout this period there was a much greater proportion of the population living in rural areas. In 1911, 43 per cent of all Australians lived in the bush. This had dropped to 31 per cent in 1945 and to 14 per cent in 1976, and the figures have remained relatively static since then.[78]

Another factor that may have influenced the increase in fire frequency in the 1920s was the post-war soldier settlement scheme, which opened more crown land for private use. Fire was used widely in land-clearing with disastrous consequences, especially in inexperienced hands. The area involved was considerable.[79] Over 2 million acres of this soldier settlement land was in fire-prone Victoria. Often ex-soldiers and their families lived in shockingly primitive conditions with no capital on unsuitable land such as in the Mallee. As historian Marilyn Lake has shown, the magnitude of the soldier settlement failure exposed the myth of the self-made yeoman farmer 'reinforcing the "town-ward tendency" of the population'.[80]

x

After the First World War, Western Australia became something of a model for the other states for it had appointed C.E. Lane-Poole as Conservator of Forests in 1916. He had tried to implement a policy of conserving forests and bringing the widespread destruction of forests by both timber interests and graziers under control, something he went some way towards achieving with the Forests Act (1919). But political forces brought about his resignation in 1921, and he was succeeded by Stephen Kessell who tried to curb overcutting and to regenerate forests. He also instituted the first regime of preventative burning in Australia, which was eventually to become the dominant form of fire control until the 1980s. Nevertheless, Western Australia still suffered fires, especially in the long summer of 1918–19. These outbreaks parallel the return of ex-servicemen.

The 1920s saw an increase in fire frequency all over Australia. This followed poor rainfall over the Murray–Darling Basin, western Victoria, south-eastern South Australia and Tasmania. In 1921–2 there were fires across Victoria from the Western District to Gippsland. Much of the central zone of New South Wales had already been cleared by this time, although there were still large areas of remnant forest. These fires would have been deliberately lit in order to clear the forests. They got out of control and spread to the grass and wheat-lands. In the Northern Territory following very heavy rain, 'Fires in 1920–21 and the next few years probably burnt out an area as large as NSW'.[81] The fire season of 1922–3 was even worse. In January–February there were massive and widespread fires across Victoria and in the Riverina. Between December 1922 and May 1923, around 219,000 hectares of forest and agricultural land in Victoria were burnt out in a total of 265 fires. For three summers in a row (1923–5) Western Australia suffered widespread fires, and in October 1923 there was extensive damage done to sugarcane fields at South Ballina, New South Wales, from burning-off operations that got out of control.

Then came the 1926 fire season. It followed very low rainfalls across eastern and central Australia. The fires were the worst since 1851, and rank with 1939 and 2002–3 as the most extreme in the European history of Australia. The trouble began in early January 1926 with widespread outbreaks in New South Wales and the Australian Capital Territory which reached their peak in early February. The fires were to the west and south of the small city of Canberra, reaching right across the Brindabella Ranges. There were also immense fires around Wagga Wagga and Albury, and Rydal to the west of Lithgow, with 85,000 hectares of forest and farming country lost. In mid-February the fire focus shifted to Newcastle. The causes were drought, high temperatures, low humidity and high-velocity north to north-west winds.

At the same time in Victoria 426 fires destroyed at least 395,000 hectares of the state. The fires came in waves and the worst day was Sunday 14 February, 'Black Sunday'. The Victorian disaster began in late January 1926 with fires at Violet Town near Benalla, all around Melbourne, in the Dandenong Ranges, in South Gippsland, and for five days around Healesville. The temperature in Melbourne reached 40°C on 31 January.

In the first two days of February there was a break in the Victorian fires, and New South Wales seemed to be bearing the brunt, especially around Albury. However, the break was short-lived and throughout early February the weather see-sawed. The *Age* of 2 February reported that 'the scorching north wind which was responsible for the worst fires on Sunday had given place to a southerly wind yesterday, but as the day advanced the wind increased in velocity and set the fires travelling in different directions . . . unfortunately many smouldering fires were fanned into new life by the south-easterly wind and sent along new paths that had not been touched while the north wind was blowing'. Fires exploded again in the mountain ash forests, and a fierce conflagration broke out at Cockatoo on the south side of the Dandenong Ranges where houses were destroyed.

Meanwhile the situation in the Upper Murray east of Albury was horrendous with exhausted firefighters, tinder-dry country and 'fires still burning fiercely in the hilly country between the Murray and Tumbarumba and in the hills beyond in the direction of Batlow and Kosciusko', according to the *Age* of 3 February. It seemed that the whole of Victoria was alight. The fires surrounded the south-east of Melbourne and stretched across the state to Ballarat, Warburton, Moe, Bendigo, Bright, Ararat, Bairnsdale, Colac, Whitfield, Garfield, Sale, Wonthaggi, Traralgon, Milawa, Rochester, Tynong, Mirboo North, Kyneton, Euroa, Dromana and Avoca.

The *Age* of 2 February 1926 reported police investigations into the causes of the fires:

> Detective Byrne said yesterday the cause of the main fires around Healesville had been traced to some settlers on the Chum Creek. Fearing the approach of other fires to their properties they had endeavoured to burn a fire break around their settlements, but their fires got out of control in a strong wind . . . A more serious view is taken by police in regard to the big fire which is now raging on Mount Riddell (south-east of Healesville). It was reported yesterday morning that a man had been seen alighting from a motor car in the vicinity of the place where the fire was discovered, and retreating hastily afterwards.

Campers were also blamed. In one case a work-boy was caught lighting an open fire to heat water for his washing! The Victorian Minister for Forests, Horace Richardson, said that 'he had every reason to conclude that most of the fires [in Gippsland] had been deliberately started by holders of grazing licences for Crown lands'.[82] They were after green pick.

By the end of the first week of February the feeling was that the fires 'were under control'. But the break was short-lived. On 14 February, Black Sunday, the fires broke out again with renewed intensity. The temperature in Melbourne reached 40°C. The strong north-westerly wind revived the smouldering fires and the whole of the mountain country around Melbourne and central Victoria was soon ablaze. Reports were coming in from everywhere: the Dandenong Ranges were alight, there were fires right across Gippsland and the Otways. Many towns were threatened, as were outer Melbourne suburbs.

By Tuesday 16 February the full extent of the tragedy was known. The fires had raged across the whole of Victoria, especially in the mountains. The logging township of Noojee was destroyed. The fire entered the town at about 3.00 pm on Sunday afternoon. At first townspeople sheltered at the railway station, but when it burnt, they fled to the Latrobe River where about 50 people took refuge beneath the bridge and sheltered in the water until midnight. Others took refuge in a ploughed potato field where they covered themselves in wet blankets. Five people died in the Noojee area, four of them from the Olsen family who had lived just outside the town for twenty years on a selection cleared out of virgin bush. It was said that the fire which destroyed Noojee spread 'faster than a horse could gallop'. The trestle railway bridge was destroyed.

A vivid description of what happened comes from John Rankin, a survivor from the Fumina district:

> The wind was blowing a hurricane. The trees were being uprooted everywhere in the gale and the air was thick with smoke . . . My father, brother-in-law and I stayed behind to save the house. . . . We had placed a few buckets of water at vantage points . . . Soon, very soon, we saw the fire rushing up from the

Science and technology

Also in this section

86 How siestas help memory

87 The search for ET

87 Nuclear forensics

Tech.view, our online column on personal technology, appears on Economist.com on Fridays. The columns can be viewed at Economist.com/techview

Climate and combustion

Fired up

in areas that might, on the basis of their soils and ambient temperatures, have

With this, she proposes to create a "pyrogeography" of the Earth—a spatial analysis of fire across the planet.

Fires, she explained to the meeting, need three things: fuel, which is related to the productivity of local vegetation; a source of ignition such as a lightning strike or a volcanic eruption; and a lack of moisture to make the fuel combustible, something that is linked to the seasons and the climate. These three factors can be combined to allow a fire "niche" to be mapped across the Earth's land surface. Dr Krawchuk has been looking at how this niche changes in response to shifts in the climate predicted by various models of the future.

Although the different models predict different details, they agree that the fire risk will change for about a quarter of the land surface. Some areas, notably parts of California and Australia, should expect to become more fire-prone. Land to the north of California, by contrast, is likely to see less fire because the amount of rain it receives will increase.

What is likely is that a changing distribution of risk will increase fire's impact on both humanity and wildlife. From the human point of view, some places will become more hostile, others less so. Work by people like Dr Krawchuk may allow these changes to be planned for. The same, mi-

How siestas help memory
Sleepy heads

team wanted to test was that the ability to form new episodic memories deteriorates with accrued wakefulness, and that sleep thus restores the brain's capacity for efficient learning.

They asked a group of 39 people to take part in two learning sessions, one at noon and one at 6pm. On each occasion the participants tried to memorise and recall 100 combinations of pictures and names. After the first session they were assigned randomly to either a control group, which remained awake, or a nap group, which had 100 minutes of monitored sleep.

Those who remained awake throughout the day became worse at learning. Those who napped, by contrast, actually improved their capacity to learn, doing better in the evening than they had at noon. These findings suggest that sleep is clearing the brain's short-term memory and making way for new information.

It is already well known that fact-based memories are stored temporarily in an area called the hippocampus, a structure in the centre of the brain. But they do not stay there long. Instead, they are sent to the prefrontal cortex for longer-term storage. Electroencephalograms, which measure electrical activity in the brain, have shown that this memory-refreshing capacity is related to a specific type of sleep called Stage 2

SAN DIEGO
Researchers say an afternoon nap prepares the brain to learn

MAD dogs and Englishmen, so the song has it, go out in the midday sun. And the business practices of England's lineal descendant, America, will have you in the office from nine in the morning to five in the evening, if not longer. Much of the world, though, prefers to take a siesta. And research presented to the AAAS meeting in San Diego suggests it may be right to do so. It has already been established that those who siesta are less likely to die of heart disease. Now, Matthew Walker and his colleagues at the University of California, Berkeley, have found that they probably have better memory, too. A post-prandial snooze, Dr Walker has discovered, sets the brain up for learning.

The role of sleep in consolidating memories that have already been created has been understood for some time. Dr Walker has been trying to extend this understanding by looking at sleep's role in preparing the brain for the formation of memories in the first place. He was particularly interested in a type of memory called episodic memory, which relates to specific events, places and times. This contrasts with procedural memory, of the skills required to perform some sort of mechanical task, such as driving. The theory he and his motor performance. Then comes 30 minutes of stage 2 sleep, which refreshes the hippocampus. After this, between 60 and 90 minutes into the nap, comes rapid-eye-movement, or REM, sleep, during which dreaming happens. This, research suggests, is the time when the brain makes connections between the new memories that have just been "downloaded" from the hippocampus and those that already exist—thus making new experiences relevant in a wider context.

The benefits to memory of a nap, says Dr Walker, are so great that they can equal an entire night's sleep. He warns, however, that napping must not be done too late in the day or it will interfere with night-time sleep. Moreover, not everyone awakens refreshed from a siesta.

The grogginess that results from an unrefreshing siesta is termed "sleep inertia". This happens when the brain is woken from a deep sleep with its cells still firing at a slow rhythm and its temperature and blood flow decreased. Sara Mednick, from the University of California, San Diego, suggests that non-habitual nappers suffer from this more often than those who siesta regularly. It may be that those who have a tendency to wake up groggy are choosing not to siesta in the first place. Perhaps, though, as in so many things, it is practice that makes perfect. ■

its. Dr Krawchuk has looked at how nature reserves around the world will fare. Like the land in general, she reckons a quarter will have to cope with big adjustments.

Of course the presence of humans themselves also changes the odds of fire, as people are a frequent source of ignition—both accidental and deliberate. The invasion of mankind into places where fires could thrive but do not, often leads to a dramatic increase. Clearing woodland by burning it is a trick almost as old as humanity itself, but the presence of man also increases a forest's susceptibility to fire. The evergreen forests of the Amazon and Congo basins, for example, are generally fire-free because the closed canopy maintains a level of humidity that inhibits burning. Road-building and the clearance of patches of land by foresters and homesteaders breaks up the canopy's continuity, permits drying and encourages the spread of fire over and above the extra risk of ignition that humans, the one species to master fire, bring with them.

What some master, though, enslaves others. The human fascination with fire has a Manichean quality. Last year's Los Angeles wildfire was neither an accident of nature nor an act of human carelessness. It was set deliberately by an arsonist. Factoring that sort of behaviour into the models is the hardest thing of all. ■

SAN DIEGO

This year's meeting of the American Association for the Advancement of Science looked at, among other things, fire, siestas, alien life and nuclear forensics

CALIFORNIA, though regarded by some as one of the more civilised parts of the world, is prey to wildfires. Last August, for example, a fire took hold to the north of Los Angeles. Over the course of almost two months it devoured 65,000 hectares (160,000 acres) of chaparral and forest, destroyed 89 houses and claimed the lives of two firefighters. Over the past few years other parts of the American west have burned in similar fashion. Australia, too, saw serious fires last year. The question on many people's minds, therefore, is whether such fires are becoming more common, and if they are, whether that is a result of climate change.

To examine this question (and many others, in many fields of scientific endeavour), the American Association for the Advancement of Science (AAAS) held its annual meeting this year in San Diego. On February 19th a session was devoted to the environmental role of fire. It brought together palaeontologists and ecologists, as well as climate experts, because fire is a natural and useful part of both past and present ecosystems—a fact that human fire controllers have only reluctantly come to recognise in their zeal to snuff out any blaze as soon as possible.

The role of small, frequent fires in clearing out the undergrowth so that larger, tree-destroying ones will not eventually take hold is now understood. But fire is good—or, at least, necessary—in other ways. Most notably, as Jon Keeley, an ecologist at America's Geological Survey, pointed out to the meeting, the seeds of between 100 and 350 species of Californian wild flower germinate in response to smoke. They may sit dormant for hundreds of years waiting for a fire. When one has passed, they leap into action to blossom in the clearing that the fire has created.

Prometheus's gift

That fact alone—that species have adapted to fire and turned it to their advantage—shows that fire is a longstanding part of the natural environment. Just how longstanding can be seen in the rocks. Deposits of charcoal show that the chaparral-like shrub lands roamed by dinosaurs were periodically disturbed by fires. Fire also had a hand in the success of a recently evolved group of plants, the grasses. The spread of grass, starting roughly 10m years ago, was greatly aided by fires culling trees

gist at the University of Arizona—in other words, he studies the recent past through the medium of tree rings. Some of those rings are in wood that has been preserved as charcoal. Others show evidence of carbonisation in passing fires that the tree containing them has survived. Such "fire scars" in long-lived species like giant sequoias allow him to estimate the past frequency of forest fires. Dr Swetnam's work suggests that variations in temperature were well correlated with the frequency of fire during what climatologists call the medieval warm period, from 900 to 1300. More generally, he says, the incidence of large fires seems linked with warming trends and droughts in parts of Canada and the western United States, as well as Australia, Siberia and the Mediterranean.

Meg Krawchuk of the University of California, Berkeley, though, warned the meeting that such an analysis could be misleading if extrapolated planet-wide and into the present day. First, the evidence is not derived from a random global sample (it tends to come from countries with a well-established scientific tradition). Second, it represents a period when patterns of rainfall could have been very different. In an attempt at a more systematic appraisal of the global fire risk, Dr Krawchuk has been trying to build a model that explicitly includes all the factors that create fires. ▶

creek—a massive fiery flame which started at ground level and ascended to the tall tree tops, and 100 feet above. "We can't save the house, where will we go?" my brother-in-law shrieked. A few chains away from the house was my father's garden, an acre plot... We each grabbed a couple of buckets of water and a few woollen blankets and made off in haste. The fire was all around us by the time we arrived at the plot. The tall dry trees soon burst into flame and showered our garden plot with coals. We crouched down to the ground and tried to keep our blankets damp—and we prayed. It was hot... Soon the trees began to fall as they were burned down, and we had to run several times to dodge them. One tree fell within ten feet of us.[83]

Another family, the McHughs, also took refuge under wet blankets in the open. Tom McHugh and his wife and four children, including a baby, survived for three hours as the fire burnt around them on a clear patch of ground. McHugh was badly burnt on the face, neck and arms. Stranded for seventeen hours, they survived on a loaf of bread and some milk from a neighbour's cow which happened to wander by in the chaos. One woman, Mrs Mitchell, died when she ran back into her burning house to recover belongings. Most of the buildings in the district were destroyed.

That same Black Sunday afternoon 21 people were killed when fires swept down on the mills around Powelltown in the Little Yarra Valley between Yarra Junction and Noojee.[84] The first detailed news of what happened reached Melbourne when 'two scorched and blinded men', Harry King, 21, and Joseph Walker, 22, were admitted to Melbourne Hospital. King told the *Age* of 16 February, 'an epic story of a struggle through blazing scrub... stumbling almost blind and terribly burned... to the nearest habitation'. The *Age* reported that King told his story 'in gasps' and at the end 'opened one badly burned eye a trifle and whispered, "I'm dying for a smoke, dig"'. A moment later he was puffing at a cigarette with true Australian stoicism, while his face broke into the nearest approach to a smile that his injuries would permit.'

His story was horrendous. The fire descended on Worlley's mill, between Gilderoy and Powelltown. The seventeen people there tried to protect the machinery and their possessions, but they were forced by the flames to retreat to the tram track. There they were surrounded by flames. Some of the group retreated back to the mill, but the Rowe family who kept a boarding house at the mill, and Mrs Duncan who had come with her newly born child to work for them, took refuge in a creek with less than half a metre of water in it.

> Mrs Duncan's infant son became separated from the rest of the party, and was cut off by the fire. Mr Rowe forced the two women and his son Clarence down into the bottom of the race, and kept splashing them with water while the flames roared overhead . . . The little boy Duncan was seen enveloped by the fire . . . Mr Rowe had by force to restrain Mrs Duncan of entering certain death by dashing into the flames in an attempt at hopeless rescue. The child was burnt to death before its parent's eyes. (*Age*, 16/2/26)

Those back at the mill were trapped. Fourteen people were burned to death, including four members of the Walker family from Tasmania. One man's body was never found. King and Walker survived because they made a run for it. 'Taking off their coats to shield their faces, the two men made a dash at the wall of flame and smoke . . . They were only wearing sleeveless singlets leaving their bare arms exposed. For five or six chains they stumbled through the inferno of flame and smoke, their arms being terribly burned, and in spite of the protection of their coats, their eye-brows and hair were singed off and their faces badly burned.' Exhausted and practically blind they struggled into a small creek. Here they lay for several hours before they eventually emerged to find the Rowes and Mrs Duncan in the same creek. The two men then returned to discover the mill destroyed and everyone dead. King and Walker, both badly burnt, literally crawled 4 kilometres along the tramline to raise the alarm.

Twelve kilometres away in Warburton many people took refuge in the Yarra River. The Catholic church and nearby houses were set alight by spot fires. At Big Pat's Creek, 6 kilometres east of Warburton, Thomas and Mabel Donald and their three sons aged eight, six and four were trapped at Grant's mill when 'a tornado of flame caught them'. It was a tragic misjudgement, for they had apparently gone back to the mill from a safe area to recover property. Just before her death Mabel Donald had rescued a woman neighbour, and in an act of extraordinary generosity, handed her 'a roll of notes, representing practically the whole of her savings, with the remark that as her husband was working and the husband of the neighbour was seriously ill, she was more in need of the money'.

Like Mabel Donald, the women in Powelltown showed tremendous courage. A schoolteacher, Florrie Hodges, lay across the bodies of her two infant sisters. All three were badly burnt, but she saved their lives. Another report credits her with saving the lives of the schoolchildren. When the Ryan family were driven by the flames into a narrow gully, Mrs Ryan also lay across her children's bodies while her husband did all he could to protect them.

A vast area of mountain ash forest was destroyed from Warburton east to Woods Point and south-east from Yarra Junction across to Moe in north Gippsland. In the Kinglake West area a fierce and intensely hot fire raged on a 40-kilometre front. It literally rained 'balls of fire'. These were huge, blazing pieces of bark which set off more fires wherever they fell. In the Dandenong Ranges the town of Selby was wiped out, and Lilydale was threatened. There were also fires at Ballarat, where Mount Warrenheip was ablaze, in the Western District, at Horsham, Mornington, Bright, and Ararat. In the Upper Murray on the New South Wales–Victoria border, mountain fires continued to rage.

Altogether 60 people were killed in Victoria and there was widespread destruction of farms and homes. The smoke was so bad it crossed the Tasman and reached New Zealand. After the fires an *Age* editorial praised heroic firefighters, and reflected that:

> Our weather comes to us in quite discernible cycles . . . In almost regular rotation comes the period of drought with some weeks of intense concentrated heat. Then, from whatever direct cause, conflagration breaks out . . . Answers will one day have to be found . . . Our forestry system is in a most unsatisfactory state. The inadequacy and laxity of laws relating to large grazing

areas, the free-and-easy methods of burning-off to ensure a better grass crop, the unheeded masses of forest debris—these and numberless cognate matters are important; the need for giving them consideration is urgent. (17/2/26)

Tasmania was also experiencing a terrible summer, especially between February and March 1926. There were widespread fires in the north-west forests, along the west coast and across the farming and forest areas in the central north to the north-east. Simultaneously outbreaks occurred in the Huon and Derwent valleys. In South Australia in mid-February there was a very destructive fire at Mount Pleasant just west of the Mount Lofty Ranges.

xi

Another very severe fire season hit eastern Australia in the summer of 1926–7, but this time the epicentre of the fire-storm was in New South Wales and Queensland. Things were already looking bad in northern New South Wales in October 1926, with severe drought creating tinder-dry conditions. Eventually the whole region from Brisbane to Murwillumbah to Glen Innes was ablaze with areas of rainforest destroyed. There was much damage and many homes and stock were lost. Despite floods in Bourke and rain in coastal areas, on 6 December 1926 the central-west of New South Wales exploded into terrible fires, driven by strong winds and accompanied by dust storms.[85] At Forbes the wind reached almost cyclonic force. At Dubbo the temperature reached 40.56°C and Cowra was on the edge of being evacuated with children ready to be taken to the Lachlan River to shelter in the water. At least six people were killed, hundreds lost everything, and many were injured and affected by smoke blindness. Among those who lost their lives was a man who tried to escape with three mates in a car that would not start. Another was suffocated by smoke. It is estimated that more than 2 million hectares were burnt in these massive fires. Total losses in the Parkes district alone totalled over £1,000,000. The fires lasted right through to around 17 December when there was widespread rain.

At the same time as these fires there were fierce outbreaks in the Newcastle district with many people in that city collapsing from smoke inhalation. Mid-December also saw massive fires in southern New South Wales and the Riverina with much damage to buildings and stock. These were the worst fires ever experienced in the Wagga Wagga area. They intensified up to 12 December. On the south coast a north-west wind of almost cyclonic force drove a wall of flame at 65 km/h towards Pambula. Eden was also threatened. In January 1927 the worst fires ever experienced in the Junee and West Wyalong districts occurred.

Animals suffered terribly. 'In the Forbes district one farmer lost 21 horses. Maddened with heat and pain the animals dashed into the fire and were found afterwards alive but in agony. A kindly neighbour dispatched them with rifle shots. Sheep with scorched bodies and hooves roamed the stock route', reported the *Sydney Morning Herald* on 13 December 1926. Dead animals were being buried all over the countryside. Fires on this scale and intensity did not recur in the state until 1939.

Despite torrential downpours on the central and north Queensland coast, south-eastern Queensland and the area around Brisbane experienced extensive fires in October and November 1926. Fires returned in November 1927 with outbreaks in the Esk–Toogoolawah district, resulting in damage to banana plantations, stock and farms. In late October 1928 there were fierce fires at Cooroy between Nambour and Gympie, and also south of Brisbane in the Samford district.

The 1928–9 fire season in New South Wales started very badly. Driven by gale-force winds, fierce fires exploded in the Sydney area on 8 October as well as along the central coast from the Hawkesbury River to Newcastle; 62 weekend cottages were destroyed. The whole area around Lake Macquarie was on fire. Outbreaks also occurred in the Blue Mountains and at Lithgow. Up to 80 houses were destroyed at Mittagong by a fire on a 22-kilometre front. Fires exploded again in New South Wales in January 1929. There were outbreaks on the south coast and the southern tablelands, as well as around Sydney and the Wollondilly valley, in the Tumut, Adelong and Wagga Wagga districts, as well as at Pambula, Bombala, Albury, Jingellic and Bathurst. In February destructive fires occurred in and around the town of Coonamble where disaster struck and 35 shops, two banks and two hotels were destroyed.

In an interesting sidelight to the terrible fire seasons of 1926–9, there was much discussion about the use of aeroplanes in Canada for firefighting in forest areas. The Canadians used both spotter planes to pinpoint fire grounds and larger aircraft to dump water to retard the fire. However, it was pointed out that the Canadian wilderness, unlike Australia, is rich in large lakes with a plentiful supply of water. In the 1929–30 fire season the Victorian Forests Commission, in co-operation with the Royal Australian Air Force at Laverton, used a two-seater Wapiti bi-plane for fire spotting. Evidence at the 1939 Stretton Royal Commission shows that this idea was developed throughout the 1930s.[86]

xii

With 1929–30 came the Great Depression. Things hit rock bottom economically and socially and many left the cities seeking employment or, more likely, subsistence existence in the bush. Unemployed people, many with families, simply decamped on land and began clearing it and building shacks. As Australia emerged from the Depression, with roads getting better and the number of cars increasing, prosperous city folk began travelling and camping. Everyone wanted to have a summer picnic in the bush and boil the billy. Others took to bushwalking, or to buying blocks, subdividing and building weekenders, especially in attractive forested areas close to the cities, such as the Blue Mountains or Dandenong Ranges. The bush was being invaded, and more people meant more fires. This was also the period of the maximum expansion of state railways. More steam engines meant more sparks and, despite spark-arresters, a greater risk of fire. There was never a summer without a serious outbreak in one or more states.

The fire history of the decade prior to 1939 was even worse than the decades preceding it. These ten years saw a veritable onslaught of fire, almost all deliberately lit. Pyromania seemed to have infected everyone. Perhaps the reason is that reporting of outbreaks improved. Certainly the newspaper record from the 1930s onward is littered with references to bushfires every year from late spring to early autumn. The 1930s were literally a decade of fire.

For example, late January 1930 saw the worst fires in Western Australian history. Rainfall had been well below average from October onwards, and heatwave conditions occurred in late January with temperatures reaching 40 to 45°C with winds of up to 35 km/h. Outbreaks were particularly bad around Perth, Northam, and in the south-west. There was much damage in the Denmark district on the south coast near Albany and around Katanning. Twelve months later in January 1931 fierce fires broke out around Margaret River township, and the worst fires ever in the districts around Mount Barker and Denmark in the south.

There was good rain late in 1931 in the western zone of New South Wales. It led to abundant grass growth, which in turn dried out and large fires broke out around Menindee and Ivanhoe. They were a foretaste of what was to come. By mid-December huge grassfires were raging on an 80-kilometre front between Tilpa and Cobar in the north-western outback. Given the scattered nature of the population it was very hard to control these fires. In December that year there was a massive grassfire at Oatlands between Berrigan and Culcairn in the Riverina.

It was Victoria's turn again in the summer of 1931–2. Following a dry spring and summer, smouldering fires which had begun in December 1931 blew up on 11 January and spread right across the state, including uncontrolled outbreaks in the Grampians, around Melbourne, in Gippsland, and at Cowes on Phillip Island where the mutton-bird and penguin rookeries were swept by fire. The *Age* of 5 February was optimistic because the birds were in their burrows and they 'probably escaped'.

By late January the fires had spread and worse was to come. On Friday 4 February 1932, two days before the 81st anniversary of Black Thursday, appalling fires swept across the mountains and timber settlements from Warburton through to North Gippsland. The small town of Gilderoy between Yarra Junction and Powelltown was completely wiped out and nine people were killed. There were big losses to the timber industry. After a long fight to save the town, Noojee just survived because of a lucky change of wind direction. At Erica in North Gippsland six people lost their lives and three men were badly burnt. Two firefighters, George Cook, 62, and his son Ronald, fourteen, were killed near Warragul. The

Age of 6 February reported that 'the unfortunate man and his boy went to the assistance of a neighbour whose homestead was threatened and during a change in the wind the flames swept back on them and they were enveloped in smoke and fire. The older man was terribly injured and his condition was looked on as hopeless, and though the boy was not so seriously burned they both succumbed to their injuries and shock within a short time of each other.' Another hero was James Vague, 26, a bachelor schoolteacher, who rode through the blazing forest to warn mill workers in the Walhalla district about the approaching fire, only to be trapped at a mill himself and burned to death with five others. All up, twenty people died in 307 fires and 207,000 hectares of forest were destroyed. Eventually heavy rain put out most of the fires.

In October 1933 fires again broke out in North Gippsland and the mountains between Warburton and Healesville. The summer got worse in January when a terrible fire swept through the Otway Ranges from the Gellibrand River through Beech Forest to Apollo Bay on the coast. There were severe fires around Hepburn Springs, and at Ferny Creek, near Sassafras in the Dandenong Ranges. There was another bad year in 1936. Scattered fires had broken out in Victoria throughout January and February, but in March widespread and intense outbreaks occurred around Portland, the Dandenong Ranges and in Gippsland. The tiny town of Gunyah, north of Foster in remnant forest in the Strzelecki Ranges, was destroyed. There were also extensive grassfires across Victoria, with many stock killed and houses burnt. After this Victoria remained relatively fire-free, until 1939.

On the other side of the continent things were bad in Western Australia from the 1931–2 to the 1935–6 fire seasons inclusive, with outbreaks in many parts of the state. The worst season was in January 1933 with fires north of Perth and in the central south-west of the state. In early February serious fires broke out due east of Perth at Narembeen and Kellerberrin. By mid-February the fires in the south-west had become particularly intense, especially around Bridgetown, Frankland, Busselton, Pinjarra, Boyup Brook, Bunbury, Collie, and as far east as Lake Grace. In mid-February there were also fires on a 40-kilometre front at Westonia (near Merredin), north-west of Perth. In October 1934 there was a fire 600 kilometres north of Perth and south of Carnarvon and just east of the North West Coastal Highway, where 40,000 hectares of bush were destroyed. The period January to March 1935 was very bad. Fires began in mid to late January in heavy forest country in the Darling Ranges north-east of Bunbury. In late February disastrous outbreaks flared up in the south-west around Kojonup, Gnowangerup and Broomehill. March saw these fires continue in the south-west around Beverley (near Marradong), Bridgetown, Pinjarra, Kojonup and Brunswick Junction (near Bunbury).

The worst fires of the decade in South Australia occurred in 1933. In early February serious fires broke out around Mount Gambier in the south-east and at Port Lincoln on the Eyre Peninsula. Between 8 and 13 March there were fierce fires at Blackwood and Belair, now south-eastern suburbs of Adelaide, and at Mylor. Tasmanian fire seasons were mild until January–February 1934 when the state experienced extreme bushfire conditions in the Derwent Valley, the Huon and Channel districts, and Mount Wellington, as well as in the north-west and north. There was a dense smoke haze in Hobart with ashes and debris falling. Visibility dropped and people had sore eyes and difficulty breathing. The uncanny darkness created panic among some in the city. Smoke was carried across Bass Strait to Victoria.

Between 1933 and 1939 there were only two really severe fires in New South Wales. In November 1935, about 16,000 hectares of the Pilliga Scrub east of Baradine and north of Coonabarabran were burnt, and in October–November 1936 fires broke out throughout the Blue Mountains. On 6 November Glenbrook experienced its worst fire ever with many houses destroyed. In December Coffs Harbour had serious bushfires throughout the district. There were fires also around Kempsey, Taree, Narrabri and Glen Innes with 56,000 hectares burnt out in 469 forest fires.

Queensland had a break from fires from 1927 until January–February 1932 when fierce winds drove fires across the southern part of the state from Wallangarra to Inglewood and Goondiwindi, destroying grasslands, pastures and crops. In November 1935 many houses and thousands of hectares were burnt around the Inglewood–Pelican area, as well as the Maryborough, Tiaro and Gympie districts. At Augathella, north of Charleville, there was a massive fire on a 48-kilometre front. Driven by fierce westerly winds in late October 1936 fires swept over south-eastern Queensland from Gympie to the New South Wales border. By 30 October much of the east coast from Gympie to Coffs Harbour, a distance of 480 kilometres, was on fire. North Queensland was burnt again in late October 1937 with fierce fires destroying large areas of rainforest near Cairns. On 2 November a massive fire 80 kilometres wide and 160 kilometres long broke out around Charters Towers and from October to November there were widespread grassfires to the north of Hughenden.

The worst bushfire season in the European history of Australia occurred in 1938–9, affecting Victoria most, but with bad fires in New South Wales, the Australian Capital Territory and South Australia. The season began early. In March–April 1938 there was a fire in state forest at Killen-Butta near Molong, New South Wales. By late September Victoria's Mount Macedon was burnt, and by early October there were serious fires around Gisborne, Erica, Kalorama (in the Dandenong Ranges), Beaufort and Walwa, where thousands of hectares of forest were lost. Fire also swept through the Noojee area where the trestle railway bridge that had been rebuilt after the fires of 1926 was damaged again. By mid-October there was a terrible fire in the area between Ballarat, Creswick and Daylesford. The fire around Gisborne intensified. Another fire had broken out in the Plenty Ranges near Kinglake. On 19 December a large fire in the area of Mount Buffalo destroyed the chalet. Late in the month there were fires around Bairnsdale, Maffra, and the fire in the Daylesford area was still burning at nearby Eganstown.

There were also outbreaks around Mount Gambier in South Australia. In New South Wales a fire broke out around Dubbo in early December and burnt for three weeks. In the Sydney suburb of Lugarno, thirteen houses were destroyed. Queensland was not exempt. In mid to late December there was a big bushfire at Landsborough near Caloundra, a severe fire at Kingaroy on Christmas Day and a Boxing Day fire at Goondiwindi.

xiii

Black Friday was not a one-off aberration. It was the culmination not only of the fire history of the 1930s, but also the result of a form of communal madness, the white Australian addiction to fire-lighting. This is what the Stretton Royal Commission inherited and confronted. The vast majority of witnesses, both expert and otherwise, reflected the dominant view that Europeans in Australia could only control natural wildfire with controlled fire. The attitude that had driven people since settlement and which dominated the evidence was that fire was the prime instrument for clearing the continent of its natural biota and replacing it with European-style agriculture. Precious little was said that contradicted this approach. The irony was that fire demanded more fire because burning encouraged the very scrub that became fuel for the next fire. An endless round of burning led to the proliferation of the very fuel which would feed the next destructive wildfire. It was a form of collective madness—literally a pyromania—that very few in the Australian community, especially the rural community, could perceive.

There is a real sense in which Australians were fascinated by fire and addicted to it, especially as a tool for 'clearing' and providing green pick. But it was more than that. It was both a spectacle and a form of moral combat in which Australians could be tried, tested and proved. To conquer the fire, or even to be consumed when fighting it, somehow symbolised a form of human superiority and dominance over nature. There is a kind of moral, even metaphysical aspect to it all in which an evolving national identity was forged. It was an identity from which particularly rural Australians would ultimately have to retreat.

3

'BURN, BURN, BURN', 1939

i

Reading the minutes of evidence of the Stretton Royal Commission into the 1939 fires is not only a herculean task; it is mind-numbingly boring. Yet Stretton had to sit through the lot, keeping awake and alert on hot, oppressive afternoons, following a heavy lunch, often in uncomfortable circumstances, such as in the tent at Woods Point. He was not helped by Counsel-Assisting, Gregory Gowans, KC, whose approach was often confused and illogical. The 2608 pages of evidence is spread over 34 days of testimony and 200 witnesses who, with a few notable exceptions, are overwhelmingly repetitive, prosaic and pedestrian. They usually reflect a crassly utilitarian mind-set: the land existed to be exploited, the forests to be cleared or logged, the native animals to be eliminated as 'pests' and 'vermin', words regularly used in 1939. The trained foresters also reflected the exploitative ethos. Few appreciated Australian flora and fauna.

Forests, of course, have always been viewed ambivalently by humans.

Until the modern era in Europe they had been seen as dangerous, mysterious, dark places, the domicile of outlaws and strange, menacing beasts, like wolves, as well as the spirits of the dead.[1] While there were things that could be collected from them and sometimes domestic animals could graze in them, you could not grow anything in a forest. If you were a farmer you either gave them a wide berth, or you cleared them. Australian forests similarly were often unusable, so landholders either cleared them for farming or, as the timber industry developed, used them as a source of raw material. The Aborigines also kept away from dense forests. This was probably less to do with alienation than that there was no reason to utilise them when they had ample food supplies on the coasts and plains.

In the late nineteenth century in Australia, forests were only valued when there was prized timber, like mountain ash, that could be logged and sold. Trees that had no immediate economic value were 'rubbish', another word that is used regularly in the royal commission transcripts to describe non-millable timber. Frederick Leake, a farmer from Apollo Bay who lived in the Otway Ranges for 50 years, was an exception. He was asked by a rather bemused Gowens: 'What about the forest itself? Would you allow it to stand as it is and let it go on in its natural state?' To which Leake replied, 'Most decidedly I would leave it untouched. I notice that in various places the Forests Commission has gone through and knocked down *all the beauty* of the forest' [my emphasis].[2] The few witnesses like Leake who dissented from the prevailing view stand out like beacons in the dark.

The other trend that emerges clearly from the evidence is the overwhelming need to blame someone for the fires, to demonise those who are supposedly responsible. This tendency emerges after all major bushfires; the preferred scapegoat is usually an official institution or a group of people rather than an individual. But it seems that human agency in some form—even if it is only the accusation of bad management—is usually cited. What is involved is a process of shifting blame, usually from the real culprits. In 1939 the Forests Commission's policy of not 'cleaning up' the bush by regular burning was constantly blamed.

Also criticised was the appointment of young, inexperienced, city-born forest officers who might have a different approach to what forest inspector Robert Code called the 'burn, burn, burn' philosophy.³ Clarence Patterson, for instance, boarding-house keeper from the Ruoak mill in the Rubicon forest, told Stretton: 'There are some of the forestry officers who have no idea what a fire will do or what it will not do . . . Some of them are not experienced . . . Unless a man has twenty years' actual experience in the bush, I do not think he can be termed "experienced".'⁴ Peter O'Mara, timber contractor from Noojee, said that, 'Many of the officers are good men, but there are many of the others who are only city men'. They need to be 'practical bushmen, and [to have] been for many years'. In response Stretton jokingly asked him, 'You would be in favour of taking the Forests Commission out of the forest and putting in a tribe of blackfellows to look after it?' As O'Mara began to answer Stretton interrupted and said, 'I wish to accentuate that I spoke in jest'.⁵

By 1939, all of the foresters had completed the three-year diploma course at the Forests Commission school at Creswick. Yet, for many bush people, it counted for nothing. In fact, it was held against the officers by Patterson who said, 'They have a lot of theory'.⁶ In defence of his own men, Forests Commission Chairman, Alfred Vernon Galbraith, told Stretton that the forestry school at Creswick was set up in 1910 precisely to train 'good bushmen'. They were hardly 'boys' as their average age was 39, and they were called upon to supervise on average 121,000 acres [48,900 hectares] of reserved and protected forest. Their equivalents in New Zealand looked after 7900 hectares, and those in Prussia supervised 538 hectares.⁷

According to many witnesses, the problem was that the Forests Commission no longer allowed them to burn when it suited them. They claimed that fires emerged from national parks and forests, destroying private property. This attitude was summed up by William Irvine of Yarra Glen, a farmer, road contractor and councillor of the Shire of Eltham. 'Feeling is very strong against present-day forestry methods by people who have lost their all by fires coming down on them from national parks and forest reserves . . . [Undergrowth] should not be allowed to accumulate for ten years, as has been done in the past in both Forests Commission and MMBW country. The forestry men realise that it is so dense you cannot poke a stick into it.'⁸

Analysing why fires were lit, First Constable Ralph Brown of Mirboo North, the only impressive uniformed policeman to give evidence, told Stretton he had spent most of his life in the bush and said that people lit fires to get new grass, to protect their property, and because campers and people travelling in the backblocks wanted to boil the billy for tea. And there were those who were just driven by pyromania: 'A good many people light fires for no apparent reason, possibly only to get some sort of satisfaction in seeing a good blaze'. He commented that in his own area it was hard to get people to volunteer to fight fires. 'They do not take notice until it has reached a danger point . . . They are very individual, and I have noticed that amongst the farming people a great deal.'[9]

Nevertheless, the Forests Commission was still the main scapegoat for the Black Friday fires. While a few witnesses spoke up in its defence, it was ultimately left to the chairman, Galbraith, to try to exonerate his organisation. With a diploma from the Creswick school and a diploma in commerce, Galbraith, a rather arrogant, authoritarian man, had been appointed secretary of the commission in 1920, commissioner in 1925, and chairman from 1927.[10] He attempted to drown Stretton in overwhelming detail, but the evidence shows that Galbraith was in a no-win situation, and that even Stretton himself fell into the trap of looking for a scapegoat and focusing the blame for the fires on the commission. In an extraordinary attack in his final report Stretton said that Galbraith 'found himself in the embarrassing position of being the truthful sponsor of what he thought was a bad case'. Allowing he was 'a man of moral integrity', Stretton went on to comment that, 'If he were freed from the preoccupations attendant upon a life of enforced mendacity on behalf of his Department' he would be of greater value to the state and be able to give his attention to 'what should be the first consideration of every forester, the problems of fire prevention and suppression'.[11] This is a completely unfair assessment, especially given it was not within the commission's legal remit to control fires. While Stretton showed tolerance to and protected rural witnesses who were responsible for fires that had led to many deaths, he seemed almost to enjoy putting down the mighty from their thrones, while forgetting that, as a judge, he himself was one of the powerful.

The simple fact was that legally the Forests Commission was not the fire warden of Victoria, even though *de facto* most people thought it was. The real culprits in the Black Friday fires, as in almost all previous conflagrations, were the fire-lighting landholders, timber millers, farmers and graziers. Stretton acknowledges this. Analysing the causes of the fires, he conceded that these groups were the worst fire-lighters.[12] The other problem was the totally inadequate, disorganised, and unsupported firefighting response from local areas. As First Constable Brown pointed out, people were 'very individual' and so parochial that they were not interested in outbreaks that were not immediately close to their own properties, that is within a kilometre or so. It was only when their own immediate interests were threatened that they responded.

In his evidence Galbraith quotes his predecessor, Owen Jones, writing in the 1926 *Empire Forestry Journal*:

> The history of forestry in Victoria . . . has been one of vicissitude, of bitter, protracted and often unsuccessful struggles against ignorance, prejudice, greed and self-seeking. Our permanent and racial antipathy to forest progress . . . has in Victoria been intensified by the conditions prevalent in a new country, where much forest land must of necessity be cleared to permit of settlement and where in consequence the forest too frequently comes to be regarded as a hindrance and a foe. These hostile ideas, ingrained in the older people, have been handed down from generation to generation.[13]

This sums up the pre-1939 Australian attitude to forests and fire. Galbraith's experience confirmed this. 'There is not the slightest doubt', he told the royal commission, 'that any attempt at control of burning by landholders was extremely difficult and stoutly resisted. This attitude has been induced by a long period during which unrestricted or "free" burning for the clearing of debris was permitted.'[14] He was particularly critical of graziers, but pointed out that any attempt to prosecute those who broke the law was hindered by 'meagre and inadequate legislation'.[15] He proposed that the Forests Commission be legally made the fire warden of the state and that it have strong coercive powers to enforce the law.[16]

One of the other major issues with which Stretton was concerned was the construction of dugouts. After the 1932 fires the Forests Commission had made this a condition of the renewal or granting of logging licences. But not all forest officers had enforced this. Like most mill owners, some were sceptical about the effectiveness of dugouts. Investigating the Fitzpatrick mill tragedy, Stretton discovered that the inspector of the central division of the Forests Commission, Finton Gerraty, was particularly evasive on whether he had enforced the policy on dugouts. The judge asked Galbraith directly why the commission did not enforce the construction of dugouts and virtually accused Gerraty of either lying or not enforcing official policy. On this issue Stretton was like a dog with a bone. 'I don't think we have got to the bottom of it yet . . . I should like a clear statement why, having made it [the need to have dugouts] a part of the policy, you [Galbraith] did not see that it was enforced.'[17]

What emerged was that the Forests Commissioners felt dugouts were safe and wanted them, but at the local level some field officers, like Gerraty, were doubtful about their effectiveness and were very slow in enforcement. Stretton commented: 'If you [Galbraith] had insisted . . . all the Fitzpatrick people would have been saved'. Galbraith replied that, 'Some people would not stay in the dugouts and rushed outside'. Under continuing pressure from Stretton about the Fitzpatrick tragedy, he eventually conceded the area 'seemed to be secure. The Forest Officer considered they were safer than the people were elsewhere. The policy of the Commission was that if they were satisfied that other means of escape existed, we should not press for the dugout principle. The other means of escape included a running stream or a road close handy where people could be placed in safety.'[18] Stretton expressed scepticism about this, and Galbraith eventually admitted that some officers 'have been far too friendly with the millers'.[19] There was some amusement when Gowans read an opinion from the Crown Solicitor who, in an attempt to protect the Forests Commission from legal responsibility, said that perhaps dugouts should have a notice over the door: 'He who enters here does so at his own risk'.[20]

Galbraith outlined the problems encountered by the Forests Commission as a result of lack of funds, staff shortages, a dearth of clear legislation, and divided control over bushfires. He claimed that the only solution was to place everything to do with fire and forests under the control of his commission. Representing the Melbourne and Metropolitan Board of Works (MMBW), Alexander Kelso protested that the board was unwilling to surrender authority over water catchment areas. (The board, founded in 1891, drew together a multiplicity of local water authorities. It was a world pioneer in water conservation and the storage of water in pristine conditions.) Saying that Black Friday had changed everything and calling Kelso a 'pharisee', Galbraith declared, 'The desire of the Forests Commission is to act in the public interest, and the system of divided control is wrong'.[21] One can sympathise with him; he was trying to bring order into administrative chaos. In addition, giving evidence must have been a humiliating experience for a proud man like Galbraith.

ii

One group who came in for a lot of Stretton's attention were the timber millers. Most of the deaths in 1939, as in previous twentieth-century fires, occurred at milling sites in the bush. The evidence is that some millers were genuinely concerned with safety: Feiglin's no. 2 mill was a good example, where an adequate dugout saved lives.[22] However, other mills were deathtraps, such as Fitzpatrick's at Matlock. If the management had not been killed, they surely would have been charged with manslaughter. There was long discussion about what was done with 'heads'—the hacked-off logs that were abandoned in the bush prior to milling. Very few millers had a policy of burning heads in non-fire danger periods.

Other mills had sawdust heaps that burnt continuously. The Rouch mill in Healesville was situated less than 3 kilometres from the centre of town. A neighbour, Jonathan Marriott, told the commission that it had a sawdust heap 'as big as a house. . . [burning] for weeks without cessation'.[23] Nine days after Black Friday the fire in the heap spread into the surrounding bush on the afternoon of 22 January. Marriott complained to both mill workers and the local police without result. A very evasive Senior Constable James Slatter conceded that the fire was a threat to Healesville, but said he could do nothing about prosecution without permission from his superior officer, a Superintendent Green, in the far-away Melbourne suburb of Malvern! Since the mill was run by a company, Slatter claimed he was also not sure who to charge. The mill owner, Edward Rouch of Heidelberg, tried to minimise the danger of the burning sawdust and pleaded ignorance when told that he could be prosecuted.[24] Rouch's manager, Thomas Grant, pressed by Stretton, admitted that the burning sawdust heap involved 'a certain amount of danger'. He then added, 'You get so familiar with it you just do not realise it'. Stretton replied, 'That is the whole trouble with the outlook on bushfires. Familiarity breeds contempt'.[25]

iii

Some exceptional and insightful people appeared before the commission. The most influential was Alexander Edward Kelso, the engineer in charge of water supply for the MMBW.[26] A short, self-contained, precisely dressed, intelligent, kind-looking man without a hair out of place, Kelso was in the odd position of being both an advocate for the board and a witness. Born in June 1894, he was educated at Melbourne University and after war service, joined the MMBW in 1923.[27] A Master of Civil Engineering, but untrained legally, he showed forensic skill and logical toughness in dealing with complex issues of law and fact with evasive, hostile witnesses. Stretton recognised this when he said that Kelso had 'done so well it makes one wonder if legal training is of as much value as lawyers think it is'.[28] No one at the royal commission had a better opportunity to put his case. As well as questioning witnesses, he had two protracted periods in the witness box, and gave a final summation.[29] Kelso, however, was not an 'environmentalist' in the modern sense. Essentially he was an engineer who developed compaction techniques for dam building.[30]

The case that Kelso argued before Stretton was based on the conviction of the MMBW from its foundation in July 1891 that Melbourne water catchments must be kept in their natural state to preserve the pristine purity of the water and to protect them from biological contamination and water-borne diseases. There was sound reasoning behind this attempt by the board to 'lock up' the mountain ash areas close to Melbourne. They wanted the original forest preserved to prevent erosion and the silting-up of dams. They also wanted the surface litter and soil protected so that water could percolate through the ground and be purified. Another concern was the loss of old, 'overmature' trees which only took up a minimum of water. This is one of the key reasons for preserving old-growth forests in their natural state. Young eucalypts, especially growing mountain ash, soak up an enormous amount of moisture from the soil.

The MMBW gradually persuaded government to hand over control of the whole of the city's catchment, although this was continually challenged.[31] Kelso told Stretton: 'The Board recognised that the only satisfactory means of ensuring the permanency and sufficiency of water through these areas is to absolutely conserve the forests . . . because the Board knows from experience that even a small fire does destroy the mountain ash sooner or later'. The MMBW was consistently opposed by timber interests and the pro-development *Age*. Kelso acknowledged that, 'There is a clash between those who use forests for commercial purposes and those who control forests for water supply, because you cannot use the forests for commercial purposes and still have them . . . It is the policy of the Board that for water purposes the forest must not be used for industry because of the risk of fire.'[32]

At the core of Kelso's case was the statement 'you cannot use forests for commercial purposes and still have them'. What the MMBW wanted was a forest 'in its natural condition, as it was before' European settlement.[33] The MMBW argued that repeated fire eliminated forest litter which was made up of 'fallen and dead leaves, the sticks and small matter lying on the ground to a depth of a few inches. Under the litter there is the humus, passing gradually into the soil.'[34] It is through this material that water percolates. Fire destroys this top layer and the humus in the soil, and leads to the growth of bracken that is highly inflammable.

Kelso told Stretton that after several fires, 'There is a gradual increase of what you can rightly call fire bush and types of scrub that are generated, thickened and helped by continual burning . . . It is our experience that if fire is excluded that kind of bush will gradually revert to the cleaner type of bush.'[35] He argued that, 'In the natural forest there is not much scrub. I can show you forests . . . in which you can ride uninterruptedly without trouble . . . it does not carry dogwood scrub and this other kind of bush which normally follows after a fire.'[36] In these natural forests, 'scrub does not generally flourish'; they are far less fire-prone than forests that have been frequently burned. Kelso stressed that he was not suggesting that forests in their natural state are not inflammable; every forest is. But he argued that they usually only burn in terrible fire years like 1939. A sceptical Gowans asked him if this happens 'merely by the action of nature?' to which Kelso answered 'Yes'.[37]

Given the constantly over-burned and scrub-ridden state of most Australian forests in 1939, Stretton pressed him to estimate how long it would take to restore a forest to its original pre-European state. Kelso responded: 'You are asking me something I am not able to answer . . . I cannot specify and say 5 years or 50 years, but I do feel confident that . . . under this exclusion policy [it] will gradually become natural again.'[38] Ever the pragmatist and concerned with jobs and the economy, Stretton was not convinced by this notion of a natural forest ecology. Stretton only embraced environmental concerns some years later. His 1939 report promoted a more social justice-oriented view, which is that nature exists ultimately for humankind, especially for the working man, and that human needs could not be suspended until nature reached a new balance.

Kelso understood that the whole forest system is interactive, with climate, soil and vegetation intimately linked. Logging, fires, and human intervention interfered with this balance. While Australian forests have historically always been subject to intermittent fire, he contended that the mountain ash forests have been 'nearly immune. . . . The whole forest in its natural condition is remarkably immune from fire . . . fire does not readily penetrate.'[39] Natural fires occur in cycles 'possibly of a frequency of 20, 30 or 36 years. From the persistence of [mountain ash] . . . it is fairly clear that bad fires or frequent fires were certainly not part of [the] environment of the virgin forest.'[40] He pointed out that in the 1851 fires pristine areas like the South Gippsland rainforest were not penetrated by the conflagration. It was largely where the forests had been broken up that the fires were worst. The reason was because forests in their natural state retain their moisture and greenness. He argued that just because a forest had been burnt it does not mean that it is 'less inflammable than a natural forest'.[41] He also pointed out that forest litter does not accumulate endlessly, but the debris is broken down and in a natural forest stasis is eventually achieved and maintained. The growth of scrub is controlled. 'The continuity of the canopy, the shelter provided by the grown timber, and the exclusion of light by big trees do control the growth of scrub. I have no doubt that this is the main influence in keeping down the under storage of scrub in virgin forests.'[42] So if the forest is allowed to recover its natural state and fire is excluded, there is a distinct possibility of its returning to a natural stasis. This may well be an ideal to strive for today.

As human settlement increases so does the frequency of fires. Kelso was worried that Victorian forests were increasingly being penetrated by loggers and settlers. Many of them were people down on their luck after the Depression. The problem was that these people worked on the assumption that the easiest way to clear is 'to put a match into the bush . . . They feel that fire is not particularly harmful to the forest.'[43] What settlement and human penetration did was to break open the forest, interfering with its natural rhythms and making it vulnerable to wildfire.

Central to Kelso's argument was the attempt to exclude fire completely, especially in the mountain ash forests. If fires do start they must be tackled immediately. 'I suggest that if all fires were put out . . . when and where they occur, the type of fire that we cannot put out would not arise.'[44] He opposed the use of controlled fire. He argued that 'light burns' were very destructive. In order to protect both forests and settlements Kelso argued for a kind of buffer zone between the two. Its width would vary according to local conditions. He also advocated that in bad fire years the logging industry should be closed down completely, especially 'when it becomes obvious that we have reached a critical period'.[45] The US state of Oregon had already imposed this kind of closure in bad fire years on its timber industry. Kelso advocated the need to develop an experienced firefighting force trained for forest conditions. The logging industry should be charged for this service. He also wanted to stop a plethora of small mills developing in the bush; if necessary they should be closed down and consolidated.

Kelso was not alone in espousing the possibility of restoring 'natural' forests. He was supported by Charles Edward Lane-Poole (1885–1970).[46]

English-born, educated at the Ecole forestière at Nancy in France, he worked in South Africa and was later appointed Conservator of Forests in Sierra Leone. He came to Western Australia in 1916 in the same role. A tall, tough, acerbic man and an experienced horseman, who did not suffer fools gladly, he was an influential believer in protecting forests for their commercial value, and he directly opposed the kind of mania for clearing, land development and farming that infested most state governments. In 1918 he drafted a Forest Act which legislated to bring the timber industry under control and provided for a permanent forest estate. This led to a counter-attack from established timber interests who were unwilling to surrender their power, concessions and nominal fees to an Englishman who wanted to concentrate power in himself and his own department. Eventually the West Australian Government caved in to timber interests and Lane-Poole resigned in October 1921.

While also wanting to protect forests for their commercial and conservation values, his successor as Conservator in Western Australia, the Wollongong-born Stephen Kessell (1897–1979), became a strong promoter of the use of controlled burning to suppress fuels and 'clean up' the forest floor.[47] Kessell had enormous influence on forestry practice and set up the Institute of Foresters of Australia in 1936. Lane-Poole and Kessell set out the parameters of the two different approaches to fire prevention and suppression in Australia. A central question became—and still is—should regular, wide-scale controlled fire be used to prevent wildfire?

After Lane-Poole left Western Australia, the Commonwealth Government asked him to report on the forests of Papua New Guinea and his work first made Australian administrators aware of the enormous forests of PNG.[48] He was appointed Inspector-General of Forests for the Commonwealth and head of the Australian Forestry School, reorganising forestry policy and education in Australia. Through his attendances at Empire Forestry conferences Lane-Poole also placed Australia on the forestry map of the English-speaking world.

Like Kelso, he advocated the elimination of all fires in and around forests. If they did break out, he wanted immediate action to control them. He was opposed to the fire-lighting of the settlers, and to the scientific attempts of Kessell to use regular controlled burning. However, he was not a proto-environmentalist. His evidence shows he belonged in the utilitarian tradition of scientific forestry: what is the best way to preserve the forest in order to use it? He maintained that, 'the fires before . . . [European] settlement were not of the intensity nor did they occur at the frequency as they have done since the white man settled the country'.[49] Asked if it was desirable and feasible to 'clean up' the forests, Lane-Poole replied:

> That is the great question of controlled burning . . . The usual opinion is that an enormous accumulation of inflammable material occurs if the forester excludes fire. That is incorrect. In [an] area . . . [in the Brindabella Ranges near Canberra] from which fires have been excluded since 1926, the natural succession of plants was just reaching the stage when the wattles were dying out, and the ground floor of larger shrubs was becoming clearer. We were beginning to get back to those conditions which Mr Kelso has described so well which all the old people in Australia remember, that is, our forest was becoming more open through the exclusion of fire . . . The thickening up of our forests is entirely due to fire and the exclusion of fire will render them less susceptible to fire because it will get rid of an enormous amount of inflammable material.[50]

How long would it take to get the forests back to natural stability? Lane-Poole replied it depended on the species of eucalypt and how often they have been burned. Stretton asked again: but how long before forests will be safe? Lane-Poole replied: 'The answer is we have not done it yet in Australia. Fires have always come in before we have been able to reach that position where the wattles have disappeared.'[51] Lane-Poole contended that controlled burning should only be resorted to in extreme cases. His central concern was the danger of erosion, especially in steep areas. He was also adamant that there should be only one fire authority and that it should be run by scientifically trained foresters, not local councils 'because the local authority has not the vision to see anything more than the immediate grazing revenue . . . none of them look further than that'.[52]

There were only a couple of witnesses who were 'proto-environmentalists'. One of them, Sir James William Barrett (1862–1945), was without doubt the most broadly educated and influential man to appear before Stretton.[53] Trained as an ophthalmic surgeon in Melbourne and London, he was chancellor of the University of Melbourne. He was a founder and administrator with his doctor sister of the Bush Nursing Association, a student of German language and culture, a fine musician and one of the founders of the Melbourne Symphony Orchestra, and a member and chair of many boards and committees, among which was the National Parks section of the Town Planning Association. Not just a theoretician, Barrett was a person of enormous practical competence, a real 'Renaissance man'. But he was also a tall, shambling man who took himself far too seriously. He was disliked and ridiculed around Melbourne, as much from jealousy of his ability and influence, as for his pomposity. Reading the evidence, one has the feeling that Stretton shared the antipathy of most of Melbourne's nabobs to him. Perhaps his social justice bias prevented Stretton engaging with the surgeon's arguments.

Barrett was totally opposed to regular burning. He had travelled all over Victoria on Bush Nursing work and had seen an enormous amount of damage to forests. He studied three areas particularly—Apollo Bay, Mallacoota, and Wilsons Promontory where the forest had been destroyed by graziers burning for green pick. He highlighted particularly the effect on native animals and birds of the continuous fires. Describing Mallacoota, he said: 'The birds have been expelled and the natural beauty has gone'.[54] On Phillip Island the large koala colony was under threat from settlers' fires. Barrett pointed out that in the Wyperfield National Park near Hopetoun in the Mallee a strict condition of land leased to graziers was that if a fire occurred from any cause whatsoever, the lease was cancelled immediately. He commented that as a result 'the natural increase of birds and animals has been extraordinary'.[55] He commented that, 'The tourist who comes to Australia does not come here to see the city; he wants to see the country as it was with the animals and birds in their native state'.[56]

Barrett articulated a genuinely spiritual vision of nature. Without the natural world, 'I think future generations will suffer badly,' he said. 'Mankind, crowded in cities, working hard and fast, worried and stressed, turns to nature for their holidays and recreation. It is natural; it is instinctive; it is deep down. If such places as National Parks were not provided, the world would be the poorer . . . To allow [them] to be devastated would really be a sin against posterity.' He insisted that, 'we need these biological islands in every state, and these islands must not be burnt or destroyed in any way . . . what is wanted is rigid protection of the islands when they are established'.[57] He compared Australia unfavourably to Canada where, he said, people were more disciplined and fire-aware; they knew they need to protect the environment. To change Australian attitudes Barrett recommended education. He said that 'The [Canadian] forests are very carefully patrolled, and they have this advantage over Victoria, that nearly everywhere you can get water'.[58] Much of the patrolling and fire control in Canada was carried out by aircraft.

Barrett's vision was extraordinarily comprehensive for his time, but nothing of it is reflected in Stretton's report. However, that does not mean that, as Tom Griffiths shows, the judge did not eventually develop a 'forest conscience' as he called it himself.[59] But in 1939 he was convinced that public opinion was crucial; that natural ecology was superseded by social ecology.

Barrett's evidence about Wilsons Promontory was supported by that of Archibald Campbell of Kilsyth, a University Extension Board teacher. He said that Wilsons Promontory National Park was being ruined by constant burning. It was no longer 'a national park; it is a national pasturage'.[60] He did not get a very sympathetic hearing from Stretton because he suggested that graziers should be educated to stop burning and punished if they did not obey the law. Soon afterwards on that same long, hot afternoon in the Exhibition Building in Melbourne, the apiarist, Vernon Davey of Toolern Vale, north-west of Melbourne, argued that regular spring burning affected native birds. He said, 'The spring is the nesting period of birds, and we cannot afford to ignore the birds because they are a national asset'.[61] While conceding that he was not against all burning, he suggested that any grazier who lit fires indiscriminately should lose his lease. This led to an extraordinary outburst from Stretton:

People come here and advance theories about jurisprudence, crime and punishment. They should not because they do not know what they are talking about. Do you suggest because it is hard to convict a man we should make it easier to dispossess him? . . . I suggest you witnesses should keep to your own department. You should talk about bees. You do not know about human nature and what can be done under the law if it gets into the wrong hands.[62]

Davey replied with spirit: 'And leave the birds out?'[63]

Another witness with views similar to those of Barrett was Alfred Douglas Hardy, a retired drafting and botanical officer from the Forests Commission, who had a long interest in the mountain ash forests and especially in the tallest of these trees. In 1939 he was the vice-president of the Victorian branch of the Forests League.[64] As well as giving evidence, Hardy also doubled as counsel representing the interests of the league in the same way as Kelso represented the MMBW.

Hardy's evidence is interesting because of his personal contacts with people like Ferdinand von Mueller, who knew the forest before European penetration, and because of his collection of historical photographs. Hardy pointed out that thick undergrowth is a characteristic of rainforests. 'My father told me they had the greatest difficulty in 1868 . . . in getting through the forest because of the enormous amount of cutting that had to be done.'[65] Hardy's view was that when fires penetrated the forest they burnt away the natural undergrowth and allowed bracken and other weeds to penetrate. He told Stretton, 'In the western Dandenong forest, in the heart of Sherbrooke, following a fire there came a growth of bracken about eight feet high . . . These are the aliens that come in and take possession of a mountain ash floor if they get the chance.'[66]

iv

What did Stretton make out of the material presented to him and the procession of witnesses? His report is, without doubt, one of the most significant royal commission reports ever written in Australia, even if it is lacking in any sympathy with the views of those who spoke for the landscape, flora and fauna. The introduction to the report is an elegant masterpiece of English. It is Dantesque in its proportions and the force of its expression. Its power arises from the use of an objective, passive voice and prose which is aloof, almost remote, yet which forcefully recounts what happened. Its reserve is its strength and the sheer power of Stretton's writing demands attention. If Black Friday was literally an invasion of bushfires, a *Götterdämmerung*, a final destruction by the forces of evil, as fire historian Stephen Pyne calls it, then Stretton was determined to get to the bottom of it, expose it, and make practical and realisable suggestions to prevent it happening in the future.[67]

The remote causes were obvious: a long drought, a hot summer, an acute water shortage, dry winds which sucked the last drop of moisture from the ground. As a result:

> Men who had lived their lives in the bush went their ways in the shadow of dread expectancy. But though they felt the imminence of danger, they could not tell that it was to be far greater than they could imagine. They had not lived long enough. The experience of the past could not guide them to an understanding of what might, and did, happen. And so it was that, when millions of acres of forest were invaded by bushfires which were almost State-wide, there happened, because of great loss of life and property, the most disastrous forest calamity the State of Victoria has known.

And the proximate cause of this catastrophe? 'These fires were lit by the hand of man.'[68]

Seventy-one people were killed. Sixty-nine mills and millions of acres of forest were destroyed, together with a vast array of infrastructure. Hundreds of thousands of horses, sheep and cattle, were obliterated, often with great suffering, although Stretton does not mention the native animals. They were mere collateral damage. 'On that day it appeared that the whole State was alight. At midday it was as dark as night.' Few were prepared for what was coming, and some were criminally unprepared.

At one mill [Fitzpatrick's], desperate but futile efforts were made to clear of inflammable scrub the borders of the mill and the mill settlement. All but one person, at the mill, were burned to death, many of them while trying to burrow to imagined safety in the sawdust heap. Horses were found, still harnessed, in their stalls, dead, their limbs fantastically contorted. The full story of the killing of this small community is one of unpreparedness, because of apathy and ignorance and perhaps of something worse.

'The speed of the fires was appalling', Stretton said, noting that spotting occurred up to 11 kilometres ahead of the main blaze. 'Such was the force of the wind that, in many places, hundreds of trees of great size were blown clear of the earth, tons of soil, with embedded masses of rock still adhering to the roots; for mile upon mile the former forest monarchs were laid in confusion, burnt, torn from the earth, and piled one upon another as matches strewn by a giant hand.' The fires were so intense that the soil itself was burnt. These were the worst fires since 1851.

Stretton was blunt. 'The truth was hard to find . . . Much of the evidence was coloured by self-interest. Much of it was quite false. Little of it was wholly truthful.' Most timber workers were too afraid to give evidence, fearing that they would not be employed again, and while the forest officers were 'in the main, youngish men of very good character . . . they were afraid that if they were too outspoken their future advance in the Forests Commission's employ would be endangered. Some of them had become too friendly with the millers whose activities they were set to direct and check.'

The report denounced departmental rivalry and attempted 'to expose and scotch the foolish enmities which mar the management of the forests by public departments who, being our servants, have become so much our masters that in some respects they lose sight of our interests in the promotion of their mutual animosities'. Stretton's view was that eventually some form of consensus would have to be reached which involved a recognition that, 'No person or department can be allowed to use the forest in such a way as to create a state of danger to others'. If people cannot be convinced of this they need to be cajoled, which would involve 'deprivation of rights, rather than, but not to the exclusion of, fine or imprisonment'.

What Stretton revealed was a situation worthy of *Yes, Minister*. There was internecine feuding between departments and authorities. One of the most contentious issues was whether the Forests Commission was primarily responsible for fire prevention and suppression. The act did not specifically lay this duty on the commission, but did so 'by implication' in that the commission had to report to parliament regarding 'the protection of state forests from (*inter alia*) fire'.[69] Stretton highlighted the ludicrous position regarding 'protected forests'. In these the Forests Commission managed the trees and the Lands Department the ground cover; that is, the grass! The Lands Department had no fire prevention policy and the Forests Commission could not activate theirs because that would almost certainly damage the ground growth. Nevertheless, the Lands Department could grant—without consultation—leases and licences for grazing in protected forests 'to any person, however bad may be his reputation for the illegal and dangerous use of fire. Although the grazing licences . . . contain a condition to the effect that the licensee shall protect the leased area from fire and extinguish any fire that may break out, and that breach of this condition may lead to forfeiture of the lease or licence, the condition is a nullity, in reality, as there is no supervision of his conduct in this respect.'[70] In fact Stretton found that there had been no prosecution for five years. He described the Lands Department's attitude to the royal commission as 'tactical' and their behaviour 'to be deprecated'.[71] When the secretary of the department, William McIlroy, appeared before Stretton he was almost monosyllabic and played a straight bat to every question. His mantra was, 'We feel we cannot interfere' with the activities of the Forests Commission.[72]

There was also 'a long-standing feud between forestry officers and officers of the MMBW, and between officers of these bodies and those of other water authorities, on the one hand, and farmers and settlers, on the other, who consider their properties are endangered by the state of nature which the water conservator and the forester consider to be the best attainable state in the forest areas under their control'.[73]

As a result:

the policy (if any) of prevention and suppression of fire has in each case been determined by and subjugated to what each has considered to be its major interest and has, as in the case of the Lands Department, been non-existent . . . Each excuses the unsafe condition of his own territory by protesting that his own undertaking is of vast importance, and that he must aim at the perfection of production or supply . . . Both [Forests Commission and MMBW] have ignored the advice and supplications . . . of the private landholder whose interests have for years past been placed in jeopardy by the refusal of these bodies to protect him against the danger which they have brought to his door. Both have, in turn, been exposed to the danger which the landholder has caused by the illegal measures of self-help [i.e. fire-lighting] which have been forced upon him by the inflexibility of the law.[74]

v

Stretton is clear about causes. 'The major over-riding cause, which comprises all others, is the indifference with which forest fires . . . have been regarded. They have been considered to be matters of individual interest.' Perhaps the greatest attitudinal change that Stretton's report ultimately achieved is that rural people have gradually come to realise that bushfires are a social responsibility. He clearly heard some of the expert witnesses. He agreed that in pre-European times there were 'clean' forests because:

> the forests had not been scourged by fire. They were in their natural state. Their canopies had prevented the growth of scrub and bracken . . . They were open and traversable . . . Compared with their present condition, they were safe. But the white man introduced fire to the forests. They burned the floor to promote the growth of grass and to clear it of scrub . . . [and] the balance of nature had broken down. The fire stimulated grass growth but it encouraged scrub growth far more. Thus was begun the cycle of destruction which cannot be arrested in our day. The scrub

grew and flourished, fire was used to clear it, the scrub grew faster and thicker, bush fires, caused by the careless or designing hand of man, ravaged the forests; the canopy was impaired, more scrub grew and prospered, and again the cleansing agent, fire, was used.

As a result in the once open forest 'the wombat and the wallaby are hard put to it to find passage through the bush'.[75] Kelso, the MMBW, the Forests League, Lane-Poole and Hardy would beg to differ that the cycle of destruction 'cannot be arrested' by natural means.

Who were the principal fire-lighters, according to Stretton's findings? 'Settlers, miners and graziers are the most prolific fire-causing agents. The percentage of fires caused by them far exceeds that of any other class. Their firing is generally deliberate. All other firing is, generally, due to carelessness.'[76] Stretton found that the law failed 'because it is not fitting for the widely diverse conditions and circumstances which obtain in Victoria'. The problem was that having a proclaimed period when fires cannot be lit (November to March) did not take into account the diversity of geographical circumstances across the state. As a result 'the settler decides to burn in defiance of the law and, not wishing to be detected in the act, leaves the fire untended, either to die out or to rage across the countryside'. Further 'the law is so notoriously unpopular . . . that there is no public opinion to check an intending lawbreaker'. Even police and forest officers were unwilling to enforce it.[77]

There was also widespread apathy about fire prevention. When fires came the Forests Commission, the MMBW and the almost nonexistent Bush and Country Fire Brigades were expected to deal with them. But 'their powers only arise when a fire has begun'.[78] No one had a duty of prevention, nor the power to enforce preventative measures. 'The law', Stretton reported, 'was inadequate, ineffective and flouted'. People would often refuse to support the Forests Commission officers when they asked for volunteers. This was not true universally, but there was certainly bad blood between officers and locals in some places, especially central and east Gippsland, and in the Otways. Neal Oldham, the forest officer at Beech Forest, told Stretton that even on Black Friday he could get no volunteers to deal with a fire that eventually took a fortnight to control: 'The men would not go in on the bad days' because they wanted to protect their own homes.[79] So forest officers often had to use their own resources and staff to deal with fires.

Stretton also argued that the Forests Commission had failed in its duty of care to landholders close to the forests. It was very slow in developing a co-ordinated fire policy, which was not produced until 1935, despite the terrible fires of 1926 and 1932. A chief fire officer was appointed only fourteen months before Black Friday. Stretton said that the controlled burning that had been carried out was 'ridiculously inadequate'. He was critical of the commission's failure to protect forest margins, establish trails, tracks and roads, and to construct small dams to conserve water for firefighting. All this only began just before the 1939 fires. The commission also failed to force millers to dispose of heads and mill waste, nor did it prevent burning by graziers. However, he conceded that the commission's staffing and funding had been 'ludicrously inadequate'. He praised the commission for doing 'all in its power, having regard for the disabilities under which it labours, to suppress fires which break out on or near its areas'. He also acknowledged forestry officers and field staff who 'risked their lives in the performance of acts of courage in their attempts to stem the spread of the fires'.[80]

Turning to the MMBW, Stretton commented that water catchment areas under the board's control were dangerous because the board refused to do any preventative burning. 'It appears that a large part of the Board's policy of prevention of outbreak and spread of fires is to be left to Nature. Nature, however, in another department of its working, sends the abnormal season which encourages the major fire which consumes the forest.' This is a trivialising of the arguments put to him by Kelso on behalf of the board, whose policy nowadays is seen as extraordinarily far-sighted. Stretton concluded that, 'the condition of the Board's areas assisted the spread of fires', a view which is not justified by the evidence presented to him.[81]

He devoted much attention to safety at timber mills because so many workers had been killed. He was again critical of the Forests Commission, stating that it was 'absurd' for the commission to say that the word 'sawmills' only included the mill buildings and not accommodation for staff, thus excusing it from insisting on clearing trees and vegetation around the workers' dwellings as well as the mill buildings. 'It is true that the mill workers, for the greater part, did nothing to protect themselves. But that . . . [is] irrelevant to the question of the conduct of the Forests Commission in this matter.'[82] He was equally critical of the conduct of the commission regarding dugouts. After the 1932 fires the commission had decided to impose the construction of dugouts on millers as a condition of their licence. But 'Having made its considered decision, the Commission at no time thereafter took any steps to compel the observance of the condition'. Galbraith's excuse to Stretton had been that while the commission was convinced about dugouts, it 'feared it might be liable at law if people were asphyxiated in them' even though the Crown Solicitor said they were not.[83] 'So far', said Stretton, 'it is a sorry story. The conclusion is worse.' The 'conclusion' was the commission's failure to demand a dugout at Fitzpatrick's mill at Matlock with the resulting deaths. He also criticised 'the contumacious conduct of some of the millers' and their refusal to maintain even basic safety standards at mill sites.

vi

So what was Victoria to do about preventing fires, and dealing with them when they did occur? Firstly, Stretton recommended the establishment of a state fire authority, as well as local fire authorities. 'It is strongly recommended that no public department or possible combination of public departments interested in forests should be permitted to control this authority'.[84] The state authority should consist of nominees from the Bush Fire Brigades Association, the Country Fire Brigades Board (which provided regular fire protection for country towns), the Forests Commission, and local shires. The state authority should be concerned with general policy concerning fire prevention, suppression, protection of life and property outside areas under the control of the Forests Commission and the recruiting and organisation of local volunteer brigades. The MMBW should maintain authority over areas entrusted to it.

The focal organisation in Stretton's reforms was the Bush Fire Brigades Association (BFBA), a voluntary organisation that had been formed as a result of the 1926 fires. But it had little money and was scandalously under-resourced. The insurance industry, for instance, contributed just £100 to the organisation in 1938–9! Colin Campbell, treasurer of the BFBA, told Stretton that 'we have had no financial assistance whatsoever' from any level of government until they threatened to close the organisation down completely.[85] Their credit balance in February 1939 was £120. Few of the local brigades had any money.

Stretton recommended that there be a closer association between bush fire brigades and shire councils. 'For preventative and protective measures . . . the municipal engineer, acting with the advice of the brigade should determine the local policy.'[86] To some extent this follows the evidence of Joseph Cook, a Healesville guesthouse owner and local councillor, who wanted shires to take on responsibility for fire control and argued that some men should be paid for their services. One of the difficulties, according to Cook, was that 'volunteers are usually men who have to drop their occupations to fight fires and, while in an emergency the whole population does this, it is nobody's business to look beyond the task of getting the present outbreak under control. When once it has been checked, men should be employed to control the fire and make it safe.'[87]

Stretton recommended that where there was no volunteer fire brigade in a fire-prone zone, the shire should be obliged to form one. But, he said:

> No money is provided by the government . . . In the last ten years the brigades have received from the Forests Commission equipment to the value of about £700, i.e. less than £1–10–0 per week. Such gifts were made only to brigades situated in the vicinity of State Forests or Crown lands and who give an undertaking to assist in the suppression of fires in such forests or lands . . . These brigades have in the past saved many thousands of pounds worth of property from destruction . . . Where these brigades have come into existence there has been a lessening not only of destructive spread of bushfires, but also of outbreaks.[88]

While nothing was done immediately, what was to become the Country Fire Authority gradually emerged in the 1940s after more disastrous bushfires.[89] These ideas also began to permeate the other states.

Stretton also made other practical suggestions. Among the important ones were that land occupiers and lessees ought be obliged to keep their lands cleared of combustible material. The local fire authority should be empowered to instruct them to clear their land, or charge them to do it if they failed to comply. A failure to clear in dangerous, fire-prone areas should be made an offence at law. These same principles should apply to the MMBW, the Forests Commission and public authorities, who must keep the perimeters of their lands clear. Local brigades should be able to enter private land without permission in order to put out fires, and landholders should pay for this service. He also recommended that the prohibited period be decided and proclaimed by the state fire authority and that this should vary from zone to zone and period to period depending on conditions at the time. There should be 'absolutely' no fires in the open during the proclaimed period. However, local brigades should be allowed to do protective back-burning in an emergency during the proclaimed period. Stretton pointed out that an acute fire period occurred every six to ten years in south-eastern Australia and that these periods could be predicted.[90] He suggested the adoption of an idea presented in evidence that when the state fire authority proclaims a prohibited period all milling operations should cease and that the mill workers become firefighters, paid by government.

One of his interesting suggestions was the importation into fire legislation of the legal notion of 'reasonable suspicion'. He saw this as a useful weapon to curb and convict arsonists and other fire-lighters. What it amounts to is that if there is a reasonable suspicion that a person had lit a fire, but that it cannot be proved because it was done furtively, and there were no eyewitnesses and no confession, then a magistrate could still consider the possibility of finding the defendant guilty if there was a reasonable suspicion that he had lit the fire. This idea still has practical applicability today.

There is a long section in the report on suggested reforms of the Forests Commission. Stretton was particularly critical of the lack of government funding. The best evidence of this is his comment that field staff numbers were 'ludicrously inadequate. The fact of their numbers in relation to the multiplicity of duties which devolves upon them and to the area of the forests which they are expected to maintain and protect, calls up the recollection of Lewis Carroll's "forty maids with forty mops" '![91] Detailed safety precautions for mills were also set out, including the disposal of heads under the supervision of forest officers, the clearing of mill sites, the provision of dugouts, providing a reliable water supply, and the forced removal of mills in extremely dangerous forest areas to safer places in open country.

vii

Stretton's report was submitted to Lieutenant-Governor Frederick Mann on 16 May 1939. The report, important as it was, was ignored even after it was tabled on 28 June by the Premier. It was not until 4 July that the Minister of Forests and Deputy Premier, Albert Lind, began tearing into Stretton. He treated the Royal Commissioner with sheer bastardry. The minister was presenting a bill concerned with the salvaging of as much of the burnt timber as was possible, as well as plans for replanting.[92] Lind turned to the royal commission report just before the suspension of parliament for dinner, and he invited honourable members to have 'a field day on this report'.[93]

After a no-doubt ample three-course dinner and liquid refreshment, Lind returned, fortified, to the fray.

> I had hoped that when the [royal commission] report was presented it would be found to contain a good deal that would be helpful . . . I regret, however, that it contains too much generalisation and too little particularisation . . . I challenge any honourable member to find anything in the file [the minutes of evidence] to support some of the innuendoes contained in the Report of the Royal Commission . . . The Royal Commissioner has indulged in too much generalisation.[94]

This is nonsense. The report is full of practical suggestions. But facts did not trouble Lind. When UAP member Thomas Maltby asked: 'Is there any question put forward by the government that has not been answered by the Royal Commissioner?' Lind shifted his ground and attempted a defence of the witnesses. Quoting Stretton saying, 'The truth was hard to find', Lind indulged in a broad-brush whitewash of witnesses:

> I do not believe that those witnesses referred to told untruths when examined. I know many of them and they are reputable citizens. Some are leaders in various industries. Many enterprising people are associated with the milling industry . . . the bush workers are some of nature's very own. So far as concerns the forest officers, I say it was unfair for the Royal Commissioner to state that those lads told untruths in the witness box.

Challenged by Maltby again that Stretton's hard words were for the Forests Commission and not its employees, Lind said: 'The reflection was more general than that. The statement of the Commissioner is a reflection on dozens of witnesses representing every class in the community'—no doubt especially witnesses from Lind's own electorate in East Gippsland![95]

Perhaps the principal reason for Lind's antagonism was an *obiter dictum* that Stretton made in the report when he suggested that the Forests Commission's failure to act might be the result of political pressure. There was no imputation that Lind had acted improperly, but he was the Country Party member for the East Gippsland seat of Bairnsdale with its logging and milling interests, and there was no doubt in Stretton's mind that there had been a weakening of the Forests Commission's power 'to apply sanctions to those whom it wishes to discipline'; that is, mill owners.[96]

Part of the problem was the irregular and insufficient budget allocations given to the Forests Commission. Stretton referred specifically to 'a large amount of such moneys [being] . . . removed from the control of the Commission, to the Minister of Forests, so that in addition to irregularity there is complete deprivation of control of a substantial part of such moneys' in the Forests Commission's budget.[97]

Lind spent some time trying to explain this away, and then he argued that the fires were not caused 'by the hand of man' but by 'intense heat'.

He continued: 'I constantly preached the adoption of systematic and controlled burning of our forests. I was strongly criticised in certain Melbourne newspapers, one stating that I was a menace to Victoria because I advocated the burning of our forests. The trouble is some people do not know the difference between the intelligent and stupid application of fire. The only way to save our timber, as any practical bushman knows, is by controlled burning at that season of the year when it will do no harm. I have preached that policy in season and out of season.'[98]

Lind's contributions to the parliamentary debate give the impression that he was not across his portfolio. As the acerbic Thomas Maltby commented in passing, 'You are beyond help'![99] Eventually Sir Stanley Argyle, the Leader of the Opposition, moved a vote of no confidence in the Dunstan Government on the grounds of 'maladministration . . . in its forests policy' as exposed by Stretton.[100] This motion was defeated along party lines but it did at least highlight the points that the royal commission was trying to make.

The *Age*, at least, looked to the future rather than to Lind's 'merely raking over dead ashes'. In an editorial on 6 July 1939 it said that 'the obvious next stage is to devise as early as possible a well-considered administrative and legal system in place of past laxity, unsuitable law and chaos of inter-departmental feuds and jealousies . . . indicted by Judge Stretton . . . An effective system of bush fire prevention . . . remains a cardinal responsibility of Cabinet and Parliament.' At least Stretton had got that message through. Over the next decade in Australia attitudes to bushfire were to change radically. For that we have to thank Leonard Stretton. He had succeeded in his herculean task.

PART 2

AFTER STRETTON

BLACK DAYS, 1939-66

i

It was two days before Christmas, 1943, and in a paddock at Tarrawingee near Wangaratta they were destroying hundreds of sheep that had been badly burnt in grassfires. A group of thin, hardworking, older-looking men with broad-brimmed hats and rifles waited as injured sheep were herded into pens and shot. The men looked tired and shell-shocked. They had been fighting an 8000-hectare grassfire for over 36 hours. And they knew that at least seven firefighters were dead and others were critically injured in Wangaratta hospital. The death toll was to rise to ten with three of the injured men dying. Over 1000 men fought this fire, including air force and army contingents. There was an acute shortage of manpower as the Second World War dragged on.

The fire had begun on the Hume Highway near the Bowser railway station, probably started by a cigarette butt thrown from a car. It was a dry, hot day with a temperature of 34.44°C with a strong north-easterly wind. The fire took off eastwards, burning tinder-dry grass across the flat, open country towards the town of Tarrawingee. A fire truck with volunteers from Wangaratta and the surrounding area was heading out to fight the outbreak along the Ovens Highway (now the Great Alpine Road) when they were trapped. George McLaughlin, a 48-year-old mercer from Wangaratta, was driving. He tried to turn around and retreat, but 'blinded by smoke and fumes he drove the truck too far to one side of the road and it was jammed in a ditch', according to the *Herald* report of 24 December. The truck caught fire. Panic set in and men started 'running in and out of the flames blindly'. Angus Simmonds, a carrier, kept calm and crouched behind a burning tree. He heard someone calling for help. Despite the fact that his trousers and hat caught fire, he dragged two men to safety: a young farmer, John Allen, from Everton, and an older man, George Spencer, a technical school teacher from Wangaratta, who later died in hospital. Two of the others who died, Claude Hill and Kevin Dunkley, were fourteen-year-old boys. Twelve others were seriously injured. The fire was out by Christmas Day. This was the first fire of the season, and much worse was to come.

As a result of this grassfire, fifteen children under fourteen years lost their fathers and seven women became widows. The sum of £1500 was paid in compensation to the victims' families by the Bush Fire Brigades Association (BFBA). Despite the recommendations of the 1939 royal commission, the BFBA received very little funding and equipment from government.[1] The Victorian Government, which had been asked by the BFBA to provide each brigade with at least a truck, a tank and a power pump, was very slow to respond.

Out across Australia drought still dominated the landscape. There were rains after the terrible fires of 1939, but by the end of the year a pattern of weather conditions had begun that signalled the onset of dry or even drought conditions in eastern Australia. By August 1940 one of Sydney's supply dams, the Nepean, was empty. There were water restrictions in Brisbane, and Perth experienced one of the driest years on record. In Queensland in July and August 1941 some 120,000 hectares around Richmond and Julia Creek were burnt.

Mid-1942 saw good rain, but drought returned in 1943–4. Some areas along the Murray and in the Wimmera were between 25 per cent and 40 per cent below average rainfall. It was also cold. Mean temperatures were well below normal for the greater part of the year. However, there had been enough rain to support a plentiful growth of grass, and most of the bad fires of the 1943–4 season were grassfires. One of the worries was the danger to the immature mountain ash forests which were just beginning to recover from 1939, but which were too underdeveloped to produce seed. Another fire through them would have been disastrous.

ii

When the bone-dry summer of 1943–4 hit, the stolid, parsimonious, Victorian Country Party government of Albert Dunstan and his do-nothing Minister for Forests, Alfred Lind, still refused to implement Stretton's recommendations. They were much given to issuing warnings about fire safety, but these were largely a public relations exercise. Nothing practical had been done to lessen the bushfire danger.

By early January the 1944 summer was beginning to shape up as another 1939. Most of the state was a proclaimed bushfire area and it was illegal to light fires in the open. Despite this, on 4 January 1944 there were outbreaks at Mount Eliza which threatened the homes of the wealthy, including that of newspaper magnate Sir Keith Murdoch, father of Rupert. The strenuous efforts of 200 firefighters, no doubt mainly of the working class, saved the homes of the well-to-do.

On Saturday afternoon, 8 January 1944, a fire driven by a strong wind swept down from the wooded hills around Mount Cameron and Talbot, 15 kilometres south of Maryborough towards the town of Clunes. The 40-bed hospital and fifteen houses in Clunes were destroyed. Firefighters just saved the primary school. Grave fears were also held for the town of Creswick, 18 kilometres north of Ballarat. All through that Saturday afternoon the fire bell was rung in Ballarat calling for volunteers and by day's end over 1000 men were fighting the fires, including 350 air force men. Eventually the wind then turned the fire back on itself, but not before 12,000 hectares of farmland were burnt.

But the worst outbreak that weekend was just to the north of the pre-Christmas fire around Tarrawingee. This broke out in the hills of the Mount Pilot Range near Byawatha and Eldorado, in red stringybark and box-ironbark forest. The fire moved with incredible speed, covering almost 13 kilometres in twenty minutes. Eighteen farms were burnt as the fire moved along the northern edge of the pre-Christmas outbreak, reaching as far as the town of Everton which lost eleven houses including that of the stationmaster, a hotel, the railway signal box, and two railway bridges. Beechworth also was ringed by fire on three sides and 1500 firefighters were deployed.

Outbreaks continued and by Wednesday 12 January the position was fraught. There was a sense that fire could break out any time especially, the Weather Bureau warned, if strong northerly winds eventuated. To this point all the fires had been in grass and scrubland. Forests Minister Lind called for a controlling body to co-ordinate fire prevention and suppression. He created a sense that much was being done by government. 'Extra pumps and radio equipment had been acquired . . . dugouts had been built', reported the *Age* on 12 January. He said that £5000 had been given the previous year to the bush fire brigades in contrast to the £700 over ten years prior to 1939 and that there was 'close co-operation' between everybody involved; again, this was largely rhetoric.

On Friday, 14 January, exactly five years and one day after Black Friday, Victoria was hit with another 'perfect' fire day with high temperatures, low humidity and a hot, northerly wind averaging 35–55 km/h. Blistering wind gusts swept through Melbourne and the temperature reached 39.5°C.[2] The fires that day caused an extraordinary amount of damage in upmarket Melbourne suburbs. At about 11.30 am a raging north-easterly wind whipped up the flames through 280 hectares of coastal scrub and tea-tree overlooking Port Phillip Bay. In Beaumaris, Black Rock and Mentone 58 houses, many of them brick, were destroyed. 'Formerly one of the beauty spots of the inner bay-side, [Beaumaris] presents a haggard prospect with the remains of fashionable houses and beautiful gardens and the sweep of once shapely tea-tree, now black and naked . . . One remarkable feature of the fire was that in the midst of widespread ruin of brick houses, some weatherboard houses were still standing', reported the *Age* on 17 January.

A week later the first ever scientific study of house resistance to external fires was undertaken by G.J. Barrow of the Council for Scientific and Industrial Research, forerunner of the CSIRO.[3] This was based on a careful study of seventeen Beaumaris houses. Barrow found that simple precautions such as enclosing underfloor spaces, keeping shrubs and trees clear of walls, covering ventilators and eaves with mesh and not stacking wood beside the house, were more effective deterrents than the building material used. Bushfire researchers Justin Leonard and Neville McArthur describe him as 'the first person to scientifically identify the ignition mechanisms of bushfire attack on houses, and to document the fact that houses tended to burn down from inside. His work attacked existing myths about the destruction of houses in bushfires, and gave clear guidelines to improve the performance of houses in bushfire prone areas.'[4] His conclusions were ignored.

In the Western District a huge grassfire swept across the flat, lake-studded landscape on 14 January. The conflagration ran eastwards from near Hamilton along a north-east line to Dunkeld, Lake Bolac and Skipton, and south-east through Mortlake to Lismore and Cressy. Hamilton lost 40 houses, Dunkeld 35 and fifteen more in the district were destroyed. Three small towns were wiped out—Derrinallum, with 76 houses, Mount Bute and Berrybank. As the fire raced eastwards it destroyed twenty houses in Lismore. Cressy was saved by the local creek which acted as a fire-break, but 50 homes were destroyed within a 24-kilometre radius. Skipton was threatened and hospital patients were moved to Ballarat. At the height of the fire near Rokewood, reported the *Herald* on 15 January, 'the heat caused immense whirlwinds. At one stage six were sweeping forward, one behind the other like ocean waves, carrying the flames hundreds of feet high.'

Because of their speed and the heat they generate, grassfires are often more dangerous than forest fires. That is why seventeen people were killed and many more injured in the Western District.[5] These grassfires caused enormous damage. At least 356 houses and homesteads were destroyed, a minimum of 435,000 sheep and cattle and horses were either killed or had to be put down later, lush pastures were lost, bridges and fencing burnt out. The area destroyed was about 140 kilometres west to east on a front up to 45 kilometres wide. The damage bill was estimated to be between £750,000 and £1,000,000.

There were also grassfires around Geelong and south of Colac in some of the best farmland in Victoria. The outbreaks ran in a south-westerly direction diagonally across the Hamilton–Skipton–Cressy fires on a 40-kilometre front driven by a northerly gale. The fire was a massive, unstable wall of flame that took everything before it. 'Sweeping through tall grass and flax fields, the flames were 20 feet high. They leapt fire-breaks, roadways, and creek beds . . . and carried on destruction until the wind abated', which was not until about midnight, according to the *Herald* of 15 January. First Constable William Cain said he had been in the forest fires in Powelltown in 1936, 'but that was nothing to the ferocity of the grassfire yesterday. It travelled at many miles an hour and at times the wall of flame was 20 feet high. Nothing could live against it. It was terrific and the desolation left is indescribable.' Even the Ford motor works in North Geelong were threatened when a grassfire swept down on the factory. All 2000 workers turned out to defend their workplace.

Central Victoria was also badly hit. A wide area of forested hill country with Kyneton at the southern end, Heathcote in the east, Castlemaine in the west, and Bendigo at the northern end, was burnt in a series of fires with the loss of 30 houses and two churches, as well as stock, fencing and pastures. The town of Glenlyon had a lucky escape when most of it was saved because it was surrounded by potato fields.

Destroying injured stock was a major problem after the fires, and the federal government had to provide much of the ammunition: 11,000 rounds were issued by police in Hamilton alone. Eventually military convoys were used to put animals down in the Western District. 'Reports indicate that thousands of cows are wandering around devastated areas with burned feet, unmilked and bewildered. Thousands of sheep have been blinded and burned and must be shot as soon as possible', reported the *Age* on 18 January 1944. The fires were followed by dust storms.

Victoria was not alone. In New South Wales there were extensive fires covering 10,000 hectares with one person killed and eleven injured. The fires centred on Harden, and Murrumburrah and further south at Junee. In south-western Queensland two days before Christmas a massive grassfire broke out in the Dirranbandi and Culgoa River districts. Temperatures reached 44°C and there were flames 12 metres high. Forty-five thousand hectares were burnt. In the same season South Australia had a mild summer, but prolific growth of spring grass meant that 75 fires were reported in open country and thirteen in forest reserves which were badly damaged.

Eventually widespread light rain fell in Victoria, easing the situation. Deputy Premier Hollway claimed that the 1943–4 fires were worse than 1939. He said that besides the usual careless lighting of fires, one of the real problems was the use of gas producers on cars and, to add some spice to his comments, he referred to the work of 'Fifth Columnists' (spies and traitors) who had taken advantage of the extreme fire conditions to create subversion and confusion.

There were many complaints about the use of gas producers to replace petrol, which was strictly rationed due to the war. Australia was a pioneer in developing them.[6] The charcoal was burnt in a small hopper with air ducted through a tuyere producing carbon monoxide which was used to power petrol engines. In an editorial on 17 January 1944 the *Age* argued that many cars were shedding 'glowing embers, to be fanned by the wind into a sea of flame along country roads'. Jim Freeman, a mechanic who worked on them, says they were 'horrible devices' full of filthy soot. But they could not light external fires since 'the burning charcoal was entirely enclosed and the gas caused . . . was ducted directly to the filter and then to the engine'.[7]

The real cause of bushfires was more prosaic, as the *Age* editorial admitted: Australian pyromania. The main culprits were 'the farmer and grazier burning off without making sure that the fire is not liable to spread, the heedless camper and picnicker leaving embers of billy fires, the smoker throwing away glowing matches, pipe ash and cigarette ends'. Arsonists also played their part.

It was not until early February that the figures were tallied for these fires. The *Age* editorial of 3 February 1944 reported that, 'The death roll from the recent fire has mounted to 25; 500 houses were destroyed, and 1,250,000 livestock lost. In monetary values the disaster is destined to have cost Victoria about £5,000,000. Every visitor to the devastated areas agrees that never before has the State suffered such a damaging blow in the scale of material losses and setbacks in productive activities . . . [the fires] were much more serious [than 1939], especially when the emphasis is upon food production.' About one million hectares of grassland were destroyed and almost 160,000 hectares of forest had been burnt in 286 fires.[8] It would have been much more except that little was left to burn after 1939. The *Age* editorial was also critical of the parsimonious attitude of the federal and state governments to bushfire victims. It pointed out that the War Damage Commission had already collected £14,000,000 in compulsory insurance from property-holders and said: 'It would be reasonable to divert a proportion of this fund to ensure adequate compensation and rehabilitation to sufferers from the fires, many of whom in country areas are uninsured'. Given the acute manpower shortage due to the war the editorial called on government to release men and material to restore farms and homes.

By the end of January things seemed to have settled down. Then on Monday 14 February 1944, a classic fire day with hot, dry, north to north-westerly winds gusting to 125 km/h, there was a series of outbreaks in central Gippsland from Mirboo North and Moe in the west, to Yarram in the south, and Morwell, Yallourn and Traralgon to Sale in the east. The Yallourn open-cut mine caught fire. Thirteen people died and more than 200 homes were destroyed. For a period on Monday communications broke down completely with only one private State Electricity Commission (SEC) phone line open between the Latrobe Valley and D24, the police communications centre in Russell Street, Melbourne. One witness, quoted in the *Herald* of 17 February, said that after the fire the view from the 750-metre lookout at Balook on the Grand Ridge Road in the Strzelecki Ranges revealed a completely blackened, desert-like landscape 'dotted with the remains of homes, dead cattle, sheep, horses, pigs, kangaroos, possums and fowls'.

By Friday 18 February the fires were largely out, including the Yallourn fire, but power cuts in Victoria were widespread. Much of the damage had been done in and around Morwell, where eight people had been killed. A long-term effect was a severe milk shortage for Victoria. Many dairy cattle perished and, because of lack of feed, the remaining cows had to be dried out.

The Dunstan Government brought back Justice Leonard Stretton to investigate the fire in the Yallourn open-cut brown coal mine. They had treated him shabbily after the 1939 royal commission, but no one was more experienced in investigating bushfires. In this instance, his remit was narrow and his report was submitted five and a half weeks later.[9] As in the 1939 report, his prose is elegant, sparse and concise and his recommendations are succinct and achievable.

The open-cut mine had commenced operation in the Latrobe Valley in 1916, and Yallourn was built in the 1920s to house staff and families. It was the complete company town. Yallourn and its surrounding 3200 hectares were dominated by the SEC, a Victorian Government instrumentality. The SEC produced electric power and brown coal briquettes for industrial and domestic use. 'The source of the material . . . is the deposit of brown coal which lies beneath an overburden of soil. The coal is won ingeniously by a method of open-cut mining', Stretton explained. Just over 2000 people were employed by the SEC, with a town population of 4000. The Latrobe River was close and its water used in the power station cooling towers. Stretton stressed that there was no civic life in Yallourn; no democracy. It was a kind of perverse utopia in which everything was provided, but no one had any rights. The air was constantly polluted with coal dust, and people constantly smelt 'the nauseating stench which comes from the neighbouring paper mill'.[10] Dissatisfaction was rife.

The fire began in timbered country on the border of SEC territory, about 5 kilometres from town. The cause was a farmer burning off scrub, probably illegally. 'The fires on the open cut were caused by air-borne burning material from the bushfire igniting the coal dust in the workings.'[11] Stretton found that there was no plan, organisation, or person within the SEC with overall authority to respond to a fire in the mine, even though this was predictable. Some preventative work had been done, but it was inadequate and had not taken into account the conclusions of the 1939 royal commission. He found complete confusion, internal jealousies, and managers with an inflated sense of personal importance. 'Had not the word "bureaucracy" recently taken upon itself an offensive connotation it might well have been used by your Commissioner to describe more succinctly [the SEC's behaviour].'[12]

There was no protective fire-break around the town or mine. Firefighting was unplanned and completely disorganised.[13] Only the local Yallourn Country Fire Brigade acquitted itself well because it was organised, trained and equipped. Much of the land to the north of the SEC enclave was leased by the Lands Department to graziers, some of whom were 'persons notorious as fire-law breakers'.[14] Stretton also discovered that 'unburned coal, known as "char", was deposited across the surrounding countryside and it added considerably to the intensity of fires once they started'. He noted that no protective burning had been carried out to insulate the town. 'In many places timber and scrub grew up to the fences of the houses built on the boundary of the town.'[15]

What happened on the afternoon of the fire was that everyone thought that it would pass south of the town. It did, but a northerly spur developed. 'It is most probable that the mine caught fire from flying embers blown from the northern thrust of the fire.'[16] The water pressure to fight the fire was inadequate. Stretton found that the neighbouring power station and the briquette factory were reasonably adequately protected, and that the briquette factory 'is built of fire-resisting material', asbestos.[17]

Stretton's report makes it clear that the lessons of 1939 had not been learned. The SEC had not offered fire education to the rural people who lived beside its enclave. 'In all years of fire in many parts of the country', Stretton commented, 'settlers cause bushfires by their illegal or inefficient methods of burning off their lands. It has long been recognised . . . that one of the fundamental safeguards resides in the creation of a state of enlightenment and goodwill in the minds of our rural populace'.[18] Probably because it was run by engineers, Stretton said, the SEC had ignored the conclusions of the previous royal commission and fire prevention was not taken seriously.

iii

Despite the failures of the SEC, the 1943–4 fires show that the war was, to some extent, forcing people to organise themselves better, and army and air force personnel were there not only to give a hand as firefighters, but also to provide an example of organisation. There was still little leadership from state governments, except in Western Australia, but the war brought about a national consciousness that bushfires were an Australia-wide problem and that organisation was needed to deal with them. In Victoria it was public anger over the 1943–4 fire season that finally forced the Dunstan Government to act, despite the fact that it had ignored Stretton's 1939 report and had let local firefighting brigades languish. After protracted negotiations the government eventually passed the legislation that led to the formation of the Country Fire Authority (CFA), which came into being on 1 January 1945.[19]

But the early history of the CFA was difficult. According to CFA historians, Robert Murray and Kate Wright, it 'unleashed enormous passions among the bush fire brigades, many of which initially resented what they saw as a "takeover" by the urban brigades'. It took many years to reach 'a reasonably harmonious working relationship'.[20] Victoria was leading the way in eastern Australia. But Western Australia had begun before the war to rationalise the dates when burning was prohibited, and in 1939 set up a Bush Fires Advisory Council which evolved into the Bush Fires Board in 1954.

Stretton's work was not finished. He chaired another royal commission, set up in 1946 by the Cain Labor Government, to inquire into forest grazing.[21] There was convincing evidence, brought into focus by the 1939 royal commission, that graziers were responsible for many of the fires that plagued Australia. For instance they caused 13 per cent of fires in the Australian Capital Territory between 1931 and 1944, 12 per cent in New South Wales in 1942–3, 25 per cent in Queensland between 1938–43, and 38 per cent in Victoria in 1940–1 and 1943–4. In contrast in Victoria in 1943–4 other land owners caused 14 per cent of fires, campers, tourists and fishermen 10 per cent and lightning only 3.9 per cent of known causes.[22] The purpose of this new royal commission was to ascertain the effects of cattle grazing on grasslands and forests, as well as to document the impact of grazing on erosion, water catchment and forest fires.

Stretton's report is comprehensive, and he used the same methodology as in 1939: he travelled right across the state, taking evidence locally and seeing the situation for himself. His report is more 'environmental'. He had come to understand that there was an intimate interrelationship between 'an inseparable trinity—Forest, Soil and Water. None of them stand alone . . . No one of them, without the others, can prosper.'[23] Destroy one, you destroy them all.

Stretton's report must have been a nightmare for government and bureaucrats, used to narrow contexts and immediate, practical implementations. He did not mince words. He ranged over human history and showed how past civilisations destroyed themselves through the exploitation and destruction of their forests. 'Civilizations have perished, leaving only the monuments of man's pretentiousness to mock their memory, because in ignorance or wantonness man's impious hand has disturbed the delicate balance which nature would maintain between forest, soil and water.'[24] There are resonances here of Percy Bysshe Shelley's poem, *Ozymandias*. Left alone forest, soil and water interact to maintain nature. It is 'the behaviour of mankind' that throws them out of balance.

Stretton, in common with many at the time, saw erosion as a core problem. This preoccupation was deepened by the destruction wrought by the 1939 wildfires. Again it was human interference that was the cause. 'Natural erosion', he said, 'is controlled by those forces which control the physical world and which, left to themselves, preserve rich and fertile places'.[25] He showed how dams were silting up because they interfered with the natural flow of rivers. The silt was the result of human interference with the upland forests and vegetation. People were allowed to clear upland because of 'tenderness towards the rights of the individual—that is his right to make money by almost any means— ... his anti-social activities will no doubt continue upon a course of destruction which cannot be abated until those activities are controlled'.[26] This could be a modern 'greenie' speaking. He cites the example of the Hume Weir. 'The Kosciusko area is a major part of the Murray catchment. Your Commissioner is convinced that that area is being severely over-grazed and burned and, therefore, eroded. If this be so, the Hume Weir is endangered.'[27]

Turning to the question of grazing and fire, Stretton stated unequivocally, 'that grazing is causally related to the occurrence of forest fires is true beyond the barest possibility of doubt'.[28] Graziers burn, he said, 'with an untroubled conscience ... [they] do not believe and cannot be convinced that burning is harmful to the forest'.[29] Their actions were viewed as acceptable as a result of tradition and folk wisdom. The report examined fire outbreaks in state forests by geographical region between the seasons of 1939–40 and 1945–6. Stretton noted: 'Orbost enjoys the infamous distinction of having to its discredit a total of 169 fires. For the 1941–2 season alone the forest division which includes Orbost is to be debited with 91 fires to which score Orbost contributed an approximately disgraceful 57 . . . The Western district is discredited with 247 fires since the 1939 fires'. In contrast, Omeo was held up as a paragon of virtue: 'although a district in which there is much grazing, [it] has consistently failed to score'.[30] He recalled the distinction between those who undertook protective burning to stop greater wildfires and those who believed, like the MMBW and Kelso, that left to nature the forests would rid themselves of scrub growth and attain stasis within 40 to 70 years. But he still believed that this 'idealistic', natural solution was not the way to go, and that the 'realists' who believed in protective burning were right.

While there had been some decrease in the number of forest fires, Stretton felt the pressure had to be kept on graziers, particularly 'fear of monetary loss' and fear of 'deprivation of their forest grazing rights'. Were they afraid that they would be thrown out of public forests, then they would obey the law and the number of fires would continue to decrease. But 'the number is still unforgivably large. The improvement is seen by comparison with the extent of the utter untrammelled lawlessness of the period prior to 1939.'[31] Very much the judge, Stretton emphasised that to be allowed to graze cattle in the forest was a privilege, not a right, and that should graziers fail to preserve the public property entrusted to them, this right ought not be renewed.[32]

Stretton is the man who put bushfires on the public agenda. Fire historian Stephen Pyne says that, 'His stature placed the issue of bushfires beyond the confines of professional forestry no less than outside rural folklore'.[33] His three royal commissions are semi-continuous. He kept hammering home the same issues: controlling wildfire through periodic burning, the need to clarify responsibility and organise prevention and suppression. The Royal Commission on Forest Grazing led to the Cain Labor Government's important Soil Conservation and Land Utilisation Bill of 26 June 1947 which set up the Soil Conservation Authority.[34] Stretton's new environmentalism reflected what was happening to thoughtful, better-informed people in Australia after the war. The national parks movement began to have an impact, the word 'conservation' entered the lexicon, bushwalkers started exploring the hinterland, and there was a new interest in Australian flora and fauna. As Pyne succinctly says: 'The tourist replaced the swagman as the archetypical Australian nomad'.[35] The 'Save the Forests' campaign began in 1944 in Victoria and quickly built up its membership. What emerged from all of this activity was a campaign to develop what Stretton called a 'forest conscience', whereby the public would maintain pressure on government to protect the state's woodlands.[36] This kind of ecological consciousness flowed through into the later environmental movement.

iv

Things were changing as the Second World War ended. Post-war migration began to provide a work force for the kinds of developments envisaged by the Curtin–Chifley Labor Government. Much of the impetus came from the Department of Post-War Reconstruction led by H.C. 'Nugget' Coombs and John Crawford. The department had begun in the war years when the Commonwealth gained increasing power over many aspects of life. Before 1939 the state governments tended to be the overarching authority in people's lives, but during the war Commonwealth authority over many areas increased, the most important of which was the power to levy income tax. When John Curtin became prime minister in 1941 he established the Production Executive which controlled agricultural and industrial production. His vision of an efficient, centralised government was to continue beyond the war into modern Australia.[37]

The Snowy Mountains Scheme, deeply flawed as it was environmentally, was a paradigm of centralised patronage by the Commonwealth. The scheme needed water. But there was widespread erosion in the mountain catchment areas from which the diverted rivers flowed. This resulted from uncontrolled burning and grazing. From the late 1940s onwards extensive study of the Snowy–Murray rivers watershed was undertaken in order to tackle the problems resulting from human activity in the high country. But the tragic irony was that while successfully protecting the headwaters of the rivers, the scheme actually destroyed the iconic Snowy River. Only now, as the massive dam wall at Jindabyne is lowered to obtain an environmental flow, are we beginning to realise the ecological destruction wrought by this 'engineering marvel'.

However, the scheme did lead to the establishment of a fire-fighting force in the Snowy Mountains. This in turn influenced attitudes to bushfire control. There was a distinct move away from localised 'folk' attitudes to a more scientific and organised approach. As foresters began to study the dynamics of fire, they realised that this was a national issue with Australia-wide implications. They were increasingly influenced by North American approaches and firefighting techniques. In the United States and Canada firefighting 'resembled a paramilitary activity', according to Pyne.[38] These ideas permeated back to Australia and were disseminated through interstate conferences. American articles were often reprinted in the periodical *Australian Forestry*. But foresters still viewed forests as a resource to be exploited. While many were aware of the unique nature of our flora and fauna, an environmental consciousness was only beginning to emerge.[39]

Fire science was developing through the work of R.H. 'Harry' Luke and Alan McArthur. Luke spent all of his working life until retirement in 1972 with the New South Wales Forestry Commission as fire officer and served for three decades on the New South Wales Bushfire Council. Together with Alan McArthur he wrote what was seen as the fundamental fire textbook, *Bushfires in Australia*.[40] Luke's *An Outline of Forest Fire Control Principles for the Information of NSW Foresters* (1947) became a working manual for foresters and firefighters.[41] The manual gradually evolved into *Bush Fire Control in Australia*, a popular work designed for non-professionals.[42] It sets out the basic scientific principles of how bushfires ignite and develop, and how firefighters and other landholders can respond to them. The American influence is clear, but Luke was also a great believer in trusting the experience of locals. He inherited a regime of indiscriminate burning and turned it into the art of 'fuel' (forest undergrowth) control and reduction through regular burning. He argued that because conditions differed across the country and local volunteer firefighters knew these conditions best, ultimate authority and responsibility for firefighting and prevention should be in their hands. Good land management began, he argued, with controlled fire, and locals knew how to do that best.

His colleague was Alan McArthur, born in 1923 in the Western District of Victoria.[43] McArthur lived and breathed bushfires. A blunt, pragmatic man who delighted in practical research, he had no patience with theorists, environmentalists, scientists, or critics. His values were absolutely clear. He stood for rural people, for the men and women who had to fight fires. Tracks had to be pushed into wilderness areas so that controlled burning could occur. It was the only way to prevent wildfires. For him protecting rural people, lands and production was the only value. And that could only be done by controlled burning. He lectured on fire control at the Australian Forestry School in the Canberra suburb of Yarralumla and he taught a whole generation that fire had to be used to control fire. 'It can be done as long as you work systematically, you can wear it down and beat it in the end', he said.[44] He advised investigations that followed major fires: the Rodger Commission in Western Australia in 1961, the Tasmanian Government after the destructive 1967 fires around Hobart, and the investigation into the western Victorian fires of 1977. The views of McArthur and Luke became the received doctrine right up until the advent of the contemporary environmental movement.

Another contributor to knowledge of the dynamics of fire was James Charles Foley (1892–1967) in his *A Study of Meteorological Conditions associated with Bush and Grass Fires and Fire Protection Strategy in Australia* (1947).[45] This pioneering work, which was inspired by the 1939 royal commission, examined the intimate, causative relationship between fire and weather. While this connection was known for many years, Foley was the first to draw together all of the scientific evidence, so that a scale could be drawn up indicating the type of weather that indicated fire danger and its severity. There is a 140-page historical section in his *Study* in which he attempts to detail by state every recorded fire from the late nineteenth century until 1946. His main sources were newspapers, still probably the most reliable historical starting point for the period before 1945.[46] With every fire cited in the twentieth century he gives detailed local weather information including temperature, relative humidity and rainfall.

v

Between 1944 and 1951 there was a lessening in fire activity across Australia. The droughts that had afflicted the country during the war and up to 1947 were broken and significant general rain fell. For four years the weather was generally more benign.

But by 1951 dangerous conditions were beginning to emerge again. In January 1951 Queensland was simultaneously subjected to both flood and fire. In mid-January there were record floods in Longreach.[47] There was flooding right across northern Queensland from the Northern Territory border to the coast. North of Longreach, at Muttaburra, the local policeman, Sergeant Simpson, said water surrounded the town 'as far as the eye could see'. There were reports of £3.5 million in flood damage, with graziers facing stock losses of £1 million and cane farmers £1.2 million. Swollen rivers flooded coastal towns.

But further south there were massive grass and scrub fires. These had been burning since 14 January around Charleville, 490 kilometres south-south-east of Longreach. There were fires to the west of the town north of the Quilpie road, and south of Charleville towards Wyandra with flames reaching 9 metres in height. Between spring and late January these fires eventually burnt around 3 million hectares.[48] As a result Queenslanders began to realise that they had to organise bushfire brigades.

A similar fire followed record rains in the Pilliga Scrub north of Coonabarabran, New South Wales, in 1951. The outbreak was started by a truck backfiring. Eventually 350,000 hectares were burnt—60 per cent of the Pilliga Scrub—over a period of a fortnight.[49]

The 1951–2 bushfire season had been preceded by what we now call the El Niño effect. This resulted in the northern two-thirds of Australia experiencing very low average rainfall.[50] The Northern Territory was particularly affected as well as the northern areas of New South Wales, South Australia and Western Australia. In the Northern Territory 1951 was one of the driest years on record: permanent waterholes were drying up, and packs of dingos were attacking weakened cattle. There was hot humid weather in Sydney and water shortages across the state. Sydney sweltered in 95 per cent humidity in the first days of 1952, and the temperature at Walgett reached 44°C. The worst drought for

40 years hit northern New South Wales from Grafton to Glen Innes, with extreme temperatures, high winds and low rainfall. Grafton received only 51 millimetres instead of its average of 406. The drought led to a meat shortage.

The fires had begun early in the season in November 1951, mainly in the north-western region of New South Wales where over 300,000 hectares of forest were burnt. Conservation Minister Weir said that of 158 fires, 58 began from burning off on private properties. In December 1951 there were fires around Sydney and in the Blue Mountains. High winds fanned a huge fire on a 16-kilometre front from Richmond to St Marys. Homes were destroyed around Londonderry. Other fires were burning around Springwood, Faulconbridge, Blackheath and Lithgow. There were more fires in the north of the state around Lismore. Some of these fires were still smouldering in early January.

South Australia—like Victoria—had heavy rainfall between April and August 1951, so summer brought a prolific growth of grass. In early December 1951 there was a heatwave across South Australia. Widespread grassfires broke around Adelaide. There were also grassfires at Maitland on the Yorke Peninsula and at Wilmington, south-east of Port Augusta. On 12 December 1951 a grass and scrub fire started on the Lower Eyre Peninsula between Tumby Bay, Koppio and Port Lincoln. Losses of stock and grain amounted to £250,000.

In Victoria fires began on New Year's Day 1952 with an amusing incident, although perhaps not so funny for the victim. Reporting it on 3 January, the *Benalla Ensign*'s front page headline read: 'Burning Trousers Start Bushfire?' 'When the occupant of a car threw his burning trousers away after a car had ignited his clothing, a big bushfire broke out in the Thoona district . . . on New Year's Day.' What had happened was that a car with several passengers from Dookie had broken down at Thoona, west of the Warby Range and Wangaratta. As they tried to get the vehicle started it backfired and 'ignited the trousers belonging to an occupant'—one wonders precisely how. Tearing them off, he threw them aside. They set lopped branches and leaves alight that had been left on the roadside by PMG workers, and from this the grass caught fire. Nothing is reported about injuries to the occupant of the trousers!

By Wednesday 23 January it was clear that south-eastern Australia faced a major bushfire crisis. There was a heatwave right across New South Wales and Victoria. Water shortages were widespread.[51] Not one town in New South Wales west of the Great Divide had a maximum temperature of less than 38°C. The temperature in Melbourne almost attained Black Friday proportions when it got to 43°C. In Sydney it was 41. There were hot, northerly winds and low humidity. In Frenchs Forest a firefighter reported seeing a bird whose feathers were alight flying into thick scrub. A few seconds later he saw smoke and fire from the area where the bird had disappeared.

Many of the 1952 fires were grassfires because there was luxuriant growth, due to a combination of pasture improvement and, ironically, myxomatosis. Farmers had worked hard to improve their grasses with superphosphates, better grass strains and aerial crop dusting. The rabbit population had also been hit hard by 'myxo'. Despite drought there was rich grass growth. In contrast to the rest of south-eastern Australia, the Western District of Victoria was deluged by rain; on 25 January 114 mm fell in Warrnambool.

The fires exploded on Friday 25 January. The chairman of the New South Wales Bushfire Council compared the situation to 1939 and warned in the *Sydney Morning Herald* of 26 January that 'the danger is extreme'. A massive fire broke out in southern New South Wales that spread across the border into northern Victoria. By Sunday the farmland and bush around Humula, Carabost and Rosewood, east of the Hume Highway and south of Tarcutta, was ablaze. The fires were so bad that the locals called Monday 28 January 'Black Monday'.

Many grazing families in the Mangoplah district had taken refuge in local dams, including Mrs Eunice Killalea and her infant son Christopher. She and her husband had abandoned their home and driven to his father's place when the fire approached. As it turned out their own farm was spared, but her father-in-law's was destroyed. Mrs Killalea told the *Daily Advertiser* of 28 January that when 'the fire caught the house, I rushed into a blazing bathroom to protect the baby'. With other family members she then 'rushed down to the dam on the property . . . We were waist-high in the dam as the fire burned all round us. At times we had to wet ourselves completely to stop from catching alight from the flying embers. We did not know when to make a break because of the terrific heat.' So strong was their community spirit that almost immediately after escaping the dam all the women started helping with catering and all the men with firefighting. No time was wasted on self-pity. Police investigated suspicious circumstances surrounding several of these fires. By weekend's end three people had been burned to death.

The situation in the Riverina on Tuesday was 'static' according to the *Daily Advertiser* of 30 January 1952. Fires continued to burn, but the focus of danger moved south to the New South Wales–Victoria border, to the rich farmland in the Upper Murray district east from Wodonga to Corryong. Tintaldra was destroyed, and only a few houses survived in Walwa. Flames 30 metres high leapt across the Murray River and 40 women and children had to jump into the river at Tintaldra to save their lives. Seven men were burnt, five of them critically. Blood plasma was flown to Walwa Hospital which was just saved from the fires. There was widespread damage to properties in the area. The mountain and forest country west of Cudgewa and north of the Murray Valley Highway was ablaze. Police worried that if these fires were not brought under control up to 2000 people might have to be evacuated from the area.

Meanwhile on Friday 25 January a highly destructive fire had hit the far south coast of New South Wales, particularly the Bega, Cobargo and Tilba Tilba districts.[52] Five people were killed. Bega Hospital was just saved, and the town was in dire danger because many of the buildings were of timber construction.

Typical were the tragic deaths of the Otton girls: Marie, sixteen, and Jennifer, fourteen, from Upper Brogo in the coastal range between Bega and Cobargo. Above the coastal strip north of Bega there are rows of hills that rise successively higher towards a 1300-metre escarpment. This is rugged, isolated wilderness that today makes up the Wadbilliga National Park and the Brogo Wilderness. Throughout January 1952 there had been many fires along the timbered ridges and in the isolated back country. People farming around tiny settlements like Upper Brogo just west of the Princes Highway were apprehensive.

On Friday 25 January the wind changed and, the *Bega District News* reported, 'a north-west gale swept the fires out of the hills into the open. The sun glowed red through a thickening cloud of smoke and flying cinders and burnt leaves as the first word came that a mighty blaze had leapt high in the Bemboka Peaks and Upper Brogo.'[53] The fires hit the settled areas in the late morning. 'The whole countryside exploded in a horrible pattern of racing, terrifying fire.' People 'found themselves fighting for their homes and lives as the flames raced over the country in a hurricane of searing heat.' One spur of the fire came out of the mountains around Bemboka, north of the Snowy Mountains Highway, and headed down into the lush Bega Valley. Another spur emerged to the north of Puen Buen and Upper Brogo heading generally eastwards towards the Princes Highway.

Calls for help came through in the early afternoon from Fred Otton's Upper Brogo farm, 'Hillfield'. The Otton family had fought many fires over the years, and Fred first felt there was no immediate danger to life. But it soon became clear that there was little chance of saving the house and despite his trying to burn a break, the house burst into flames. The men who had been fighting the fire had to take refuge in the Brogo River. Mrs Otton recounted that she and her daughters had packed the lorry with household goods and 'I had even cooked a cake and an apple pie and these were packed on the lorry too'.[54] But when it became clear nothing could be done to save the house:

> The girlies and I went up to the spring on the northern side of the house about 200 yards away. [We] . . . were left at the spring with [Rudolph] Ogilvie [a 64-year-old neighbour] and his horse . . . Ogilvie must have enticed them away. The next thing I knew was the two girlies trying to mount his horse. I called out to Marie. She came over to me and I grabbed her wrist and I said to her 'I want you to stay with me and if the worst comes we'll jump into the spring'. She wrenched herself free and ran to Ogilvie's horse again. She was afraid and panicked and I did not realise it . . . All at once the wind changed to the west and blew at a terrible speed. The heat was too much for me. I waited a second—Jen was on Ogilvie's horse and off they set down the hill to the River. Marie and Ogilvie ran off together into a gully . . . I was burning now and had to jump into the spring to save myself. Marie and Ogilvie, I'm told, did run back

but up the hill above the spring and there they were found. It was 2.35 by my clock when the fire came.

Mrs Otton just survived immersed in the spring. At one stage she had to hold a tin covering over her head. After the fire, reported the *Bega District News* on 1 February, 'a search revealed the shockingly charred remains of Ogilvie and one of the girls. On the following morning the other girl's body was found under a tree, approximately 150 yards from the spring.'

The fires lasted for seven hours. Around 5.00 pm the wind turned southerly and a change came through. But the whole Bega district had been laid waste. An area 90 by 160 kilometres had been burnt out and £1,800,000 worth of damage had been done. The massive losses of cows and sheep meant that there was a shortage of milk, butter, cheese and meat. It was estimated that the January fires had cost New South Wales £5 million.

Even before January 1952 had passed a *Sydney Morning Herald* editorial was bemoaning the outbreaks as a 'large-scale disaster' and was complaining about 'the ravages of fires illegally lit or started by reckless throwers of cigarette butts' (29/1/52). It also ran a front-page headline on 30 January quoting fire scientist Harry Luke warning that, 'We are about two days off another big blow-up. Weather conditions seem to be working up to a renewal of hot, dry, north-westerly winds, which are very dangerous. If they come fires in the present fire areas may get out of control again, and we may expect fires almost anywhere in the state.' As it turned out Luke's predictions were accurate. There were bushfires that very day around the Sydney suburbs.

A day later the worst grassfire in the history of the Richmond River district broke out south of Lismore in northern New South Wales. Patients from the local hospital were loaded into trucks in case they needed to be evacuated. A large fire on the coast between Yamba and Port Macquarie almost led to the Hastings District Hospital being destroyed. North of Lithgow on a plateau in the mountains a radiata pine forest went up in flames 'like an ammunition dump exploding' and firefighters fled for their lives. A forestry foreman said 'the whole area was like a seething volcano'. Describing what happened, Police

Inspector W.E. Lind said 'A veritable hurricane roared through the forest and fire burst everywhere at once. The men had no hope of holding it. If there had been 1,800 instead of 90 men, they would have had to run . . . The whole forest has gone. It has simply been wiped out.'

Closer in to Sydney there were fires in the Wilton–Douglas Park area, and Appin was only saved by the sheer speed of the fire that raced right through its centre. The postmistress, Mrs Floyd said, 'It was blowing like a gale. The men had no chance of stopping it. It came through with a terrific rush. It was like a blowlamp being suddenly turned on the town. Then the wind carried it through and the worst was over'.[55]

In the outer Sydney suburbs, bushfires swept up from the sandstone valleys west of Hornsby. The kind of difficulty that firefighters faced was illustrated in Dural Street, Hornsby when two 'elderly women, one of whom is bed-ridden, refused to leave when their home was menaced, but a young man carried out the invalid and a pillowcase full of valuables. When the other refused to go unless she took some of her clothes with her, another man led her out with one hand and carried her clothes with the other. Then he went back for more clothes. Firemen saved the house' (*Sydney Morning Herald*, 31/1/52).

At the same time there were fires along the New South Wales south coast from Nowra, which was surrounded by flames, and many of the coastal forests were ablaze near Burrill Lake, Batemans Bay and Narooma. The town of Milton was entirely cut off by road. There were also fires near Bermagui, and new outbreaks at Cobargo and Bega. The eastern Riverina fires also reignited around Humula and Tumbarumba. There were fires burning from Tumut to the Murray River. There was particular concern that they didn't get into the rough, inaccessible country around Maragle Mountain and the western edge of the Snowy Range. However, almost as quickly as they arose the Riverina fires died down but never went out completely, and they revived briefly again on Tuesday 5 February fanned by scorching north-west winds.

In the Blue Mountains, Blackheath was just saved from a fire that roared up from the valley at Centennial Glen and burnt the southern edge of the town. Police Sergeant Scanlon said:

> About midday we thought we had it under control. Then a choppy westerly wind whipped up the flames and it just took off. It is impossible to fight a fire like that. You stand in front of it for a little while then you have to get away from the terrific heat ... We had old men and schoolboys fighting the blaze. They used everything they could lay their hands on. Some used green branches, some used sacks. But they stopped the fire. They did a wonderful job. (*Sydney Morning Herald*, 4/2/52)

Blackheath firefighters said it was like a furnace driven by a tornado.

At the same time much of northern Victoria was still in the grip of a drought that had begun in 1944. Feed for cattle and water for human consumption had almost run out in many towns. Suddenly in early February 1952 northern Victoria exploded into flames along a section of the Hume Highway from the small town of Barnawatha to Wodonga with spurs running south-eastwards through the Pilot Range to Yackandandah and eastwards to Tallangatta on the Murray River.[56]

Two people, a man and a boy, died in Barnawatha when the fire literally swept down on the town from two directions at 65 kilometres an hour, and another man was killed when he fell from a fire truck, broke a leg and was engulfed in fire near Howlong, just across the Murray in New South Wales. There were many people badly burnt. Reporters from the *Wangaratta Chronicle Dispatch* of 2 February 1952 said that:

> flames licked both sides of the Hume Highway as we followed a water truck into Barnawatha ... thick smoke hung over the still burning shops, homes and churches ... [It billowed] like the mushroom of an atomic explosion ... Local brigade men attempted to burn a break but the wind whipped the fire away and trapped them on the highway. Six men crawled to the brick Roman Catholic church to shelter in the vestry, spraying the interior with their knapsack pumps.

This Hume Highway fire was just stopped at the edge of Wodonga, which resembled a receiving station for a bombed-out area. First aid posts were set up in the main street. Trains on the Melbourne–Sydney line were delayed because of bridge burn-outs. This fire was probably started by a careless Hume Highway motorist throwing a cigarette butt into the bush.

Closer to Melbourne there were outbreaks in the Mount Macedon–Woodend and Bacchus Marsh–Gisborne districts. Douglas Hislop became trapped by a fire at Diamond Creek, the scene of the terrible McLelland family tragedy in 1851, but he jumped off a 13-metre cliff into the Plenty River and escaped with minor burns to his arms and legs. Patricia Townrow, a Canadian war bride, and her two daughters, Marjorie, four and Gloria, two, of Nullawil north of Wycheproof in the Mallee, were trapped trying to run from the family home to a dam. The children died two days later from their burns, and Mrs Townrow, who was still in a critical condition, gave birth to a stillborn son a couple of hours after the girls died.

More fires broke out in mid-February right in the heart of Ned Kelly country east of the Hume Highway. Houses and churches were destroyed and thousands of head of sheep and cattle perished around Kilfeera, Lurg, Molyullah, Winton, Glenrowan and Greta. The fire started in a sawmill near Ryans Creek. The *Benalla Ensign* of 7 February almost waxed poetic about the beauty of the fire: 'At night, the smouldering trees on far away hills looked like the lights of big cities, while the deep red-pink glows rose all around'. But the reality on the ground was different. Two people died, nearly 40 homes were destroyed and thousands of sheep and cattle were killed. A vast area of good faming land was devastated. One of those who died from burns was 43-year-old farmer, Mick Tanner.[57] With a 65-kilometre an hour wind blowing and paddocks aflame almost a kilometre and a half ahead of the main blaze, Mick and his brother-in-law, Jack Wardle, 'were rushing sheep and cattle to safety. Their wives, who are sisters, were back at Mick's home preparing for the worst. They both felt they would never come out alive, but they attacked the flames gamely and were successful in saving the house. Mick, who had helped another neighbour yarding cattle, had moved most of his own sheep to safety, when he went back to save a haystack and he was caught in the approaching flames.'

Meanwhile Jack was struggling to save his own life. He had been trapped when his car caught fire. Abandoning it and 'rushing through the flames he fell to the ground in dense smoke, but managed to struggle to his feet and running along the road . . . escaped to safety. His car was completely destroyed.' He went back to search for Mick and 'found him in his car suffering badly from severe wounds on the hands and face'. He died in hospital a few days later. As a sad coda to the story the *Benalla Ensign* reports that, 'All Mick's rescue work was of no avail, as a neighbour cut the fences where the sheep were sheltering, and as they moved out most of them were mopped up by the passing inferno. Out of 1,800 sheep and 135 cattle, 1,500 sheep and 120 head of cattle were destroyed.'

There were also outbreaks around the Australian Capital Territory and nearby New South Wales. There had already been fires in Canberra including one that was thought to have been started by a magpie causing a short-circuit on a powerline.[58] Other fires on Friday 25 January 1952 broke out in Yarralumla and Red Hill and, driven by a 90 km/h wind, burnt south-eastwards for 28 kilometres and were only stopped in New South Wales at Royalla and Burra. At one stage the main street of Queanbeyan was threatened.

On 6 February the heat was stifling. A fierce fire broke out in isolated country near Wee Jasper and another fire started from lightning about 10 kilometres north-west of Mount Stromlo, close to the junction of the Molonglo and Murrumbidgee rivers. In an almost exact foretaste of what was to happen in January 2003, the fire was driven by almost gale-force westerly winds along Uriarra Road towards the Australian National University's Mount Stromlo Observatory, which at that stage was rather isolated from Canberra.

The fire approached the mountain on a kilometre-and-a-half front through the radiata pine plantation established in 1915. Here it gained enormous velocity. 'At one stage the Minister for the Interior (Wilfred Kent Hughes) who visited the fire, ordered firefighters to fall back from the main centre of the fire, and take up positions to protect the city area', reported the *Canberra Times* on 6 February.[59] The blaze headed up the slopes of Stromlo with 'willie willies' of flame rising more than 30 metres. The workshop with essential machinery and equipment was destroyed. The observatory was saved by a small group of scientists turned firefighters. 'As scientists battled against the fire exploding gas cylinders in the workshop scattered huge pieces of metal through the walls past the firefighters . . . Only the efforts of a small group and a sudden change of wind saved the main section of the Observatory, which contained equipment valued at more than £1,000,000.' The Director of Mount Stromlo from 1939–55, Dr Richard van der Riet Woolley, had already made 'certain recommendations', presumably pointing out the danger of pine forests close to the observatory, but nothing was done by the federal government that then administered Canberra.[60]

Reflecting on these fires the *Canberra Times* of 6 February 1952 editorialised that while there were many volunteers, the ACT's firefighting organisation 'proved insufficient for the demands made upon it, the equipment necessary for fighting fires was not actually in readiness for the outbreaks, and that safeguards against the spread of fire had not been taken on an appropriate scale . . . Fire breaks had not been provided . . . and property owners had not taken elementary precautions to remove fire hazards or to protect their property.' It sounded very much like the post-mortem on the January 2003 fires. The next day the *Canberra Times* was calling for a national approach to firefighting and fire prevention. Fire officers needed greater powers to force property owners to protect themselves and their neighbours. 'If a model were to be created to show how a regional organisation could function Canberra could well provide the centre.'

Ever so slowly the fires died down. But the destruction had been massive. In New South Wales alone, eleven people had been killed and 4 million hectares of land destroyed. In Victoria there were ten deaths and much of the state had been burnt, especially in the north-central and north-east. The summer of 1952 had been terrible, almost as bad as 1939.

vi

Between 1952 and 1957–8 rainfall was reasonably steady and there were few significant fires. The only exception was on 'Black Sunday', 2 January 1955, when the Mount Lofty Ranges east of Adelaide were devastated by fires. The Adelaide temperature reached a maximum of 43°C about 1.00 pm. Wind gusts of 100 kilometres per hour were recorded, with higher velocities in the hills. The fire was particularly bad near Upper Sturt where the railway station and houses were destroyed, as well as at Marble Hill where the governor's summer residence was burnt down, and at Crafers, Aldgate and Loftia Park resort. Two firefighters were killed and 40,000 hectares were burnt out. The damage bill was £4 million.

Throughout 1957 most of the southern half of the continent was drier than usual with parts of central New South Wales and Western Australia recording their lowest rainfall on record from March to December. By summer Sydney had strict water restrictions; there was a total ban on hoses and sprinklers. Warragamba Dam was being built, but had not yet come on line. The first bushfire in New South Wales was on Saturday 30 November 1957 when four boys were killed in the Grose Valley's Blue Gum Forest, a thick stand of *Eucalyptus deanii*, in the Blue Mountains. The shortest way in is by a very steep descent of almost 600 metres from Perrys Lookdown near Blackheath to the valley floor. A group of nine boys, aged from twelve to 21 from Saint Alban's Church of England, Belmore, walked into the valley on Friday evening and visited the Blue Gum Forest on Saturday for lunch.

About 2.00 pm they saw smoke about 2 kilometres away. Fearing fire, they headed back quickly toward the ascent up to Perrys Lookdown. Guidebooks comment on how steep it is; on average it is a two-hour climb up from the valley floor. At about 2.20 pm, just as they were beginning the ascent, the boys were surrounded by fire. They were in two groups. The five boys in the lead group tried to outrun the fire up the steep slope. All but one were quickly overwhelmed and were later found below a cliff face, burnt to death. The sole survivor of the leading group was carrying a pack and fell back and was eventually found by the second group.

The second group was saved by one boy, Barry Carter, seventeen, taking the initiative. The oldest of the group, Peter Warby, 21, told the *Sydney Morning Herald* on 2 December:

> We were going quite well but the fire suddenly burst about 30 yards from us and a large area of bush suddenly shot into flames . . . Barry shouted out 'Ditch packs . . . Get down to the bottom' several times, but the party in front didn't listen to him

and ran up the hill with the fire chasing them. We got behind the nearest rock we could find. Barry led the devotions there for a few minutes. The fire was coming up at us on all sides and we got badly singed. The heat was terrific . . . [Then] Barry [ran] through a wall of fire and I could see him in the clear shouting 'Come down' . . . We ran down, but we couldn't [find] Barry. We stayed on a ledge for a while, but it was too hot there and then we crawled into a cave. The cave was just big enough for the three of us and it was all wet inside with water trickling out around it . . . the fire passed right over the top of us.

After emerging from the cave they met the sole survivor of the first group. Warby says they eventually made their way 'to the floor of the valley and took shelter around the pools of water there. About an hour later the fire swept down on us again and we had to jump into a watercourse until it had passed us. We just had our faces out of the water.' Carter meanwhile climbed out of the valley to raise the alarm. The others were rescued by police the next day. The *Herald* reported, 'They said that but for [Barry] shouting advice over the roar of the flames they too might have attempted to run from the fire instead of into it'.

This was a terrible overture. On Monday 2 December, a day of extreme fire danger, outbreaks occurred across the Blue Mountains, and 133 homes were destroyed in Leura and 25 in Wentworth Falls. Fifteen other buildings—shops, churches, schools and a 50-room guesthouse—were lost. The damage bill was in excess of £1 million. In an act of bravery a young mother, Betty Gunter, used her own body to shelter her two young boys, Warren, seventeen months, and Trevor, three. With her house on the main Western Highway threatened and unable to drive the car, she grabbed the children, ran barefoot across burning grass and lay them in the gutter. She said that:

As I crouched in the gutter, I prayed to God that my babies would be saved. Even as I prayed I felt the pain on my legs and the flames licked over me'. Mrs Gunter, badly burnt, picked up her youngest child and ran again over burning grass in the hope she could find some safer place. While she was stumbling through the smoke, a man on a motorcycle appeared. 'Please go back and bring my other little boy', Mrs Gunter said. He drove back through the smoke and found the child. He helped Mrs Gunter and her two children get to safety. (*Sydney Morning Herald*, 3/12/57)

The Blue Mountains fires continued to burn in the Megalong Valley and in isolated country south of Lithgow and north of Jenolan Caves.

In the aftermath of these fires there was a prescient article in the *Sydney Morning Herald* on 5 December setting out the relative arguments for and against preventative burning and examining its consequences for 'ecology'—the actual word was used in the article. The main protagonist was Miss Minard Fannie Crommelin, popularly known as 'Crommy', Woy Woy's first postmistress. She had bought a property at Pearl Beach which she donated to the Botany Department of Sydney University.[61] The *Herald* article contrasts the conservationist-postmistress with 'practical' men like local fire captain Harry Small and Constable Bob Sawers, who almost seemed to be salivating for a 'burn'. For ecological reasons Crommy was totally opposed to any form of deliberate burning. 'The theory that burning-off is a protective safeguard has long been exploded by scientific investigation', she wrote in a letter to the *Herald*. '. . . Breaks should be cut, not burnt. The moist humus on the forest floor is its most valuable asset, and one of its protections. Each fire brings a drier condition, and blady grass and bracken—the great fire carriers.' A.E. Kelso would have been proud of her.

vii

In January 1958 there had been some fires around South Australia, but generally it had been a mild season. However, in April that year in a small (1370 hectares) but intense fire in a radiata pine plantation at Wandilo, north of Mount Gambier, eight firefighters died as a result of exposure to intense radiation. What happened was that a sudden and dramatic increase in the wind created a fire-storm and three fire trucks became trapped. Three men survived with moderate burns, but eight other men were killed as they ran from the trucks which were destroyed.[62] Luke and McArthur comment that, 'The main part of the fire season had been relatively mild though the autumn was dry, and one would not have predicted such an extremely serious fire.'[63]

For a decade Western Australia had been relatively bushfire free, but in January 1961 that changed. Spring and summer were dry and throughout the first three weeks of January there had been a heatwave in and around Perth. There were 33 major fires around the state, some of which were quite persistent, particularly north-east of Geraldton and north-west of Northam. In late January there were several days with temperatures reaching 44°C in Perth. By Tuesday 24 January smoke covered the metropolitan area and reduced visibility to less than a kilometre well out in the Indian Ocean. The *West Australian* reported that, 'Most of the smoke came from a huge fire burning in the hills near Keysbrook [about 55 kilometres south of Perth]. It has burnt through more than 40,000 hectares of state forest in the past four days. About 500 men fought the blaze yesterday . . . while hundreds of others battled against smaller fires burning in the hills near Perth and deeper into the forest country.'

On the evenings of 19 and 20 January 1961 there was dry lightning along the Darling Range. Humidity was low, the temperature reached 40°C and many spot fires broke out extending about 35 kilometres from Serpentine to Pinjarra.

By Tuesday 24 January the fires were driven together into a massive conflagration. Impelled by strong north-westerly winds with near-cyclonic gusts up to 120 km/h, the southern flank of the fire made a run through the isolated timber villages of Holyoake and Nanga Brook. The entire population of the area was evacuated to Dwellingup. But at 7.30 pm the Forestry Department told Pinjarra police that a completely uncontrolled fire was approaching Dwellingup. 'The last report received at the Pinjarra Post Office was that the garage at Dwellingup had exploded. The line then went out of order. All roads were cut by both flames and fallen trees and telephone and telegraph wires were brought down . . . Flames were reported to be leaping in five kilometre jumps from tree top to tree top in the path of the strong wind', reported the *West Australian* of 25 January.

Eventually radio contact was re-established. More than 50 homes, the Catholic church, the hospital, 60 cars, the sawmill and the forests department building were destroyed. There were reports of flames up to 100 metres high. Miraculously no one was killed and injuries were confined to minor burns and smoke blindness. People took refuge in nearby Pinjarra which, according to the *West Australian* of 26 January, 'resembled a wartime evacuation centre. Its streets were crowded with vehicles, many of them severely heat-blistered and smoke-blackened.' Around 145,000 hectares of bush had been burnt. By Wednesday night there was steady rain and the fires were gradually controlled.

The fire was followed by a royal commission, chaired by G.J. Rodger, a man with wide experience of forestry in eastern Australia. Among the advisers was Alan McArthur. Western Australia had been a model of controlled burning, so how could this conflagration have occurred? Could it be that the theory was not working?

The royal commission's answer was that this was a completely exceptional event, a one-in-100-year phenomenon. It was pointed out that nearly half of the fires had been ignited almost simultaneously by lightning, and this meant that it was impossible for the authorities to extinguish them all at the same time. The fires got away and once they merged they were beyond the control of anyone. The system had not failed and Luke and McArthur commented that, 'The main lesson to be drawn from [this] fire was that control over a number of simultaneous outbreaks in severe weather conditions is not possible unless fuel has been reduced to manageable proportions'.[64]

For a decade after 1952 Victoria had been largely spared from major fires, but on 10 January 1962 there were minor fires in the Little Desert near the South Australian border north of Dimboola, in the Grampians and at Casterton. Many of these fires were due to lightning strikes, but two pre-Christmas grassfires near Somerville in the northern Mornington Peninsula were lit by an arsonist. The state was tinder-dry and the fire danger extremely high. All Melbourne's dams were low, the result of a long, dry spell.

The Dandenong Ranges and the mountains around Melbourne exploded into fire on Sunday 14 January 1962.[65] The worst fires were in the Sassafras–Olinda area and around Healesville. Both were deliberately started: the Sassafras–Olinda fire began at The Basin in a backyard incinerator. Police later discovered that a sixteen-year-old youth had lit the fire and then reported it to his volunteer fire brigade and attempted to impress people as a firefighter. But the fire got out of control and spread across the Dandenong Ranges.

At Healesville the fire began in a woodshed lit by two young rabbiters. There were also fires in the Christmas Hills near Yarra Glen also lit by three rabbiters. At Lancefield another fire was started by a youth smoking a ferret out of a hole. Yallourn was again threatened. The *Age* reported that a 'new Australian' saved the town by repeatedly driving a bulldozer into the advancing fire. On late Monday fire authorities were optimistic that things were under control. Firefighters had their first rest in 48 hours. Many collapsed from heat exhaustion and smoke inhalation. Some were bitten by snakes fleeing the flames.

The key issue on Tuesday was the wind. It turned out to be a day of gusty northerly winds with a temperature of 40°C. The fires burst out anew and came within 18 kilometres of Melbourne. There were mass evacuations from the Dandenong Ranges around Kallista, Sassafras, Olinda and Kalorama. At one stage Olinda was totally surrounded, and 300 firefighters were trapped in St Andrews, just south of Kinglake, when the town was hit by a 10-metre wall of flame for two hours. Near Warrandyte at least 50 homes were destroyed in an hour. Fern Tree Gully was just saved by a wind change as 14-metre flames raced through the nearby national park to within 80 metres of the main shopping centre.

But the saddest story of the 1962 fires came from near Woori Yallock. Leslie Ockwell, 63, and his wife Linda, 60, were looking after their grandsons, Ronald, fourteen, and Geoffrey, ten. Leslie had been a hero of Black Friday when he led 37 firefighters through the flames to safety. But in 1962 he made a tragic error. When their home was threatened the four of them made a run for it in a utility truck along the Healesville road. But the wind changed direction and, blinded by smoke, Leslie drove the utility into a tree about 280 metres from their house. Like the Kerslakes in 1939, they abandoned the vehicle and started to run. But they headed straight into a cauldron of flame. Their bodies were found on the side of the road about 25 metres from the burnt-out shell of the utility. Their house was unharmed by the fire.

Twenty-five kilometres away near Panton Hill Iris Robins led her six children, aged between sixteen and eleven months, another girl aged ten, a horse, two dogs, two kittens in a basket and a canary in a cage over 3 kilometres to the main road when their house was threatened with fire. The *Age* of 17 January reported that they were picked up by a young man from Essendon in a car he had literally grabbed from the side of the road as the fire approached. After picking up another old couple they sped down the Kangaroo Ground Road. All in the car survived. The horse was left behind.

By Wednesday the situation had eased following rain and mild weather. The situation intensified on Thursday, but rain again brought relief. However, four more people had died. Looters moved into the Dandenong Ranges, robbing houses before residents returned, and an arsonist, 'whose emotions [police believed] were aroused by the Dandenong Hills blaze', said the *Age* of 18 January, unsuccessfully tried to burn down Melbourne's government house.

Three days of fires, almost all deliberately lit by what the *Age* editorial on 19 February called 'lunatics and half-wits', had left eight people dead and 454 houses and buildings destroyed. But as the newspaper's editorial pointed out, the real problem was the Melbourne suburban sprawl penetrating 'into the heavily timbered country of the Dandenongs . . . Since it is impossible to ensure the tinder dry forests against the spark of a match the main task of the government is to organise its fire-fighting services so that every outbreak can be extinguished before it has run riot.' While organisation was important, the real issue was—and still is—the consequences of the spread of suburban sprawl into mountainous forest areas.

viii

The years 1963–4 saw south-eastern Australia experience a seven-month drought.[66] The result: bushfires in the 1964–5 season in New South Wales and Victoria.

In New South Wales right through January to March there were scattered destructive fires in the central west near Wellington and inland at Coonabarabran and Inverell. While stock and homesteads were safe, these fires caused heavy losses among native animals including koalas. There was also a deliberately lit fire in the Royal National Park, south of Sydney, which was intense and dangerous. It was eventually brought under control but the consequences for the park were bad. In an article in the *Sydney Morning Herald* on 22 January 1965 the nature writer and environmentalist, Alec H. Chisholm, said the park had 'been swept by fires of such magnitude that a great part of the area is simply wasteland'. It resembled 'a blasted heath, almost gruesome in its stark desolation'. The only comment from authorities came from 'a Park employee', presumably quite junior, who said 'in a trivial and uninformed statement' that 'little harm had been done to fauna and that burnt plants would re-establish themselves in time'.

By mid-January things had settled down in New South Wales, but the fires blew up again on 7 February 1965. Driven by a strong wind, a deliberately lit grassfire near Campbelltown led to a child welfare home being threatened and ten babies aged from a few months to two years were rescued. The fires then settled down in New South Wales until 6 March when the whole state was declared a fire-risk area. There were outbreaks at Nowra and Cessnock, as well as near Canberra, and in the Kosciuszko National Park (or 'State Park' as it was then called) between Kiandra and Cabramurra. There were also fires around the northern Sydney suburbs close to what is now the Berowra Valley Regional Park.

The worst-hit area in New South Wales was the southern tablelands and a large tract of the Morton National Park, between Goulburn, Moss Vale, Taralga and Nowra, was burnt. Described, exaggeratedly, by the *Sydney Morning Herald* of 8 March as 'one of the most disastrous bushfires in NSW history', it also destroyed at least 100 homes and a motor repair garage on the Highland Way at Wingello and Tallong between Bundanoon and Marulan. One man, Isaac Mills, 55, of Tallong was killed fighting these fires and fifteen were injured. Only one house survived in the main street of Wingello. At one stage the Hume Highway, then a two-lane road, was enveloped by fire. 'Motorists who made frantic dashes to safety along the Hume Highway told of sheep dying in agony in paddocks along the road . . . "the whole highway seemed to be ablaze for up to 10 miles at a stretch", one motorist said', reported the *Sun-Herald* of 7 March 1965. The fires continued right through until Wednesday 10 March, by which time the danger zone had shifted to the north of Sydney and along the Pacific Highway around Mount Ku-ring-gai. Firefighters concentrated on saving houses as the fires again came up out of the valleys and surrounding bush. Light aircraft were used to 'spot' the centre of the blaze.

But while 1965 was a bad season in New South Wales, the worst fires were in Victoria. In mid-January 1965 there were grass and scrub fires right across central Victoria. A 70-year-old man died fighting the fires west of Bendigo, and a 29-year-old, Alan McKean, was killed near Inglewood. But the worst tragedy was the death of seven women and children just outside the small town of Longwood on the Hume Highway, 13 kilometres south of Euroa. The *Age* of 18 January reported that fire had swept into the area and they were fleeing from a threatened farmhouse in a Holden station sedan, which was the second car of a three-car convoy. It was driven by Dorothy Oxenbury. She had her three children, aged twelve, four and one, her mother-in-law, Daisy Oxenbury, and two of her nieces aged nine and eight in the car. They were killed when they ran off the road in heavy smoke and crashed into a tree. In the confusion the other two drivers did not notice the loss of the middle car until they got into Longwood. By the time they got back it was too late. Six bodies were found in the car by Dorothy Oxenbury's husband, and the body of a seventh child was found in a gully about 200 metres away.

However, the most serious Victorian outbreaks were in the mountainous wilderness of what is now the Alpine National Park. There is a parallel between these fires and those of January 2003: both were substantially alpine fires. In 1965 they began north of Maffra at Valencia Creek and Boisdale and burnt northward through isolated wilderness toward the Dargo High Plains. There was a second fire to the north-east between Omeo and Benambra. The fires continued to grow gradually during the week and the fear was that a northerly or westerly wind would drive them towards each other. Another fire broke out in the Mitchell River Wilderness through spotting from the Dargo fire.

By Wednesday 25 February the burnt-out area was 39,000 hectares. The Forests Commission was reported by the *Age* of 25 February as advising 'there was little hope of containing either fire until rain fell'. Frank Halls from the State Film Commission had a very lucky escape from the Mitchell River fire. Trapped by a wall of flame on the narrow Dargo Road, 16 kilometres south of Cobbannah, he could not turn back because he had a trailer behind the car. The *Age* of 27 February reported that he thought of

> abandoning the car, but decided he would be better off with it. 'Winding up all the windows of the station wagon, I planted my foot on the accelerator and went straight for the wall of flame. Heat from the flames poured through the closed windows, but I kept my eyes on a bank of earth to stay on the road. I thought I was finished. All I could see was smoke and flames. I thought I'd never make it'. When he broke out from the 50 feet of flame and fire that covered the road, Mr Halls said the fire sounded like a train as it raced down the mountain slopes.

The situation held until the middle of the next week. On Tuesday 2 March a minor fire fanned by a northerly wind took on major proportions in the Buckland Valley near Harrietville. By Wednesday 4 March all hell had broken loose with the two Gippsland fires uniting to become the biggest Victorian bushfire since 1939. The timber town of Bullumwaal north of Bairnsdale was evacuated. The two fires covered an area of 80,000 hectares in inaccessible country with deep valleys and steep bush-clad hills which made it very difficult to control. This was a crown fire that spotted kilometres ahead of the main blaze. The *Age* of 4 March quoted the Forests Commission as saying that these fires 'were hopelessly out of control' and only rain would put them out.

By Friday 5 March driven by strong, hot, north-westerly winds, the fires had reached massive proportions. They extended almost 100 kilometres north of the Princes Highway from Bruthen in the east to Briagolong, north of Sale in the west. At one stage there were fears for the town of Bairnsdale. The *Age* of 6 March reported that Police Inspector Warren of Sale said, 'the position was critical and would remain that way until rain fell'. A whole armada of supplies, equipment and firefighters, including 190 soldiers, were massing to deal with the conflagration and water-trains were being brought from Melbourne. The epicentres of the fires were around what is now the Mitchell River National Park and another between Tambo Crossing, Ensay and Swifts Creek on the Bairnsdale to Omeo road. Eventually rain began falling on Monday night, 7 March, and by Wednesday the fires were more or less under control.

After nearly three weeks of fires Victorians had become resigned to living with the danger. A firefighter, John East, 65, told an *Age* reporter on 8 March: ' "We've stopped thinking about it" . . . "It's just something we've got used to . . . All the blokes have been working about 18 hours a day and so have a lot of the kids. They do their share too . . . We're all very tired" '.

Group Captain Cornish from the RAAF base at Sale told the *Age* of 9 March that flying over the Gippsland fires ' "was like approaching Dante's inferno—more than 1800 square miles of country is ablaze" '. Two-thirds of New Zealand, 1900 kilometres away, was covered by smoke from these fires. 'Airline pilots said the smoke carried across the Tasman Sea on a north-west airstream extending up to about 19,000 feet. In Wellington the smoke caused a copper-coloured sunrise and tinged the sun a deep red' reported the *Age* on 9 March. During these fires one of only two brush-tailed rock wallaby colonies in Victoria was wiped out in the isolated Buchan River Valley.[67]

But these weren't the only fires in Victoria. Three men were killed on 2 March in a fire at Eltham when they ran towards the flames in a gully rather than away from them. Fifteen houses were destroyed and several others badly damaged. On 8 March an arsonist re-ignited fires near Lorne. These were driven by winds that gusted, according to the *Age* of 9 March, 'up to 80 miles an hour'. Three days later a fire in the small town of Toolern Vale, 6 kilometres north of Melton and just west of Melbourne, resulted in the loss of fifteen houses, as well as the state school, the public hall and a local church. There were also fires in the Eltham–Warrandyte district, where ten houses were burned, including one with paintings valued at £8000.

By Saturday 13 March 1965 the Victorian fires were over. Paradoxically, in mid-July 1965 there was heavy snow on the Great Dividing Range including the Blue Mountains and as far north as the Liverpool

Range, north-west of Scone.[68] The last time this happened was 5 July 1900 when there was a heavy snowfall on the Blue Mountains and the western plains.

Things settled down on the mainland until 1968, but it was a different story in Tasmania. On 7 February 1967, following four days of extreme fire danger, one of the worst single bushfire events and the worst civil disaster in the history of Australia occurred in and around the city of Hobart. The results were horrendous.

ABLAZE, SOUTHERN TASMANIA, 1967

i

Originally used in nautical parlance, the word 'snug' is defined in *The New Shorter Oxford English Dictionary* as 'properly prepared for or protected from bad weather; shipshape, compact', or being 'cosily protected from the weather or cold'. Another meaning was a safe place for ships to lie at anchor. Apparently named by a nineteenth-century ship's captain whose sick sailors recovered in the restful surroundings, the town of Snug in southern Tasmania quickly became a holiday destination for nearby Hobart. Wistful holiday-makers wrote to family and friends saying they were 'snug as a bug in a rug at Snug'.

But by late afternoon on 'Black Tuesday', 7 February 1967, Snug belied its name. It lay in ruins. Eighty of its 120 houses were destroyed with only chimneys left standing. Also destroyed were two churches, two shops and the modern section of the local school. Eleven people were dead. More than half the town's population took shelter in the sea. Among them was Lavinia Millhouse, a young mother and her four children, aged from seven to three. They had been visiting her mother Mrs Freeman, an invalid, who was looking after the family shop. Her husband was away firefighting. Suddenly, the fire swept down on Snug and Lavinia realised she had to save her children.

> 'We ran into a wall of flame, and I was dragging the kids behind me. I went back for Mum, but I couldn't get through the fire.' Mrs Millhouse then led her children the short distance to the beach. 'I got all the kids into the water, and we got in as deep as we could . . . There were other people who had arrived, and they drove their cars through the sand right to the water's edge and then they got in the sea as far as they could. We all cuddled a child apiece.' Forty adults and children stayed in the sea for over two hours. One reported 'The waves were pretty heavy, and outside the water it was jet black, just like night. All you could see were red sparks flying through the air with the wind.' Mrs Millhouse said simply: 'If the water hadn't been there nobody would have lived' . . . [She] and the others lost everything they owned in the world. (*Age*, 9/2/67)

Another woman had taken a few possessions and all her money with her, but she had to drop everything in order to cling to her baby in the waves. Fortunately Mrs Freeman was saved by an uncle who arrived just before the shop and all the houses in the street were destroyed.

The district's 300 school-age children were at Snug Area School. It was the first day back after the Christmas holidays.[1] At approximately 1.30 pm all were taken from classrooms and wooden buildings to the school assembly hall. The headmaster, Mr Bowles, reported that, 'At this time visibility was poor and in some cases children clasped hands in order to keep together'. In the hall they sat on the floor and curtains were drawn 'so that as little as possible of what was occurring outside could be seen by the children'. The music teacher, Mr Hawsley, led the children in community singing. The *Age* of 8 February reported that they sang 'My old man's a dustman' and 'She'll be coming round the mountain'. Another report says that the school librarian opened the library and read aloud to the younger children, while the older ones read their favourite books. According to the headmaster, 'By this time parents with younger children had also sought safety in the Assembly Hall and the corridor of the main school'. Reports indicate that some injured people came to the school. This was a fairly common occurrence at schools in the area on 'Black Tuesday' and 'several headmasters reported that the adults were more inclined to lose self-control than the groups of school children'.[2]

At about 2.00 pm fire broke out in the tower next to the assembly hall. Fortunately, it didn't spread, but they couldn't extinguish it because of a lack of water pressure and firefighting equipment. It was not until late afternoon that the children were either sent home or evacuated to Kingston 'by any practicable means available'. Across the road Saint Mary's Catholic church, a timber building dating from 1858, burnt down. After all the children had left the school, the fire in the tower spread to the modern, brick part of the building which was left a burnt-out shell. The assembly hall survived.

Just to the north of Snug is Electrona and the Australian Commonwealth Carbide Company factory which made batteries. It burnt to the ground as did many houses. 'Outside the carbide works were the shells of almost 20 burnt-out cars of employees and a fog of stifling black smoke still lay over the works. Around the factory only chimneys remained of many houses—five houses in a row were only smouldering ruins—and the strong winds had spread roofing iron over nearby paddocks', reported the *Mercury* of 8 February. Many of Snug's population worked there, so some lost both house and job.

Snug was typical of what was happening all across southern Tasmania that terrible afternoon. Between about 11.00 am and 6.00 pm an area of 265,000 hectares was burnt, 62 people died, 900 people were injured, 7000 were left homeless and 35,000 were affected. Fifteen hundred motor vehicles were destroyed and 62,000 livestock and 24,000 chickens died.[3] A total of 1451 buildings were burnt down, including seventeen churches (five Catholic and twelve Anglican), 1085 houses, 233 cottages or 'shacks' as they are called in Tasmania, 116 factories, shops, and other structures.[4] Eighty road and rail bridges were damaged or wrecked. About 20 per cent of the state's fruit crop was lost. In the Huon Valley:

> orchardists were readying themselves for the 1967 export season . . . Packing sheds were destroyed, along with graders, spray mixtures, tools and orcharding requirements . . . About 2,000,000 super feet of racked seasoned timber for fruit cases and building supplies [as well as] sheds with saws and equipment were wiped out. It is inevitable that this will throw men out of employment. (*Mercury*, 9/2/67)

The all-up insured cost in contemporary terms was about $101,000,000.[5]

For several hours during the height of the fires all communications were down to the rest of Australia. There were four radio stations in Hobart and three were off-air due to technical problems, including the ABC. Only commercial station 7HT remained broadcasting and they maintained a service throughout the crisis relaying messages and stopping panic. Journalists hired planes to fly from Launceston and Melbourne, but they were unable to land at Hobart Airport which was closed due to smoke and high winds. A reporter from the *Sydney Morning Herald* said that in the evening of Black Tuesday

> Hobart was a terrifying sight from the air . . . Smoke as thick as cumulus cloud was rising to 8,000 feet for 50 miles around the city. It was impossible to make out Mount Wellington . . . Several planes flew over the city during the day trying to take pictures, but each found the task hopeless. Over Hobart we picked up radio station 7HT continually broadcasting messages for its anxious listeners. The city seems one of panic. (8/2/67)

Why were these fires so destructive of both life and property? And were the fires sudden and unexpected or could they have been predicted?[6]

ii

To understand Black Tuesday we need to look at the conditions that led to it. In hindsight it is easy to distinguish causes, but at the time people were simply not prepared for what happened. Bushfires are common in southern Tasmania, with serious fires in the Derwent Valley and the Channel district occurring about every six years.[7] Nevertheless, people quickly forget previous fires. Bushfire expert A.G. McArthur concluded that, 'The vast majority of people had had no experience of such a widespread disaster, and never contemplated that such a thing could occur. Official thinking was undoubtedly in line with community thinking . . . evidenced by the fact that the fire fighting services operated on a restricted budget.'[8] This despite the dangers being regularly highlighted in weather forecasts and by fire authorities.

What were the warning signs? First, heavy rainfall, then drought. The official report on the fires notes that:

> The East Coast, Midlands and South-East of Tasmania had rainfall aggregates well above normal during September and October, 1966 which ensured good vegetation growth across the whole area. From November onwards very dry conditions prevailed with pronounced rainfall deficiencies in those areas and in the Derwent Valley . . . By late January grass and litter were tinder-dry and many bushfires were in evidence.[9]

The second factor was high temperatures and strong winds which contributed to curing the grass and undergrowth. After ignition the fires were driven forward as the winds increased.

Temperatures remained mild until Saturday 4 February. A high pressure system moved slowly eastwards across Tasmania from the Great Australian Bight, passing east of the island on 3 February. Slow-moving or stationary highs are a danger sign, and 'warnings of rising temperatures, freshening northerlies and the development of a very high to extreme fire danger were included in weather notes and forecasts' for the next few days.[10] So it is hard to understand why so little attempt was made to control fires already burning—and this brings us to the fourth element that contributed to Black Tuesday: indiscriminate fire-lighting. On both 6 and 7 February the headlines in the *Mercury* warned 'Fire danger at its peak in the south tomorrow' and 'Bush fires menacing state: danger today could be critical'. Nevertheless, people were completely blasé about the dangers. They had heard it all before and continued to light fires with or without permission for all sorts of purposes—burning off, fire breaks, boiling the billy, or malice. Obviously Stretton's recommendations had not penetrated southern Tasmania. The chair of the Rural Fires Board and Forestry Commission, Alexander Herbert Crane, prophetically warned in the *Mercury* of 6 February that 'fires still burning on Tuesday morning will be almost impossible to hold during the day. They will burn freely and travel long distances due to spots lighting ahead of the main fire.'

While people were indifferent to all the warnings before Black Tuesday, afterwards they quickly declared that the events of that day were 'unique', never before experienced. It was psychological denial. McArthur has shown that 'Black Tuesday was not a one-off. Similar conditions have occurred on three or four occasions during the past 70 years' in southern Tasmania. Using his own Forest Fire Danger Index (FFDI), he showed that in 1927 (91 on FFDI) and 1934 (94 on FFDI) conditions were close to those of Black Tuesday, and that in 1897 (62 on FFDI), 1912 (77 on FFDI) and 1960 (67 on FFDI) there were also dangerous fire days. The FFDI combines temperature, relative humidity, wind speed, rainfall, unstable atmospheric conditions and abundant combustible fuel. 'At an index figure of 1 fires are self-extinguishing, whilst at . . . 100 fires will burn so rapidly and intensely that control is virtually impossible.'[11]

Black Tuesday was a 'perfect' fire day, 100 on the FFDI. The maximum temperature was 40°C, the minimum relative humidity 11 per cent, the average wind velocity was around 60 km/h with maximum gusts in Hobart to 120 km/h (the highest on record), and the drought factor was scored at 9 out of 10. McArthur commented that 'the February 1967 conditions have undoubtedly been the severest recorded in the past 70 years'.[12]

So what happened? Within a 56-kilometre radius of Hobart 110 separate fires were identified. How did the fires begin? McArthur says:

> Only 22 of the 100 fires identified can be said to have started accidentally. Of these, 8 resulted from long distance spot fires . . . and the remaining 14 were due to a variety of causes

> such as escapes from incinerators, sparks ... breakaways from rubbish dumps, broken transmission lines ... 88 or 80% of all fires occurring in the Hobart region on 7 February were deliberately lit, mostly in contravention of the provisions of the Rural Fires Act. The motivation for most of these 88 fires is obscure. Many were lit for grazing purposes, mainly to bring on nutritious new growth before winter. A secondary reason ... was to remove inflammable debris ... Quite a large number appeared to be lit for no clear management objective and it may be suspected that they could have been started for malicious purposes. However it is doubted if this applied in more than a very few instances.[13]

Although prosecutions were threatened, few occurred.

Many of these fires were burning for several weeks prior to 7 February. Outside the Hobart area, 89 fires were burning on 6 February.[14] Close to Hobart a deliberately lit fire had been burning slowly northwards from Bonnet Hill between Taroona and Kingston along the foothills of Mount Nelson. There were other fires nearby at the Mount Nelson Signal Station. These fires were dealt with on three occasions by the Hobart Fire Brigade. Other fires in the Turnip Fields district on Huon Road indicated that 'there appeared to be deliberate fire lighting in this area'.[15] A similar situation existed in the northern suburbs around Glenorchy.

Another fire that broke out on 4 February to the south of Bothwell on the Lake highway spread south-eastwards for about 20 kilometres on 7 February towards Kempton, burning over almost 12,000 hectares of bush and grasslands. The result, according to the official report, was that most of the destruction caused on 7 February resulted from fires that had been allowed:

> to 'burn free' or were thought to be under control, but were not. The practice had grown up of permitting fires in bush and scrub country to 'burn free' so long as they were not menacing lives or property. However, under the extreme conditions of the morning of the 7 February, most of the free burning fires spread rapidly and could not be controlled by the time they

were threatening properties and the fires which were thought to be under control flared up again and became conflagrations before any resources could combat them.[16]

iii

There were several fire complexes.[17] The first was in the New Norfolk area, 25 kilometres west of Hobart. There were ten fires reported to the west, south-west and north of New Norfolk for several days prior to Black Tuesday. On 3 February a woman lit a fire in an incinerator at Dromedary on the north bank of the Derwent River which got away, and next day was out of control. By 6 February it was still only partially controlled. On Black Tuesday it destroyed two homes and a church and then took off with extraordinary speed in two directions, the northern spur moving eastward across the Midland Highway at Rogerville, where houses, the sawmill and the petrol station were destroyed. The southern spur burnt along the northern bank of the Derwent towards Bridgewater. 'Six houses, a bookmaker's premises, a barn and several sheds and cars were razed' according to the *Mercury* of 8 February. This was a massive, fast-moving fire with flames more than 20 metres high.

Another fire which had started three days earlier near Hayes Prison Farm flared up again on Black Tuesday and, driven by a strong wind, moved towards Magra, where eight houses were lost, and then on to New Norfolk where, despite the town being surrounded, there was minimal damage due to efficient local firefighters. A woman died when her 120-year-old house burnt down. At Boyer the fire hit the Australian Newsprint Mill, threatening the main factory. The *Mercury* reported on 8 February that, 'The log depot went up and massive girders twisted in the white-hot heat, two big stores disappeared, and the new lines of single accommodation crumpled to debris . . . Thousands of tons of material, imported for newsprint, caught alight . . . a tug blasted thousands of gallons of water onto the burning wharf area . . . A train of logs in a siding was still burning last night.' One of the main problems in the New Norfolk area was the constant spotting of fires due to flying twigs and sparks. The official report of the fires noted that, 'The main factor hampering the proper control of fires in this area was the fact that many people had not removed scrub particularly that close to roads . . . and the lack of equipment and knowledge of fire-fighting by the volunteers. On farms surrounded by heavy growth there were few if any preventative measures taken.'[18]

Serious fires also broke out south of New Norfolk in the valley at the back of the Wellington Range. Molesworth, Collins Cap and Collinsvale bore the brunt of these fires. Collinsvale lost 21 houses and the community hall which had taken volunteers five years to build. The whole of the Wellington Range went up in seventeen separate wildfires.[19] Driven by gale-force winds from the north-west the fires 'roared out of the timbered hills on a dozen fronts'. The *Mercury* of 9 February reported that since most of Collinsvale's men were in Hobart working, there were very few firefighters at home available to fight the blaze that afternoon. A dozen children, stranded at the local school, were taken to the general store when the school was surrounded by flame. When the store itself was threatened, the storekeeper, Mrs Betty Burr, helped by Mrs Kathleen Barnard, 'put the children in a deep culvert and covered them with wet bags'.

The second fire complex centered on Colebrook, 35 kilometres north of Hobart. This fire was deliberately lit by a group of hunters boiling a billy early on the morning of Black Tuesday near the junction of the Midland Highway and the Colebrook Road. It was the largest outbreak of the Black Tuesday fires, burning 15,300 hectares.[20] Driven by winds gusting up to 90–100 km/h, it burnt southwards down the Colebrook Road, climbed Mount Mercer, and then ignited the whole valley to the north-west of Colebrook, a small, prosperous farming town on the main railway line to Launceston and situated in prime sheep and cattle country. The fire swept into the town from the hills on a 5-kilometre front in the early afternoon, and within ten minutes it had destroyed 26 of the 44 houses, the hotel, the school, the post office, the station house, two shops, and part of the railway line. Fortunately, Saint Patrick's Catholic church on the hill, designed by the great English neo-gothic architect, Augustus Welby Pugin, escaped.[21] Two people were killed.

A Melbourne *Herald* reporter, John Sorell, was the first outsider to reach Colebrook. He reported on 8 February that as the school burned, 'Terrified schoolchildren were herded . . . to the safety of a ploughed paddock and they watched the flames speed past them . . . It was the quick thinking of a policeman's wife that saved the children. She pushed them into one of the few remaining cars and drove them down to the large paddock . . . They stood in safety in the middle of the paddock behind a row of parked cars.' Another report says they spent the time face-downwards with towels over their heads to escape the flames and dense smoke. The fire continued south-eastwards towards Brown Mountain, killing three more people and destroying farms.

Just to the south-east of the Colebrook was the third complex, the Campania–Richmond–Sorell fires. The fire began in Richmond, possibly lit by children on 5 February. The fire blew up again in the mid-morning of Black Tuesday. It jumped the Coal River out of control and travelling at 18–20 km/h and headed toward Penna and Sorell, just to the east of Hobart Airport; there are other reports of it moving at up to 80 km/h. It was joined by another fire which had begun near Campania the day before. It was this fire that burnt to within a metre of the oldest Catholic church in Australia, Saint John the Evangelist at Richmond, while the parish priest, Father Tom Garvey, was out with firefighters in another part of his parish at Sorell. The children from the Sorell school were taken by buses to the Sorell Causeway across Pitt Water, because the school oval was unsafe. Here they found themselves with the terrified elephants and other animals from Wirth's Circus which was visiting town.[22]

Sorell was particularly badly hit. Many houses were destroyed and five people killed. There were two more killed in Penna. Four people were killed in Richmond, and one at Orielton and one at Campania. One of the saddest stories was the death of Mrs Dorothy Wright, 50, of Campania. She had gone out to move the cows on the family dairy and was not seen again until her body was found some weeks later in the Coal River. The government pathologist, Dr Campbell Duncan, told the coroner that he thought she had climbed out on a willow branch over the river to escape the heat and smoke. The branch had broken under her weight and she had fallen into the river about 3 metres from the bank. At this point the river was about 8 metres wide and 2 metres deep. Her husband, Sefton Wright told the *Mercury* of 21 June 1967 that 'he did not think his wife could swim'.

There was a fourth complex east of Hobart extending from Warrane southwards to Rokeby and south-east towards Lauderdale. Hobart's eastern suburbs had experienced a plague of deliberately lit fires with 44 of them between 7 January and 7 February, 'many lit by very young children'.[23] Perhaps this reflected the instability that characterised this economically depressed Housing Commission area.

Prior to Black Tuesday there was a fire burning at the eastern end of the suburb of Warrane, either lit by children playing with matches, or an adult burning off. It took off southwards around midday on Black Tuesday on two fronts, both out of control, passing to the east of Howrah and Bellerive. One prong of the fire hit Rokeby and the other burnt well to the east of the town towards Lauderdale and Cremorne. At Rokeby several heritage homes were lost including one of Tasmania's oldest, Rokeby House, built in the 1820s. Eighteen other houses were lost in the district.

These fires left three people dead. One of them was Donald Jackson, 30, a cabinetmaker and boatbuilder, who despite shocking burns drove himself several kilometres from Rokeby to Howrah before a friend drove him to Royal Hobart Hospital, where he died twelve days later from toxemia. According to the *Mercury* of 5 July Jackson had gone back to his boatshed 'to get tools and the fire suddenly caught the shed while he was inside . . . He had thrown himself in a [nearby] sheep dip, but the flames burnt him while he was in the water.'

The same fire killed Anna Freeman, 88. She was at home at 'The Pines' (built in 1886) in Rokeby with her daughter-in-law, Jessie Freeman, 53 and granddaughter Josephine, seventeen, who were visiting. Josephine went out about 1.00 pm and saw the fire approaching the house. They decided to make a run for it. 'They left the house by the front door. The smoke was very thick and it was very hot outside. By the time they got to the front gate the fire had got very close to them' reported the *Mercury* of 5 July. Josephine said that her mother was trying to get 'Nanna' to come with them. The young woman told her mother to go ahead while she helped her grandmother. Her mother went ahead but stumbled into a ditch and injured her back and foot. Josephine also fell. She lay on the ground for a bit not knowing what to do. Then:

I got up and just ran', she said. She met some neighbors in a car, asked them to wait, and then she ran back down the road to help her mother. 'By the time I got there she had managed to struggle out of the ditch and I helped her to the car', Josephine said. It was impossible at that stage to help her Nanna. Where she left her Nanna was very close to where her body was found later. (*Mercury*, 5/7/67)

A fifth complex of fires was in a triangular area along the Tasman Highway between Runnymede and Buckland and north to the apex of the triangle at Levendale, where schoolchildren couldn't take refuge on a ploughed oval because down-drafts from nearby elevated ground caused sheets of flame to sweep the oval; they had to be driven in two vehicles through 350 metres of burning country to safety. On Black Tuesday there were also fires on Bruny Island. The official fire report noted that, 'The general opinion of residents is that some of the fires . . . were caused by burning debris from the mainland fires.'[24]

iv

The most destructive fires, besides those in the Hobart suburbs, were the sixth interconnected complex of fires across the Kingston, Huon and Channel districts and in southern Tasmania at Hastings and Southport. Around the Channel many outbreaks were connected and it is hard to sort out which was which.[25] The most northerly fires were to the south-west of the Wellington Range. These connected with outbreaks to the south and west of Huonville. There were other fires at Geeveston, further south at Hastings, and in the Cygnet–Lymington area. There was another group in the Kettering–Woodbridge area on the D'Entrecasteaux Channel across from Bruny Island. North of this was a fire at Oyster Cove. The fire that hit the town of Snug was a combination of the one that had come from the southern part of the Wellington Range with another that emerged from the hills to the north-west of Kaoota. Almost all of these fires began before Black Tuesday, were largely deliberately lit, and were enormously destructive; very few houses and structures caught in their paths survived.

In the Huon area there were fires that began possibly on 31 January or 1 February, as well as others in early February around Huonville. On the morning of Black Tuesday a fire began on a farm near Judbury which burnt along the valley of the Huon River in the direction of Ranelagh, and then turned south-eastwards across the hills towards Franklin. The biggest losses in the Huon district were suffered by orchardists. The *Mercury* headline of 9 February said it all: 'Apples cooked on trees in Huon orchards'. There was also an enormous loss of infrastructure for the apple-growing industry. There had been fires burning in the forests of the Arve Valley, west of Geeveston, prior to 7 February. About noon on Black Tuesday a hot, strong westerly wind drove them in the direction of the town. A sudden wind change saved Geeveston and the fire swung in a north-easterly direction, burning itself out near Castle Forbes Bay, but not before it had destroyed a number of houses.

The southernmost settlement in Australia is Southport. Nowadays it is a collection of holiday shacks with a few permanent homes. In 1967 there was a shop and a thriving community. On Black Tuesday disaster struck and by 2.00 pm the township was completely destroyed.[26] The fish factory, which employed many of the town's permanent population, as well as 90 per cent of the houses were burnt. There had been fires burning in the forests to the west and south as far as Cockle Creek before Black Tuesday. Between 11.30 am and 2.00 pm the fire hit Southport 'completely out of control'. It burnt right to the beach. To the north there were fires throughout the district between Dover, Strathblane and Hastings, all of which lost some houses. The road to Geeveston was cut in several places.

There were terrible crown fires in the surrounding forests. Warren Seabourne, seventeen, was trapped by a wall of flames when he was working in the bush near Strathblane. 'He spent two hours lying in a small dam with green boughs over his head. When the fire moved on he walked to Dover where he received treatment for burns on the shoulder and neck', the *Mercury* reported on 8 February.

One of the worst-hit areas was what the *Mercury* of 9 February called 'the fire-blasted Channel', the district south of Kingston and bordered in the west by the Huon River estuary, and in the south and east by the

D'Entrecasteaux Channel. A range of hills up to 830 metres runs down the core of the district. Many of the fires began in these hills in the days before 7 February. A back-burn very early on Black Tuesday morning got out of control and there were other fires that had already broken out in the hills to the north-west. It seems that all the fires in the area then combined to sweep through Woodbridge to the sea.[27] The smoke was thick and people had to lie on the ground and breathe through wet cloth. The jetty was destroyed and even seaweed on the beach was burnt.

Something similar happened around the Oyster Cove area. There had been fires smouldering since early February which were more or less under control. But by midday of Black Tuesday, these fires combined with another that came from the 800-metre Snug Tiers to the north. By 12.30 pm the town of Oyster Cove was badly damaged and the death toll eventually reached four. Flowerpot, Middleton and Gordon all suffered house losses. An elderly couple died when they ran from their holiday home at Gordon, which was not damaged. Just east of Snug two people were killed at Conningham, a rather upmarket holiday area, and further south at Kettering a woman was burned to death in her home when it exploded in flame.

One of those who died at Oyster Cove was a farmer, John Palmer, 66, who lost his life trying to save his scooter bike. His wife, Esma Palmer, said that with fires threatening they had decided to abandon their farmhouse when their daughter arrived in her car.

> We left to go to my daughter's home, my husband declined to get in the car. He said he had just finished paying for the scooter and didn't want to lose it in the fire'. Mrs Palmer said she did not see her husband alive again. When the fire was at its worst, she said, it was as dark as night. 'The wind was howling and the heat and smoke were over-powering, but my daughter drove through it to her home.'(*Mercury*, 23/6/67)

Meanwhile, John Palmer had been caught in the open by the fires on the Channel Highway where his body and burnt scooter were found. Fifteen kilometres further south at Garden Island Creek Leslie Bolton tried to evacuate his wife and retarded son from their house. The

Mercury of 8 July told of how when he got them into the car, he found 'he had not enough petrol. He got out to replenish the tank. "My son then got out of the car and ran into the house. My wife also got out and ran after my son . . . I realised they were in great danger and went back after them but the heat drove me back".' He made several efforts to rescue them. At Lymington, on Port Cygnet, the village was virtually destroyed when twenty houses were burned down.

A hydro-electricity worker, Alex Bugg, 30, married of Huonville, was killed on the road between Nicholls Rivulet and Oyster Cove when he and his workmate, Geoffrey Bond of Cygnet, became trapped. They were repairing power poles. Their truck stalled, surrounded by flames. They made a run for it, became trapped, and then separated. After returning to the truck, Bond found Bugg. 'I found his clothes were burning , so I emptied my flask on him.' They lay down together for a while trying to recover and then Bond, even though he was also badly burned himself, dragged Bugg several hundred metres out of danger, before they were eventually rescued by a local, Jackie Smith, who turned up in his utility. Bugg died of his burns, and Bond spent six weeks in hospital with severe burns to his hands, arms, face, right leg and right hip. He eventually recovered and told *The Mercury* on 23 June 1967: 'I have never seen anything before in my life like that day—smoke, flames and intense heat—and I reckon that is what hell would be like'.

By the end of the day all along the Channel Highway between Kingston in the north to Gordon in the south at least 200 homes were burned down, and more than a thousand people were homeless. For some time afterwards two to four families were sharing the surviving houses and it was a fortnight before electricity was restored. The worst-hit town was Snug. Kingston and Blackmans Bay largely escaped the fires, although eleven houses and the Masonic Hall were burnt. Most of the population—and their cars—were on the beach or in the water sheltering from the heat and flames.

The day after the fires, the *Mercury* reported that:

> All along the Channel Highway there is death, devastation and despair. A policeman estimated that 80% of the houses south

of Snug have been wiped out. Tens of thousands of bushels of apples are lying on the ground, or hanging scorched among the blackened branches of the orchards. Poultry have been wiped out by the thousand. Sheep are lying dead in the paddocks. Somewhere near Electrona 100 head of prime beef cattle have been killed. As the horror of the fire recedes, it is dark and cold in the houses that are left.

The long-term results of the fires were that the whole social structure of the area changed. Not only did people lose their homes and livelihoods, but the old sense of rural community disappeared. Many families with longstanding connections left the Channel and patterns of employment changed. 'Many orchards were ripped out and not replanted . . . It was a time of population shift, of change in emphasis in local industry and employment.'[28] Increasingly the area became a dormitory for Hobart.

v

The biggest impact of the fires was around Hobart, which in 1967 stretched from Claremont, 14 kilometres to the north-west of the city, to Taroona, 8 kilometres to the south. The elongated shape of Hobart, never much more than 2 kilometres wide and strung mainly along the west bank of the Derwent River, and on the foothills of Mount Wellington, contributed considerably to the penetration of the fires into urban areas. In three separate places the fires reached to within 1.6 kilometres of the GPO. The outer, semi-rural areas of Melbourne and Sydney had previously been impacted by fire, but Hobart in 1967 was the first time bushfires actually penetrated into strictly urban areas. For years people seeking better views had been building in the foothills and gullies of Mount Wellington in areas such as Tolosa Street, Glenorchy, Springfield, West Moonah, Mount Stuart, South Hobart, the Cascades and Mount Nelson. There were also the semi-rural settings of Fern Tree and along Strickland Avenue, as well as Lower Sandy Bay and Taroona. The official report of the fires noted that, 'There was thick native timber and scrub growing amongst the homes in Taroona and

Strickland Avenue whilst in other parts of the area such vegetation commenced on the western edge of the developed areas of the city and ran for miles over the foothills and towards the top of Mount Wellington.'[29] The gullies that ran down the mountain created a tunnel effect on Black Tuesday with winds coming out of some of them at an extraordinary 140 km/h. The new northern suburbs were also increasingly penetrating into the fire-prone grasslands and rolling hills of the north-west bank of the Derwent.

Some think that the city of Hobart could have been completely overwhelmed if a change in wind direction had not occurred at about 2.50 pm. Hobart's hilly geography, and the breakdown in communications meant that it was hard for anyone to get an overall picture of what was happening that afternoon. People not immediately impacted by the fires tended to be unaware of nearby events that were unfolding, sometimes as close as just across the hill behind them. Part of the problem was that there had been fires around Hobart for days. Most of them were lit by people burning off, or arsonists, or uncontrolled children.

The best way to understand what happened is to examine the fires one by one from the northern suburbs right through the city to the southern suburban extremity at Taroona. All of these fires were ultimately connected to Mount Wellington, which by the middle of the day had become a mountain of flame.

There had been fires burning freely on Mount Faulkner west of the northern suburbs for several days. One of these lit grassfires in Chigwell early on Black Tuesday morning. Another fire burnt across Mount Faulkner from east of New Norfolk and penetrated the suburbs by about 12.30 pm. It was burning fiercely on a broad front and caused havoc in the northern suburbs of Claremont, Berriedale, Chigwell, Rosetta and Glenorchy. Houses burst into flame as the fires exploded out of the bush. The *Mercury* of 8 February reported that one couple in Berriedale were sitting quietly in their lounge 'watching the fire burn down from the hills a mile away. "Next thing another fire came out of the gully behind us. We got the car out...but that's about all", said the man. Next door a woman rushed into the street clutching her baby in her arms. That was all she saved for within minutes sheets of flame were breaking over the road like waves. The two brick houses were heaps of rubble and red hot iron.'

Several people were killed in the northern suburbs. David 'Les' Cordwell, 60, of Pitcairn Street, Glenorchy died when he and his neighbour, Senior Constable Clark, tried to protect their homes from a grassfire behind their back fences. The policeman told the Coroner 'When the fire was about halfway down the paddock the wind suddenly changed to a south-westerly direction and brought flames 20 to 30 feet high straight at us'. Constable Clark jumped the fence and ran to his police radio car to call for help. 'I could hear Cordwell calling out. I went to the rear of my home and found him running down through my yard, a mass of flames. His clothing and gum boots were burning' (*Mercury*, 17/6/67). Clark tried to tear his burning clothes from him with his bare hands. After helping Cordwell and leaving him with his wife, Clark also managed to save their house, his own, and those of several neighbours. No ambulance was available, so Cordwell was taken to Royal Hobart Hospital in a police car. He managed to survive for a fortnight in a terrible state, but died on 21 February. Clark was recommended for an award for bravery.

Another man who suffered a similar fate was Ronald Williams, 38, who literally ran out of the fire with his clothes alight to Marys Hope Road, Berriedale. Williams and some mates were trying to save a house, but became trapped in high grass when the fire cut around behind them. 'The wind had whipped up while he was fire-fighting and the flames had engulfed him. He fell into the flames, then got up and ran.' When he got out to the road his shirt had been burnt from his body and his trousers were on fire. His two mates who escaped unhurt said that 'the smoke was intense from long burning grass and they had a job to breathe' according to the *Mercury* of 24 June 1967. Williams died a week after Black Tuesday. Two others were killed in the northern suburbs. In the Glenorchy municipality between Granton and Moonah 61 houses were destroyed.

Several schools were threatened.[30] At Rosetta Primary School 200 students were evacuated with the fire racing down towards the school, but staff and local residents just saved the buildings by beating out the fire with wet sacks. The children were moved to 'safe' houses, or to the river bank at Glenorchy by the Methodist minister. At North Chigwell Primary School, according to a *Mercury* report of 8 February, there was panic when an orderly evacuation of several hundred children, arranged by the head teacher and staff about 1.45 pm, became something of a debacle when outsiders interfered. The children, 'shepherded in groups by teachers, moved out of the smoke-besieged building into the blast furnace heat of the open. Big brothers and sisters clung grimly to the hands of younger ones, many of them sobbing despite the efforts of teachers to keep them calm.' Many children were picked up by passing motorists, and simultaneously alarmed parents were arriving at the school on foot and in cars to pick them up. Eventually no one knew where anyone was. 'The organized evacuation was soon disorganized. Many children were taken to the relative safety of the nearby Chigwell school, about 20 more were taken to a private home, another group was taken to a shopping centre, and many children were picked up by passing cars, or attached themselves to other families they could see moving to safety.' Those parents who arrived after the children had dispersed 'raced in rising panic from place to place seeking lost sons and daughters'.

Further south just before 7.30 am on Black Tuesday a fire began in the Limekiln Gully Catchment Reserve at the top of Tolosa Street, Glenorchy. It is unclear as to how it started. The official report states that, 'The possibility exists that it was deliberately lit on that morning but this cannot be proved'.[31] Nothing was done about it until after 10.00 am when three Glenorchy Council employees equipped only with beaters and rakes tried to control it. At one stage they were assisted by the Hobart Fire Brigade. The council workers claimed that before midday it was under control, but then the wind picked up to gale force and drove the fire in a southerly and south-easterly direction. They 'considered it too dangerous for them to stay there . . . [They] left the area some time before 1.00 pm and the fire was then travelling over the hills towards Lenah Valley.'[32] Other firefighters were so busy at this time in the northern suburbs that they were unable to supply men to control this fire. 'Flying debris was falling ahead of the main fire and was aggravating the situation with widespread spot fires.'[33] Eventually fifteen houses were destroyed at the top of Tolosa Street and Barossa Road. The Limekiln fire became the most destructive in Hobart. It caused widespread damage in Lenah Valley and right across the western edge of the city bordering Mount Stuart, West Hobart, the Cascades, South Hobart, Mount Nelson and Fern Tree.

By 1.00 pm the fire reached the hill between Lenah Valley Road and Brushy Creek Road.[34] Despite being warned of the serious danger by a group of firefighters in the area, Harold Goodwin, 51, and his son William, 23, both of Bellerive, had gone up the top end of Lenah Valley Road where they had a fruit and raspberry patch. Their brother-in-law thought they had gone to protect their almost mature crop. They were right in the path of the fire and their bodies were later found, badly burnt, about a kilometre up the road.[35] The fire then burnt through the top end of Pottery Road, across a large rubbish tip which, the *Mercury* noted on 9 February, locals claimed 'had been smouldering for days', into the Knocklofty Reserve and on to McRobies Road where twelve houses were destroyed. By 1.30 pm there were spot fires at the lower, inhabited end of Old Farm Road and on the ridge to the north of Marlyn Road, which lost fifteen houses.

The fire was moving at about 3.2 kilometres per hour. By 1.45 pm it had left Marlyn Road and reached Strickland Avenue. 'Very heavy crown fires were burning at the eastern end of Old Farm Road resulting from massive spot fires which were being drawn upslope . . . During the next fifteen minutes, these intense crown fires continued and threw masses of burning embers into the Cascade Brewery, setting it on fire.'[36] The *Mercury* of 8 February reported that 'the century-old Brewery was destroyed in a spectacular blaze that was punctuated by cannon-loud explosions as fire reached gas cylinders and other explosive material'. By 4.00 pm the building was still burning furiously and efforts to control it were useless. The shell was still alight that night.

In the Cascades area, reported the *Sydney Morning Herald* of 10 February, one extended family of British migrants was particularly badly hit. Two brothers, Brian Webb, 28, of South Hobart, and Bernard Webb, 30, of West Hobart, were policemen. Both were fighting fires elsewhere when their homes were destroyed. Their sister was married to Ken Plumstead. This couple's home was also destroyed, as well as that of Ken's father-in-law, Arthur Miller.

A story of resurrection centred on the South Hobart Baptist Church in Macquarie Street. Despite the building being burnt down in ten minutes to its convict-quarried foundations, the congregation of 50 still attended in the ruins the following Sunday to begin the service with the hymn 'O God Our Help in Ages Past'.

By the time the fire crossed Huon Road at about 2.00 pm it was up to 2.4 kilometres wide. There were now houses alight in Lenah Valley, Mount Stuart, West Hobart, South Hobart and the Cascades. The fire began to move more quickly and by 2.15 pm its eastern flank had reached Waterworks Road and moved into the edges of Dynnyrne and Mount Nelson. Side fires were breaking out well behind the main front. On Waterworks Road one of those disasters occurred which happen in every major tragedy. The Hobart City Council had cut off the water supply that morning to reconnect a relocated water main. 'When the fire threatened, temporary arrangements were made for the water to be turned on. However, the valve could not be fully opened due to the risk of a cap blowing off. There was full pressure of water available at the top end of Waterworks Road but there was no water available at the lower end during the height of the fires in that area.'[37] As a result 29 houses were lost.

Tragic also was the death of Anna Petrenko, 61, pensioner, of South Hobart. At the inquest it emerged that she had become very frightened while her husband, Alexis, 55, tried to fight the fire at their house.

> His nose started to bleed and his wife panicked. He told her to go into the house, but she said she would 'go to Pritchett's'. He threw buckets of water on the burning steps of the house and then noticed his wife had gone. 'I never saw which way my wife went. There were flames all round me', he said. 'I sent my Alsatian dogs to look for her. They returned with the scarf and cardigan she was wearing' . . . Mrs Bessie Isobel Pritchett said she was fighting the fire when somebody called out. She saw Mrs Petrenko lying on a chair. She had badly burned arms and legs. She attended to her wounds as best she could and made her comfortable. Mrs Pritchett said Mrs Petrenko who appeared to be petrified said: 'Alex told me to stay in the house with the dogs, but I got frightened and came to you to ring for a taxi'.
> (*Mercury*, 23/6/67)

Mrs Petrenko later died of her injuries.

Other 'stay-put' residents paid with their lives. Two of them, Mrs Marjorie McCullum and Mrs Kathleen Hall, lived on McRobies Road. A witness, Mrs Lydia Sydlowski, tried to persuade the two women to leave, 'but Mrs Hall had said the fire would soon be over and went back inside her house. Mrs Sydlowski said she had warned Mrs McCullum who lived a few doors away, but Mrs McCullum swore and said she wanted to save her house. She said Mrs McCullum had a hose in her hand, but there was no water running through it', reported the *Mercury* on 14 July. The two women were good friends and often visited each other. Their remains were found in Mrs Hall's house. There was little the authorities could do to assist residents: everything was out of control and resources inadequate. At one stage eleven houses were burning at the same time in Wentworth Street, South Hobart, within a couple of hundred metres of each other, as well as the Cascade Brewery. At the same time 25 homes and the 'Top of the World' swimming club house in the Waterworks Road vicinity were burning when an inferno swept down the ravine carrying the Sandy Bay Rut into Dynnyrne. Side fires were still burning down gullies and destroying houses well behind the main fire-front. The Lenah Valley fire had burnt across the Knocklofty Reserve and merged with the fire around McRobies Gully. 'This merger created an extremely hot junction zone showering burning embers deep into South Hobart and Dynnyrne, igniting houses in Forest Road, Liverpool Crescent, Lynton Avenue, Greenlands Avenue and Davey Street.'[38] This area was virtually inner-city Hobart. Strickland Avenue is a continuation of Macquarie Street, one of the main inner-city streets. It snakes its way past the Cascade Brewery and around the hilly contours of the foothills of Mount Wellington, more or less following the Hobart Rut until it turns south-east to join the Huon Highway just below Fern Tree. Sixty-one houses were destroyed on Strickland Avenue.[39]

By 2.30 pm the western fire-front was travelling up Proctors Road towards where the Mount Nelson campus of Hobart College now stands. It then headed down Proctors Road more or less parallel with what is now the Southern Outlet Road. The eastern side of the fire had burnt into the Mount Nelson area. Showers of embers were causing spot fires in the area south of Taroona about 4.4 kilometres ahead of the main front. This fire was spreading north-west against the wind, probably 'due to the topography of the area causing a back-draft'.[40] It was threatening Taroona, which was also in danger from another fire which had started about 10.00 am in Churchill Road.

Many women and children took shelter at the Mount Nelson Signal Station. A fire two years earlier had burnt out vegetation and prevented the spread of the main fire from the north-west. Nevertheless, houses on Mount Nelson were hit. When her house was threatened, a very athletic grandmother and a pioneer of the area, Mrs Broughton:

> ran cross-country from Mount Nelson to Red Chapel Ave., Sandy Bay, a distance a fit young man could cover in about nine minutes. Most of the way she ran barefoot . . . Another woman, alone in a house with six children, did not realize until too late just how close the fire was. The intense heat drove her back inside, but miraculously the wind changed and the house was spared. (*Mercury*, 9/2/67)

By 3.00 pm the climax had been reached. Telephone communications throughout the city and with the outside world had largely collapsed. At about midday the Hobart–Launceston telephone trunk route was destroyed at Bridgewater. Launceston was Tasmania's portal to the rest of Australia. Operators tried to re-route calls via the east coast to Hobart, but eventually they were beaten as more lines were lost. 'Though the telephone system never completely failed, many suburban and outlying areas were cut off, much equipment was damaged and heavy use of phones by the public hampered their use by organizations engaged in firefighting and rescue.'[41]

At the same time a potentially disastrous situation was developing 450 metres up Mount Wellington on the Huon Highway near the Fern Tree Hotel and Post Office, where about 250 people had congregated around the junction of Summerleas Road, which lost 58 of its 74 houses—78 per cent.[42] Four fires were converging on this area. People from the whole area around Fern Tree congregated at the hotel and road junction. Some were sheltering in a small green field.[43] Others were caught in cars that police had stopped because there was fire across the road. One

Powelltown, Victoria, 14 February 1926. Mr and Mrs G. Vennell and their baby daughter took refuge in Powelltown after a six-kilometre walk to safety. During the blaze, their home was destroyed and they were forced to take refuge under an iron sheet behind a rock.

Noojee, Victoria, Black Friday, 1939. A photograph from the Melbourne *Herald* shows Noojee in ruins. Apart from the hotel and two other buildings, the town was wiped out.

Noojee, Victoria, Black Friday, 1939. The famous trestle railway bridge on the Noojee–Warragul line, still alight. First constructed in 1919, it was destroyed in the 1926 and 1939 bushfires, but was rebuilt and stands today, even though the line was closed in 1954.

Black Friday, 1939. A typical roughly built dugout. Most were inadequate, although many lives were saved where they were provided at mill sites. Some mills, such as Fitzpatrick's mill at Matlock, did not build any, and a heavy death toll resulted.

Black Friday, 1939. The remains of James Fitzpatrick's mill in the Matlock Forest where fifteen men were killed, including the owner and his son. The only survivor was George Sellers.

Black Friday, 1939. The famous picture of George Sellers, the only survivor from Fitzpatrick's mill at Matlock. He lay on the ground for three hours in an already burnt-out area, wrapped only in wet blankets. Clearly in shock, he refused to abandon the blanket.

Black Friday, 1939. Mill workers carry out the body of Ben Saxton from the Saxton mill at Fumina, Victoria. Saxton's wife and a mill employee also died when they took refuge in an inadequate dugout.

Black Friday, 1939. Part of the ruins of Woods Point, Victoria. Much of the town was burnt out when the fires swept through the narrow valley in which it was situated. Locals took refuge in mine shafts and in the local swimming pool.

Black Friday, 1939. The main street of Omeo, Victoria. This Melbourne *Herald* photograph shows many of the burnt-out buildings. The local hospital was also destroyed.

Black Friday, 1939. The clean-up job after the fires was massive, as this picture from the Maroondah Highway near Healesville shows. Before chainsaws, the biggest job was removing the giant mountain ash trees from across the road.

Black Friday, 1939. The remains of the Kerslake car on the Acheron Way. The photograph shows the fallen tree across the road. The car had no hope of getting around or past it. The Kerslakes and three Greek quarry workers died near the car.

Leonard Stretton (1894–1967) by Archibald Colquhoun (1894–1983). Colquhoun was a Melbourne artist and teacher who influenced many, including Sir William Dargie. The original of this portrait of Stretton was lost in the Second World War in a bombing raid in the UK.

Hobart fires, 1967. Mr and Mrs Guard of Sorell watching their house burn. Photograph courtesy of *The Mercury*, 8 February 1967.

Hobart fires, 1967. A few of the sixteen homes destroyed by fire, Forest Road, West Hobart. Photograph courtesy of *The Mercury*, 8 February 1967.

Canberra bushfires, 2003. The east face of Mount Taylor. Simon Mockler, the photographer, described the scene: 'Mount Taylor half an hour after the fire front crested the north ridge. The front is advancing slowly downhill. Strong winds were blowing north-south (from right to left in the photograph, hence the relatively slow advance of the front down slope) ... Centre left of the photograph is the glow from the forest on the south face of Mount Taylor, burning after the firestorm (tornado) had gone through.' Courtesy National Library of Australia.

Canberra bushfires, 2003. Looking north-west to Mount Stromlo. The Observatory can be seen on the summit of Mount Stromlo, surrounded by the burnt-out pine forest. The road leading up through the forest from the Cotter Road to the summit can be seen in the foreground. Photograph by Greg Power. Courtesy National Library of Australia.

William Strutt (1825–1915), *Black Thursday, February 6th 1851* [1864]. Oil on canvas, 106.5 x 343 cm. Courtesy State Library of Victoria.

William Strutt, *Bushfires in the Moorabbin district* [1854]. Watercolour, 15.3 x 28 cm. Courtesy National Library of Australia. Commentary below sketch: 'Bush fire sketched by the flare of the lightning and fires in my garden at Cheltenham, Moorabbin district—on the night of March 27th 1854. Looking toward the Dandenong Ranges, 20 miles from Melbourne, Australia. The fires which had been burning for days during hot winds were driven back by the strong south wind on to fresh ground and giving a very magnificent [spectacle?].'

On another crayon drawing Strutt commented: 'Black Thursday was characterized by the sky being very smoky, yellowish, even at early morning—the sun looking red throughout the day but increasing in redness as the day advanced and the atmosphere became more charged. At about 3 or 4 in the afternoon, it was thickest, emitting a deep, dirty yellowish hue.'

John Longstaff (1862–1941), *Gippsland, Sunday night, February 20th, 1898* [1898]. Oil on canvas, 144.8 x 198.7 cm. Courtesy National Gallery of Victoria.

Kosciuszko fires, 2003. Burnt out area near the Paupong Nature Reserve north of the Snowy River and south-east of Jindabyne. The area is on the borderline between montane and sub-alpine and is 1205 metres above sea level.

Regrowth and recovery of vegetation in exactly the same area in June 2006, almost three and a half years after the fires.

woman, Mrs Haseltine, described the scene for the *Age* newspaper on 9 February:

> Red sparks were jetting across the roadway in great gusts and then it would go black as pitch. Jammed front and back by cars flanked on either side by flames, we all had the dread upon us that we'd burn alive. Babies swaddled in wet blankets whimpered. Women went down on their knees praying. So did I. Sparks set our back seat alight. Then my hair caught fire, I leaped from the car in panic and made for a steel telephone kiosk, but a policeman ordered me back shouting 'You'll roast in that oven.' So we sat there choked, sobbing and couldn't breathe. My husband had just edged forward when the petrol tank of the car behind us exploded. Our parrot collapsed unconscious on the seat.

These people were saved by a wind change. Some time before 3.00 pm the wind velocity dropped slightly and the wind swung around from out of the west to the north-west. Alan McArthur comments, 'Although the change in wind direction was only 30–40 degrees, it did have a significant effect on fire behaviour throughout the whole region . . . The death of four people at Oyster Cove and the survival of 200–300 people in the Fern Tree area were all dependent on the exact time movement of this wind shear front. A slight delay either side of 2–3pm held the lives of many people in balance.'[44]

But the fires at Fern Tree were still destructive. The Springs Hotel further up Mount Wellington was burnt to its foundations. The Fern Tree Hotel itself was reduced to a heap of rubble in five minutes. Police then realised they had to get people through to Hobart. Motorcycle police had scouted the road ahead and found that, while ringed with fire, it was clear. The column of cars was sent through. The tar on the road was alight and some cars caught fire. Occupants were picked up by others. A Fern Tree volunteer fireman, Cliff Davis, told the *Age* of 9 February: 'The winds came from both sides of this tricky valley and rolled up the flames in loops. The heat blast numbed; you couldn't breathe the air. I remember thinking atomic war must be like this.'

One of the heroines of this area was sixteen-year-old Marie Clark of

Glen Huon. She had gone to Hobart with two friends, Kath Dunscombe and Yvonne Voss and Mrs Dunscombe's four-year-old daughter, Shelley, to shop, but on the return trip their utility became trapped on Huon Road near the junction with Strickland Avenue. They tried to return to Hobart, but had to abandon the car and run for their lives towards Fern Tree. Marie said:

> I was carrying little Shelley, and with flames licking around us I told the others to keep to the centre of the highway. Then we saw two cars coming from the Huon and stopped them. Mrs Voss and I got into one car driven by a New Australian. I only knew him as 'Ernie'. Mrs Dunscombe and Shelley got into the other car driven by Mr Bax. The fire was everywhere and it was too risky to go on, so the cars were backed into a driveway. (*Mercury*, 11/2/67)

They took shelter in an abandoned house and constantly put out spot fires. As well as comforting Shelley, reported the *Mercury* of 9 February, Marie kept everyone's morale up by 'cheerfully running with buckets of water to danger spots, chatting brightly about help coming, and never showing she was a bit frightened. "I was terrified, and quite certain we were going to die", said 30-year old assurance consultant, Mr A. Bax, who lost his Fern Tree home and new car in the fires.' Just when the smoke was worst and the house was on the point of exploding in flames, a Land Rover came along with two university students who took the whole party back to Hobart. On the way they found that Mrs Dunscombe's utility was untouched by the fire. When they returned for it the next day it was stolen, as well as £15 Marie had left in it when she grabbed Shelley.

By 4.00 pm there were still extraordinary fire whirlwinds on the mountain, but nearer the city the fires had stopped spreading, although houses were sometimes still catching fire throughout the afternoon and early evening. By about 7.00 pm the conflagration was generally under control. While there were threats of new outbreaks over the following days, the conflagration was substantially over. It had been the most destructive single day in the fire history of European Australia.

vi

Sixty-two people lost their lives, 53 directly from the fires and nine from natural causes probably precipitated by the fires. Only one person under 25 was killed. There were thirteen between 26 and 50, 26 between 51 and 75 and thirteen older than 75. The official report declares that, 'This indicates the susceptibility of older people to the effect of prolonged exposure to heat and heavy smoke'.[45] Most would have been unconscious before they died. People either died in their homes (seventeen), or within 10 metres of it (ten). The majority of these were either old, infirm, or both. Eleven died some distance from the house while escaping, eleven died fighting fires, two while travelling in an open car, and two mustering stock. Of those caught outside their homes in the open, they might have had a chance if they had stayed with their houses, but 'it must be remembered that the fires were spreading at speeds beyond the experience of most Tasmanians and in many cases their deaths were due to an error in judgement as to the location of the fire and its speed. Some were engulfed in a mass of spot fires which landed round them whilst the main fire was a mile or so distant.'[46]

According to the *Mercury*, looting was a problem. The day after the fires the paper reported that looters were taking furniture and valuables saved from the fires in Forest Road, West Hobart. '"They took everything I had ... even my bed" said an old woman sitting in the gutter', quoted in the *Mercury* of 8 February. The police commissioner, Mr Fletcher, said looters should be flogged, and a *Mercury* reporter called for the return of the birch. People reported that they returned to find their homes untouched by the fire, but everything of value stolen. The *Sydney Morning Herald* of 9 February reported, 'Police said looters disguised as firefighters robbed homes in West Hobart yesterday'. The *Mercury* returned to the story the next day. 'Scum is an ugly word to apply to a human, but it is too good a name for ghouls who cashed in on the misfortune of others in the first few days of the fire tragedy. These are the people, comparatively few fortunately, who pillaged homes while pretending to help save furniture and precious possessions.' Associated with looting were false claims for relief assistance. The police were active to catch impostors, who faced a twelve-month jail term. However, a commentator on the fires, R.L. Wettenhall, points out that fear of looting is part of 'the folklore of disaster . . . the expectation of it is often enough to produce lurid results'.[47] Journalists particularly love looting stories, which make good copy. They can wax righteous, a particular habit of the media. Hobart produced its fair share of looting stories, but in the end police could not find enough evidence to bring any charges even though people were convinced that looting took place. Only four people were charged with abusing the relief fund, one receiving the full twelve months in jail.[48] In fact, in the end Wettenhall concludes that there was 'little anti-social behavior. Panic, looting and profiteering were minimal.'[49]

More significant was traditional Australian pyromania. In words redolent of Stretton, the *Mercury* editorial of 9 February 1967 said this was not a 'natural' disaster. 'The hand of man had much to do with it.' The usual questions were asked: Why did this happen? How can we stop it happening again? The *Mercury* admitted that the natural conditions were there, but 'somebody had started the fire, deliberately or carelessly'. Why weren't these fires brought under control by the authorities? Stephen Pyne says that 80 to 88 per cent of the fires 'were the product of unregulated rural burning'.[50] They highlighted the problem of the urban–rural boundary. They also showed the total inadequacy of the Tasmanian system of fire wardens who themselves reflected the complacent and inadequate folk wisdom of their neighbours.

The person who most influenced the reforms that Tasmania introduced was Alan McArthur. Accompanying him in his investigations was the young Phil Chaney. The orthodoxy of fire prevention by burning came to Tasmania. But the real issue, especially around Hobart, was not preventative burning but the invasion of the urban fringe into the bush.

This was to become an increasing problem not only in Tasmania, but on the mainland.

6

ON THE URBAN FRONTIER, 1968–2002

i

The fires that most people recall nowadays, especially in Victoria and South Australia, are the 'Ash Wednesday' fires of 16 February 1983. That fiercely hot afternoon the Mount Lofty Ranges east of Adelaide burst into what Stephen Pyne calls 'a blizzard of flame'.[1] Broadcaster Murray Nichol, whose house was destroyed as he watched, said that the conflagration 'reared up . . . with the sound of a hundred jumbo jets—so loud you couldn't shout above it. It didn't come as a wall of flame . . . it was a huge one-hundred kilometre an hour shockwave of roaring red heat that hit us like a bomb. It was as though the very air itself was on fire. Our entire world exploded firey red as we ran for our lives to shelter.'[2] Fourteen people lost their lives that terrible afternoon. Beautiful homes were lost, among them 'Mount Lofty House' and the Anglican seminary, Saint Michael's at Crafers, with its invaluable 40,000-volume theological library. Kym and Julie Bonython lost 'Eurilla', a house full of valuable paintings and antiques and a unique collection of jazz records.[3]

That same afternoon, Uniting Church minister, writer and social justice activist, Rev. Dr Gordon Powell, was taking shelter from the fires in the sea at Lorne, Victoria. His diary records what happened.

> I had gone to Lorne to spend five days with [my] brother . . . On Wednesday morning . . . it was so hot . . . At 1pm we went for a swim, but the north wind was so hot we kept ducking under. Our faces hurt. But no sign of any fire. The house was [still] cool so we had lunch and a siesta. When we came out there was a great pall of smoke streaming out to sea. On the radio we heard Lorne was on fire and everybody was on the beach. Planes and helicopters were flying over all the time . . . In the evening we gathered on the pier with scores of others and watched flames 100 feet high behind the town. Then we saw the clubhouse at the golf course ablaze from end to end . . . about 60 houses were burning, including that of Heather . . . She had rented it to a woman and her elderly mother. When Heather heard the police warnings she went down to make sure they had got the message. The woman had gone to the shops. The old woman refused to leave. Heather forcibly pulled her out and shoved her into the car. They had gone no distance when they heard a loud explosion and the whole house went up . . . I drove back to Melbourne [next day] . . . Every house between Lorne and Anglesea was gone except for two at Aireys Inlet . . . Charlie Wright had a house at Anglesea and his son David had another nearby. When I rang, Chas in his usual cheerful voice said, 'Both gone'. Later he said 'It's a bit hard to start again when you're 70'.[4]

Ash Wednesday saw the worst fires in Victoria and South Australia since Black Friday. The large number of fatalities—75—and the massive damage happened because many of the fires occurred along that dangerous frontier where the bush adjoins urban development; a frontier that had been developing since the Second World War. Many of the worst fires of modern Australia occur along it, as is vividly illustrated over many fire seasons between 1968–9 and the present.

ii

The Blue Mountains, on the western edge of Sydney, are a classic urban–bush frontier area. This was demonstrated in the lead-up to 1968 when Sydney's rainfall was 24.55 inches, 'little more than half the average of 47.75 inches'.[5] Fires had been burning in the Grose Valley since mid-October 1968. Sudden outbursts periodically ran up the gullies and creek-lines. Three firefighters were killed near Springwood on 29 October when an unexpected wind change brought a wall of flame racing toward them up a gully. Then late in the afternoon of 25 November a fire along Springwood Creek threatened houses in Huntley Grange Road and nearby streets. This is just a kilometre from Springwood railway station. People ran from their homes as the fire raced at terrifying speeds through the tree-tops. The *Sydney Morning Herald* of 29 November reported that, 'Flames leapt over roofs and firefighters threw themselves on the ground to avoid burns . . . A wall of flames 100ft high and 400 yards wide sent residents and firefighters running for their lives . . . Those with hoses turned them upwards as the fire went overhead.'

A month later fire again swept across the lower Blue Mountains. Driven by a 100 km/h north-west wind, a small fire from Faulconbridge suddenly took on massive proportions and with terrifying speed burnt along the Great Western Highway and the railway line through Springwood, Valley Heights, Warrimoo, Blaxland, and down to Emu Plains. Part of the Blaxland shopping centre was destroyed as well as 123 houses and other buildings. The Blue Mountains remained threatened for several days from the fire in the Grose Valley. A *Sydney Morning Herald* reporter who flew over it described it on 30 November as 'an awesome fire topped by an atomic-bomb like mushroom cloud'. Many of those most affected were older people who had retired to the mountains and who found it overwhelming in their late 60s to have to think about rebuilding destroyed homes. By Tuesday 3 December rain was falling across the area. The danger had passed, for the moment.

In 1968–9 there were other fires at the urban–bush frontier. North of Sydney at St Ives, Hornsby and throughout the Ku-ring-gai areas outbreaks were burning up the gullies, although only one house was destroyed, that of Mrs Molly Jenner, who escaped her home near what is now the Garigal National Park with a cat under one arm and a dog under the other. She recalled her escape for the *Sydney Morning Herald* on 29 November:

> I saw the fire burning more than a mile away . . . and thought it would take a long while to reach my home. I was hosing sheds and bush at the rear of the house and looked up to see the fire coming down on me . . . the back of the house was on fire as I ran out the front door clutching one of my pet dogs, Coquette, under one arm and my cat Timothy under the other. I was only wearing what I stood up in. I have lost everything and my home was not insured.

Fires so close to Sydney led to the usual rash of editorials. The *Sydney Morning Herald* of 3 December 1968 fell back on conventional solutions. 'The bushfires which have plagued the Blue Mountains down through the decades have had their beginning in the rugged regions backing the townships . . . Controlled burning programs are urgently needed. The major fires which devastated areas of the Blue Mountains in 1948, 1957 and again last week all followed a consistent path. How many times do homes and people have to be burnt before the lesson is learned?' A fair question, except the real issue was not controlled burning in remote valleys, but allowing people to build houses so close to the bush. It will always burn. The variable is humankind.

A month later the focus moved to Victoria. On Wednesday 8 January 1969, a day of complete fire ban, 230 fires, 21 of them highly destructive, broke out, burning more than 324,000 hectares. The worst were on the urban–bush fringe in a semi-circle around Melbourne and several were dangerous, fast moving grass-fires. It was a classic situation: the temperature reached 38°C, the humidity was the lowest ever recorded, and north-westerly winds gusted from 40 to 125 km/h. The fire danger index was in excess of extreme; that is, worse than Hobart in 1967. A huge smoke pall hung over Melbourne.

There had been a fire the previous day near Anakie on the Geelong–

Ballan road. This had been quickly extinguished, but it broke out again early Wednesday morning, possibly from a cigarette thrown from a car. This is flat, treeless grassland south-west of Melbourne between Werribee and Geelong and extending across the Princes Highway to the western shore of Port Phillip Bay. The only prominent features are the You Yangs. In the southern sector of the plain is the town of Lara, and close by is Avalon Airfield, used for pilot training by international airlines, aircraft maintenance and as a secondary airport for Melbourne. The fire quickly picked up speed and intensity, racing eastwards across the plain. It was unstoppable and hit Lara like a tornado, destroying the beautiful 90-year-old neo-gothic Anglican church and 40 houses, leaving 150 people homeless. Hundreds fled to the swimming pool and sports ground for safety.

A kilometre east of the town the fire crossed the four-lane Princes Highway. The road was a dual carriageway and there was fire on both sides. Caught in dense smoke, a number of cars travelling towards Geelong stopped. Other cars banked up. Some people remained in their vehicles, windows wound up, as the fire-front swept through. The cars offered protection and many motorists remained in them, driving on to safety as the smoke thinned. However, others panicked and abandoned their vehicles.

Eighteen people, including several children, tried to outrun the fire. Eight were burnt to death on or near the highway, including five members of one family, two people, including a young boy, died on the way to hospital, and eight died later of their injuries. This incident became a template of what to do when trapped. 'While it is preferable to remain inside a car than to be caught outside during the passage of a bushfire, a far more successful strategy is not to be caught out driving in the vicinity of a fire.'[6] Except, of course, when you least expect it—on a main highway.

Another drama was unfolding almost simultaneously just a kilometre away. Qantas had flown two Boeing 707 jets to Avalon Airfield for pilot training. As the fire swept towards the parked aircraft, pilots rushed aboard the planes, started the engines and taxied them quickly away from the fire into the lee of a large hangar where they sat for two hours with engines running in case they needed to be moved again. The pilots would have tried to take off, but flames across the runway and an 80 km/h cross-wind prevented this. The fire burnt to the edge of the western suburbs of Melbourne and right to the boundary of the Altona petro-chemical plant which, if it had exploded, would have caused massive damage to surrounding suburbs.

Other fires quickly spread across the state, especially in the semi-circle north-west and north-east of Melbourne. The central part of Diamond Valley and Kangaroo Ground near Warrandyte were evacuated.

By day's end the fires were under control. Twenty-three people were dead, 100 injured, 230 homes and 21 buildings destroyed and 800 homeless. Countless native animals and 12,000 livestock had died.[7] Victoria Police had charged 32 people with lighting fires in the open. An *Age* editorial of 10 January 1969 claimed that 30 years after Stretton, responsible people had come to understand the real causes of bushfires.

> Nature may have provided the conditions which made the flames uncontrollable, but in many cases man provided the initial spark . . . It is already clear that carelessness and stupidity, and a refusal to take elementary precautions, contributed a great deal to many outbreaks. After all the warnings there are still people who throw lighted cigarettes and matches from car windows and who light fires in the open on days of total fire ban. There are still householders who are too lazy or irresponsible to clear the bush back a reasonable distance from their homes . . . fire safety laws must be stringently enforced. Organization, money and research are keys to curbing fire: only common sense can prevent it.

A very sensible editorial, but Victorian Premier, Henry Bolte—never known for subtlety—was still blaming nature. 'People don't seem to realize that Victoria is the worst place in the world for fires', he said. While this is true, he had apparently forgotten that the Blue Mountains and Southern California around Los Angeles, were equally dangerous in similar conditions.

Meanwhile, the *Sydney Morning Herald* of 10 January was singing Victoria's praises in an article by Malcolm Elder. 'It's clear to me', he said, 'that the CFA south of the border could teach NSW a lot'. Elder pointed out that Victoria was organised, with radio links to the 26 fire regions, each with a full-time, paid officer. 'On a day of total fire ban when conditions suggest fires are likely, the CFA can have 120,000 volunteer fire brigade members on standby . . . Each officer and organisation has its activities planned ahead—and overlapping. This does not happen at NSW fires.' In New South Wales, amendments to the Bushfires Act tried to iron out the problems resulting from so many authorities involved in controlling bushfire. However, a subsequent committee reported that Victoria also had a far from perfect organisation. There were problems with equipment, communications, staffing, training and the development of community fire prevention. It recommended legislation to compel 'all concerned to carry out basic fire prevention'.[8]

A key factor along the urban–bush frontier is the arsonist. Melbourne had a spate of deliberately lit fires around the suburbs of Lysterfield and Rowville below the Dandenong Ranges in January 1973.[9] On 19 January the temperature was 39°C and there were strong northerly winds. The fire danger was extreme. Carloads of youths were suspected by police of going around lighting fires in the tinderbox area below Mount Dandenong. That Friday over 2000 hectares of bush and residential land were burnt. Up to 2000 firefighters worked to get the fires under control.

Sydney was again surrounded by fire in early December 1976.[10] With temperatures of 37°C and winds gusting up to 50 km/h on Friday 3 December, there was a ring of about 40 fires from Pennant Hills and South Turramurra in the north, to Fairfield and St Marys in the west, to Woronora, the Royal National Park and Bundeena in the south. Railway sleepers on the line across the Harbour Bridge caught fire. There were also fires in the Blue Mountains where a spectacular back-burn on Tuesday 7 December between North Springwood and the Lower Grose River area caused near-panic even though authorities had publicly announced it. This particular back-burn nearly got out of control when it came within about 6 kilometres of Faulconbridge. There was also a petrol shortage during this period and authorities were concerned about the danger of explosions. A couple of burning houses had already exploded and this was attributed to people hoarding petrol at home.

Appearing for the first time in a prominent public role during these 1976 fires was Phil Koperberg, deputy fire co-ordinator in the Colo–Blue Mountains region. He had become a volunteer in 1967, and after recovering the remains of the three volunteers killed in North Springwood in November 1968, he determined to devote his life to firefighting. Within three years he was on the permanent staff, eventually rising to the rank of commissioner in 1985.

The threat to greater Sydney returned in 1977. Fires began on Saturday 10 December in the southern suburbs and in the Royal National Park and on Friday 16 December, driven by strong westerly winds and mid-30s temperature, the focus of the fires shifted again to the Blue Mountains. An outbreak began around 11.00 am near Bullaburra on the north-west side of Lawson and, driven by strong winds, swept along the urban–bush fringe through Hazelbrook, Linden, Faulconbridge and Springwood. Phil Koperberg described the situation as 'critical'. Evacuations were ordered. The *Sydney Morning Herald* of 17 December described the drama:

> Yesterday's searing heat, and the mountain winds intensified by the fire itself, threw the mountain towns into chaos as outbreaks swept along the gorges and raced on the ridges toward houses . . . Lawson at one stage was surrounded by fire and threatened on all sides. Opposite the railway station hundreds of parents huddled with their children, dogs and birds and what they had salvaged from their homes . . . A train was still at the platform . . . unable to go because of sleepers alight on the track ahead.

Among the buildings destroyed was the Sisters of Saint Joseph convent at Lawson that was being used as a residence for 35 refugee Vietnamese nuns. About 50 homes, a nursing home and a railway station were destroyed. One person died: Bridget Sternbauer, a fifteen-year-old girl who, in the confusion of her family's home being burnt down, became separated from her parents and brothers and died nearby on a bush track.

Fires were still burning in the valleys and over the weekend the situation blew up again with 2500 homes threatened and hundreds of people evacuated. The worst-hit area was the southern side of Springwood at Burns Road which extends along a ridge-line to Sassafras Gully Reserve and 'where houses sit like fleas on a camel's back' as a *Sydney Morning Herald* journalist put it. The fire crested the ridge-line at Burns Road after midday, jumped the road and then burnt its way down the opposite slope. Only one house was destroyed. Firefighters put this down to residents being better prepared than before. Phil Koperberg commented in the *Sydney Morning Herald* of 19 December that 'this was one of the worst fires in the Mountains' history. We had fire storm conditions up here today . . . That's where the fire goes berserk, it goes everywhere. It's the sort of inferno which doesn't need any fuel to carry it along. It just explodes.' At the same time fires threatened the town of Helensburgh on the southern edge of the Royal National Park.

By Tuesday 20 December the fires were under control. They caused damage of more than $10 million in 1977 values, and the post-mortems began. As it turned out, nature was not really to blame at all. Some of the most disastrous fires were deliberately lit. Police believed they knew the identity of one arsonist but did not have evidence to charge him. However, few asked about the wisdom of building homes along ridge-lines with bush on either side. No one wanted to confront the dangers embedded in building along the urban–bush boundary. The closest anyone came to questioning the practice was journalist Tim Dare, writing in the *Sydney Morning Herald* of 19 December about Burns Road:

> If the wind changed the blaze would sweep through the street. The residents were calm, but they knew: their cars were parked at the edge of driveways, ready to go. They are among the thousands who choose to live in the Blue Mountains despite the hazard of . . . cyclical bushfires. Disaster struck in 1957, 1968 and now 1977. Suburbia has pushed into the bush, into the tinderbox, heedless of the fact that the Blue Mountains is one of the most fire-prone settlements in Australia. Cheaper housing, and the lure of the bush which can turn against them with awesome intensity, are stronger than any doubts they might have.

In December 1979 New South Wales was the driest it had been since 1972 and the fire threat was extremely high. With beaches crowded during a heatwave, widespread fires broke out over the weekend of 1–2 December 1979 right along the coast from Nelson Bay to Wollongong. Both major roads north and south out of Sydney were closed. There were accusations that despite a total fire ban, the ALP held a fundraiser attended by federal leader, Bill Hayden, with an outside barbecue.

The fires continued to burn throughout the following week and they claimed their first victim, firefighter Colin Carter. Cut off in the Mount White–Calga area, he went back to rescue a friend. They became trapped and took refuge in a small cave as the fire passed. A tree at the entrance exploded and the clothes of both men caught fire. Despite being badly burnt himself, Colin dragged his colleague from the cave and tried to beat out the flames with his hands. He then helped his companion back to their truck which he drove 5 kilometres to Mount White. He died later in hospital with 90 per cent burns to his body.

The total fire ban continued throughout the week. On 5 December in a temperature of 39.9°C, 170 schoolgirls were rescued from Karloo Pool, a swimming hole in the Royal National Park about a kilometre east of Heathcote. In what seemed to be an ill-judged decision, three high schools sent students, many from migrant backgrounds, into the bush on a day of extreme fire danger to give them experience of the environment and to improve their physical fitness. A fire broke out near the pool. Some of the girls became hysterical and there was a danger they might have run off into the bush. The girls were all gathered by teachers and taken out to the local oval by firefighters, or airlifted by police, national parks and television-station helicopters. 'Pilots had to contend with thermal updrafts from burning trees and poor visibility through clouds of choking smoke', reported the *Sydney Morning Herald* on 6 December. Shortly after the rescue, the fires died down. A local council official was furious with the schools for taking students into the bush that day, and a government minister described it as 'an act of folly'. It turned out later that teachers had checked with fire authorities and the national parks before the walk began, they were not 'trapped', and were never more than ten minutes from the Heathcote railway station. The majority walked out and the report was somewhat of a media beat-up.

Over the weekend of 15–16 December there were seven major blazes in an arc around Sydney, the most serious at Duffys Forest on the edge of the Ku-ring-gai Chase National Park and another around the Lucas Heights nuclear reactor with several fires on roofs of buildings within the complex. Most fires were lit by arsonists and in the northern suburbs of Terry Hills, Duffys Forest, Ingleside, Belrose and Elanora Heights twenty homes were destroyed. They were right on the frontier between bush and suburban fringe. With erratic winds the fires were particularly unpredictable. They came at all angles and had weary firefighters literally chasing them. Homes and a restaurant along Morgan Road, Belrose were saved by the skill of a 65-year-old firefighter, Rodger Keiran. With 25 years' experience he not only guided residents in how to protect their homes from a fire racing up gullies from what is now the Garigal National Park, he knew exactly when to light a back-burn. He told the *Sydney Morning Herald* of 18 December, '"It's all in the timing of the back-burn", he said. "You leave it as late as possible so that the heat will take the flames away from the houses".'

As the week progressed the fires got worse, especially in the northern suburbs. At Terry Hills they were spotting up from gullies rather than forming a large front, which meant authorities could not focus their resources. Eight more houses were destroyed and twenty others badly damaged. There were also outbreaks in the lower Blue Mountains, where the highway and the railway line were cut between Warrimoo and Blaxland.

Over the weekend just before Christmas a fire east of Lithgow had gained momentum and was burning toward Mount Wilson, just to the north of Bells Line of Road at the southern end of the Wollemi National Park. On Saturday, fanned by a 50 km/h north-west wind, the fire picked up, broke control lines and headed on a 30-kilometre front towards Mount Wilson, Mount Tomah and the Grose Valley. If it got into the valley it would again threaten the towns along the ridges of the Blue Mountains. But by evening as the temperature cooled the fire was controlled. Phil Koperberg commented in the *Sydney Morning Herald* on Boxing Day: 'There is no way in the world we can control the fires deep in the Grose . . . it could break out of the Grose and all our planning is concerned with this very real threat. The only way that fire can be stopped is with torrential rain, and I don't see much chance of that at the moment.' To compound the situation, right through this period Sydney was facing a water shortage.

By the end of December the fires were under control and New South Wales was ready to face the cost: over a million hectares of land burnt with a damage bill of $10 million. Five people were dead. Even before the fires were over, editorial writers were reflecting on causes. The realisation of the folly of building in or near the bush was starting to dawn. In its 18 December editorial, the *Sydney Morning Herald* commented that, 'The severity of the blaze must also raise the question of whether local councils have been remiss in allowing houses to be built in or near thickly wooded areas. Can this be allowed to continue? If it is true, as it is, that all the precautions in the world will not hold a raging bushfire in the right conditions, surely there is a case for reviewing the location of homes in areas of patently high fire risk.'

The early 1980s brought some relief from fires around Sydney, although a tragedy occurred near Waterfall on 3 November 1980. Firefighters had responded to a fire at 5.00 am which proved peculiarly difficult to control. The major problem was lack of vehicle access to the fire-front. About 5.00 pm a 1967 Bedford 4x4 tanker truck was trapped on the Uloola fire trail by an outbreak that came from a gully in a sudden 'blowup'. There were five men aboard and they were all killed as they sheltered under the truck. They do not seem to have used their hose-lines for self-defence.[11]

Everyone remembers Ash Wednesday 1983, but in the Mount Lofty Ranges east of Adelaide there was an earlier Ash Wednesday on 20 February 1980. This fire began a fortnight earlier in garden refuse in Stirling District Council's privately operated Heathfield rubbish dump, south-south-east of Adelaide.[12] It smouldered until 20 February when hot winds stirred the fire, which blew up and burnt 8000 hectares, destroyed 51 homes, the Anglican church at Longwood and 75 farms, as well as orchards and market gardens. The damage bill was in excess of $30 million. The town of Mylor was surrounded by flames and people had to take refuge on the oval. Bridgewater and the wine-growing town of Hahndorf were also threatened. While the fire itself was over in one day, ascertaining responsibility for this fire led to an extremely protracted series of court cases that dragged on until the early 1990s.[13]

iii

And so we come to the infamous Ash Wednesday of 1983. The context was what would now be known as a strong El Niño impact on eastern Australia, possibly the worst of the twentieth century, with resulting drought conditions. According to the Bureau of Meteorology (BOM), 'Below average rainfall patterns were established in April 1982 and continued almost unabated up to and including February 1983 when southern-Australia experienced heat-wave conditions . . . the vast bulk of Victoria and the southern halves of both NSW and South Australia, together with large tracts of southern and western Queensland, recorded record low [rain] falls for this particular 11 month period.'[14] There was low humidity. The forests had dried out and any serious fire would be almost impossible to control. Vegetation was severely drought-stressed and highly flammable. All that was needed were strong winds, and this is what Ash Wednesday provided. The BOM reported that:

> On Tuesday, 15 February extremely hot, dry conditions were reported over most of South Australia and north-western Victoria with maximum temperatures varying from 42 degrees at Ouyen to 47 degrees at Cook and Eucla on the Nullabor Plain . . . [On] Wednesday 16 February . . . winds had tended north-westerly by the afternoon, becoming strong, and occasionally of gale force strength . . . Temperatures varied between the high thirties and mid-forties over most of eastern South Australia and Victoria, with relative humidities generally less than 15 per cent.[15]

In other words, the worst possible type of fire weather.

In Victoria the fire season began on 25 November 1982.[16] Between then and 16 February 1983, 854 bushfires were reported. In January two Forests Commission workers were killed when they were trapped by a deliberately lit fire just north of Ballan. On 1 February Mount Macedon suffered a devastating fire, while another fire in the Cann River forest area burnt out 120,000 hectares over a period of ten days. There was another early February fire around Mount Donna Buang. The danger of the situation was apparent.[17] It was one of the hottest, driest Februaries on record and serious preparations for the fires were initiated by the Cain Labor Government.

The weather in the early hours of Ash Wednesday was complex and did not clearly signify how the day would develop. A front separated hot, dry air coming in from the landmass to the north, from cooler air moving eastwards from the Southern Ocean. Ahead of the front was 'hot, turbulent, gale force northerly winds'. Temperatures rose in the early afternoon with Melbourne reaching 43°C, and the winds raising 'considerable dust from the drought affected surface ... when fires started to develop the drifted smoke contributed to a severe decline in visibility ... which concealed the exact whereabouts of individual fires to people downwind'.[18] On the streets of Melbourne it was like a blast furnace. From mid-morning McArthur's fire danger index was in excess of 100 in several places in Victoria and South Australia. The cool, south-westerly front moved rapidly east-north-eastwards in the early evening passing through central Melbourne about 8.40 pm with maximum wind gusts up to 102 km/h. Much of the loss of life resulted directly from this change in wind direction driving the fires on a new tack.

Early on Ash Wednesday morning there were 104 fires already burning in Victoria, most controlled or contained. It was only after 2.00 pm that the situation exploded. As the Ash Wednesday Disaster Investigation stated: 'The fires were severe ... In almost all cases they were well-alight in the day-light hours in the hottest and driest part of the day ... Except in the west of the state there were several hours before the most devastating phase of the fires when the strong, squally wind-change occurred, and within less than an hour in many cases the fire had rampaged through townships and populated areas destroying property and killing both firefighters and residents.'[19] People were overconfident and the speed of the fires caught them unawares and inadequately warned. The result was chaos and panic.

The Victorian fires formed several complexes. The complex in the Western District near Warrnambool originated from a defective private powerline. This fire burnt south-south-eastwards through closely settled farming land to the small town of Framlingham, which was destroyed. A wind change then directed the fire in an east-north-easterly direction along the Princes Highway towards Garvoc and across to Cobden where the fire was partially brought under control. Meanwhile another front had burnt south-eastwards towards Curdie Vale, and then eastwards towards Timboon. The Warrnambool *Standard* reported that 'Small settlements along the way were evacuated and farmers piled belongings on trailers and left sprinklers going near their houses in a bid to save their homes'. The fire was moving along a front 'kilometres wide and at frightening speed'. This was dairy country and many were faced with the choice of saving either their homes or their stock. 'Most did not have time to do both. They chose their stock.'

The cold front came through about 5.00 pm which swung the wind around to the south-west. The fire changed direction and headed north-north-east toward Terang. 'The Terang Hospital was evacuated about 8.00 pm and the town was on standby as the flames drew near. However, the fire by-passed the town on its way north', to be brought under control the next morning north of the Princes Highway.[20] These destructive fires left nine people dead, 50,000 hectares burnt, 157 houses destroyed and 19,300 livestock lost. Another person died in a smaller fire near Branxholme between Hamilton and Portland, which had also been caused by poorly maintained powerlines.[21]

The East Trentham–Macedon fire complex began around 2.00 pm on a property close to East Trentham, about 80 kilometres north-west of Melbourne. According to police, as reported in the *Age* of 18 February, it was caused 'when two SEC power lines, which had been blowing against a tree, touched and arced. A plastic spacer, designed to keep the lines apart, was not attached to the wires.' The resulting sparks set grassland alight which burned in a southerly direction through isolated country to the Wombat State Forest. There it reached an area previously burnt out on 8 January 1983. On the way it devastated the small settlement of Bullengarook. While temporarily stopped, it was an intense fire that was spotting as much as 13 kilometres away to the south-east.

At 8.30 pm the front came through and the wind changed direction to the south-west. The fire picked up energy again and headed north-east across the Calder Highway to Woodend on a 7.5 kilometre front driven by 100 km/h winds. It developed into a violent, unstable firestorm that shot off in unpredictable directions, especially in the hilly terrain with gullies and steep ridges. Kerry Murphy, captain of the Mount Macedon CFA said:

> The wind was extreme. There was incredible noise. You had to shout into the face of your colleagues so they could hear you . . . There was no discernible fire edge . . . It was difficult to breathe. We concentrated on warning as many people as we could in exposed areas . . . Driving was quite horrific. People were absolutely panicking . . . The fire was spotting anything from three to six kilometres ahead.[22]

By 9.15 pm it had reached the small town of Macedon. More than 250 people sheltered for an hour with their children, dogs, cats and pet birds in the Macedon Family Hotel. The building was protected by sixteen volunteer firefighters who stood bravely outside and on the roof with hoses and wet mops protecting the modern, brown-brick building. Around it the village was annihilated.[23] At 12.25 am the fire reached the monument on the summit of Mount Macedon at 1013 metres. Many heritage-listed houses were destroyed. It was finally controlled in an area already burnt out on 1 February. The wind had maintained its velocity for several hours which 'prolonged the critical period during which the fires maintained their intensity and made the saving of property very difficult. Some houses did not, in fact, burn until many hours after the main fire had passed.'[24]

Almost 500 firefighters could do nothing except try to save lives. Around 2000 people were evacuated. Everything happened so quickly that responses became chaotic. The fire was indiscriminate, destroying the mansions of the rich on Mount Macedon, as well as ordinary houses below. Seven people died, 20,000 hectares were burnt, 200 homes were destroyed in the town of Macedon, 150 on Mount Macedon, 50 in Bullengarook, and twenty in Woodend.[25] Among those who lost their homes that night was Dr Clifford Pannam, QC, a successful, horse-loving barrister and the author of a basic legal guide for equine owners, *The Horse and the Law* (1979).[26] His house, 'Huntly Burn' was one of several mansions burnt to the ground. It contained paintings by Frederick McCubbin, Sidney Nolan and Eugène von Guérard. Pannam told the *Age*: 'It was a lovely old house built in 1874 . . . [There] was a library the envy of every barrister of the Australian bar. I had the complete Victorian law reports, the complete NSW, the New Zealand and South Australian law reports . . . all gone . . . Every single personal treasure . . . is gone.'[27]

The Otways complex broke out some time between 2.55 and 3.30 pm near Deans Marsh where a man was killed and fourteen houses destroyed. The fire moved south-south-east driven by strong north-west winds. It was intense, spotting up to 10 kilometres ahead. By about 4.20 pm there were spot fires burning around the north-east side of Lorne on the Great Ocean Road, 23 kilometres from Deans Marsh. Lorne had a number of subdivisions that penetrated the forest, as did a number of the other towns hit by this fire. Again it was on the urban–bush fringe with insufficient cleared areas to protect the new developments. The fire swept along a gully right down to the sand dunes and out onto the ridges above the township. Many took refuge in the sea.

Firefighters concentrated on protecting assets, but 84 houses were burnt as well as the golf club. At 6.30 pm the wind changed to out of the south-west, swinging the fire-front north-eastward along the coast. The fire moved forward swiftly on about an 8-kilometre front parallel to the Great Ocean Road. It destroyed 32 houses in Eastern View, 177 in Fairhaven, and 87 in Moggs Creek. Those on cliff tops were particularly badly hit.

The fire was soon spotting 13 kilometres ahead into Aireys Inlet, then known as 'Fibro town'. Aireys was hit by 'a wall of flame' about 7.25 pm. Police reported that roofs were torn off houses before the fire got to them. It was described as 'a gigantic fire-storm, shrieking and howling and blowing everything to bits, a primal mass of almighty energy with a will of its own'.[28] The losses were tremendous: 217 houses, as well as a hotel, restaurant and CFA station. The owner of the Lighthouse Restaurant, 'Butch' Stern, told the Melbourne *Herald* on 17 February

1983 that '"We decided to get out of Airey's when a house next to the fire station exploded . . . The fibro started making a cracking sound and then all of a sudden there was this great explosion. The roof just seemed to hang in the air". Mr Stern's $185,000 restaurant now lies a charred heap. He escaped with only the clothes on his back.'

Among the houses destroyed was that of the famous soprano, Joan Hammond (1912–96).[29] The *Age* reported that 'Dame Joan Hammond's house was a shambles, with only its white brick walls standing. She had escaped with only her memories. It was so strange up there, seeing a torn music sheet fluttering in the rubble . . . The Steinway a heap of wire bones, the 5000 books gone, the photographs. The indoor swimming pool was congealed with black timbers.'[30] Another musician who lost everything was Rev. Dr Percy Jones, choirmaster at Saint Patrick's Cathedral from 1942–73, and lecturer in music at Melbourne University. Apart from his folk song research, he lost 'all written and recorded evidence of his life's work including his notes on early and renaissance music', which he had taken to Aireys Inlet to assemble into a book.[31]

Anglesea was next to face the fire which reached the town from the south-west about 9.00 pm on several fronts.[32] Again, developmental sprawl saw many houses on the edge of town surrounded by scrub-covered blocks. To some extent the town was protected by the golf course and the Alcoa open-cut mine, which caught fire, and the electricity-generating plant to the north. The fire swept up a ridge behind the town and encircled the residential area, as well as burning down to the beach and along the dunes. The main part of the town was saved, but 127 houses were destroyed.

Many residents took refuge in the sea. Among them were Gloria and Bryan Poynton and their children, Ruth and Simon. They spent four hours huddled in the sea draped in blankets. Their house was burnt down and they lost everything. Gloria said, 'There were cinders and smoke everywhere. You couldn't see or breathe. The fire came right down to the beach. It was like an atomic explosion.'[33] The fire burnt on through the night towards Jan Juc and Torquay, and it was eventually contained around 4.00 am near Bellbrae in cleared land.

It had moved with extraordinary speed due to spotting kilometres ahead of the main front. Firefighters could not get a fix on its movements;

the wind carried it willy-nilly, everywhere. Deans Marsh suffered two fire attacks and Lorne was subjected to three. An *Age* journalist commented:

> This was no ordinary bushfire, bad though they are . . . It was the worst. It was a gigantic fire storm . . . It sucked the life out of houses, and made them disintegrate in less than a minute. Aireys . . . went off like a bomb. The noise of both the wind and explosions was immense. To one man it seemed like 20,000 trains gone mad . . . A young Aireys fire-fighter said 'You couldn't make up a bullshit story if you wanted to. Nothing would be more outlandish than the truth'.[34]

What was extraordinary was that only three people were killed, although 578 houses and other buildings were lost and many people were left homeless.

The most destructive of the Ash Wednesday fires, especially in terms of the loss of human life, was the Upper Beaconsfield–Cockatoo complex. These fires did not cover a large area—about 11,000 hectares in total. But 27 people died; hundreds were injured; 535 houses and buildings were burnt; and many people were rendered homeless.[35] These fires were particularly destructive, being right on the urban–bush frontier in the foothills of the Dandenong Ranges where large numbers of people had settled hoping to escape the city, while still enjoying the advantages of urban life. Police and the CFA believe that both fires were deliberately lit by arsonists. They questioned a fourteen-year-old youth about the Upper Beaconsfield fire, but he was not charged. Once started, the conflagration moved in two directions: firstly southerly, and then after the change, it headed north-eastwards into the night when the deaths and worst destruction occurred.

The fire began between 3.00 and 3.30 pm in Birds Paddock, an overgrown 105-hectare nature reserve owned by the Shire of Sherbrooke at Belgrave Heights due south of Mount Dandenong. Half an hour later 10 kilometres away to the east another fire began just north of the town of Cockatoo (population: 3000). This fire remained static until late that night.

The Birds Paddock fire burnt southwards, on the western side of Cardinia Reservoir, driven by a strong northerly wind. There was a report of spotting as far away as Lang Lang, 30 kilometres south. The fire reached an area south of the Princes Highway at Officer between Berwick and Pakenham, more than 15 kilometres from its source. Just when the CFA were getting it under control, the change came through around 8.30 pm. The fire then turned east-north-east and burnt with great speed toward Upper Beaconsfield, south of Cardinia Reservoir. 'Fires followed selective routes guided by the terrain and the wind, curving around hills, approaching from unexpected quarters with further confusion arising from spotting from burning leaf and bark material.'[36] The disaster was compounded by darkness.

It was a terrible night: 21 people died. 'The winds, accentuated by the fire generated effects, drove the flames like a blow-torch through the area with great rapidity.'[37] Ivan Smith, the CFA Acting Group Officer at Upper Beaconsfield says, 'I went out the door of the command HQ [in Upper Beaconsfield] and was hit by cinders as big as cricket balls, that would explode when they hit the ground. The explosion of gas and fuel tanks was like bombs going off. It was like fireworks when the hardware store went up. I thought I was going to die. I was gripped by terror. I started to fear I would never see my wife and children again.'[38]

This was compounded at about 8.55 pm when Ivan got a terrible radio call from Dorothy Balcombe, the first woman firefighter ever to be killed. 'She said: "We are in real trouble. There are two tankers trapped". That was all and they were probably dead a minute later.'[39]

What exactly happened will never be known. There were two trucks, one from Narre Warren with six volunteers and one from Panton Hill with five, as well as a casual firefighter. They were part of a group trying to secure the eastern edge of the fire. The official CFA report said that when the wind change came through at 8.50 pm 'with tremendous ferocity ... the fire developed quickly around the immediate area'. Basically they were on a bush track connecting two sealed roads. The trucks were trapped on a wooded slope. The CFA report continues: 'It can only be presumed that, because the fire was gathering intensity behind and beside them they attempted to out-run the fire'.They were less than 50 seconds' drive from the safety of the sealed St Georges

Road. The coroner concluded that, 'Upon the arrival of the wind change, the fire-fighters and their vehicles were subjected to violent flames, air-borne coals and embers and extreme heat'.[40]

Meanwhile at about 7.30 pm the fire that had begun north of Cockatoo, 'described as the smallest but the worst of the Ash Wednesday fires' suddenly exploded.[41] This fire had been deliberately lit during the afternoon and local firefighters believed they had the fire out by late evening. But, as local supermarket owner and CFA captain, Laurie Butcher, told the Melbourne *Herald* on 17 February, it suddenly broke out again. 'We had a major fire on our hands. The winds were very fierce and within minutes we had no control . . . Another big problem was that nearly all our trucks were down helping fight the fire at Belgrave. Our fire only lasted an hour at the most, but it went right through Cockatoo.' As the fire hit the town the warning siren sounded—one local described it as like a 'banshee wailing'—and those still in town, mainly women, children and elderly, fled their homes and came to the local kindergarten. Here a 'miracle' happened. The kindergarten teacher, Mrs Iola Tilley, recalled:

> All the men were already out fighting fires and those mothers and children who had not left Cockatoo came to the kindergarten with their dogs, cats, budgies and goats . . . The children were magnificent. All of them were brave little things. They never made a sound, they never panicked, they trusted us completely. They placed wet towels over their heads and lay on the floor and never complained despite the air being thick with smoke . . . why the kindergarten never went we shall never know. Everything around us was burnt to the ground when huge fireballs exploded overhead.

Two heroic men remained on the roof of the kindergarten right through the fire-storm, keeping the building wet. One of them was David Adam. He said that,

> The wind was at least 100 mph—it was ripping trees out of the ground and bodily picking up firemen . . . with the wind came

the fire—scores of fireballs each of them at least as big as a soccer ball. It was the most frightening thing I have ever seen . . . All the time the wind and flames were roaring around us, the fireballs and trees were exploding. How the children never panicked I don't know. How they weren't all killed either by the flames or by the lack of air when it was all sucked out of the kindergarten by the fire, has to be a miracle. (*Herald*, 17/2/83)

Within an hour six people were dead including a firefighter, 307 homes and buildings destroyed and 1,800 hectares burnt.

The last of the Victorian fires was around Warburton in the foothills of the mountain ash forests. The fire began at Millgrove about 7.30 pm and quickly spread between the Warburton Highway and the Yarra Junction–Noojee Road. Warburton was threatened and several houses burnt. With long experience of fires, 200 people gathered at the Powelltown oval, and in the McMahons Creek–Reefton area 83 people, mainly women and children, sheltered in a narrow underground tunnel between two water pipes from Upper Yarra Dam to Melbourne. Thirty houses were destroyed. 'Police and CFA men say the Warburton fire is particularly bad. One reason is that much of it is burning in steep, inaccessible country. Another is that it creates its own winds in bush that is very dry' reported the *Age* on 18 February.

All up, in Victoria on Ash Wednesday 47 people died, almost 150,000 hectares were burnt, 1620 houses and more than 1500 other buildings were destroyed, and about 32,400 livestock killed.[42]

iv

On Ash Wednesday morning there was little in the *Advertiser* to indicate the danger South Australia faced except that 42°C was the expected temperature and the Weather Bureau predicted a hot to very hot day with a strengthening northerly, and a gale warning for the central and south-eastern districts. In other words, a day of extreme fire danger.

In the morning the winds brought red dust from central Australia which covered the city. Driven by a north-westerly, the outbreaks began between 11.30 am and 1.00 pm. Many began suddenly and police suspected arsonists. By noon major fires were burning in a south-easterly direction in at least four places in the Adelaide Hills, in the Clare Valley wine-growing area 120 kilometres north of Adelaide, and in the south-east of the state around Mount Gambier. By 1.00 pm all of these fires were out of control and by 3.00 pm the situation was extreme. Then the cold front came out of the south-west with winds at 90 to 100 km/h. The sheer intensity of the fires created their own weather patterns. Things became so bad that it was reported on British radio that 400 had died and half of Adelaide was destroyed. Australia House in London was inundated with calls. A state of emergency was declared around 4.00 pm for the first time in the state's history.

While intense, the disaster was short-lived. By 9.00 pm rain started to fall in the Adelaide Hills and the emergency was essentially over. The fires in the south-east near Mount Gambier took longer to control. The final toll was 28 people dead, including three CFS volunteer firefighters, more than 1500 injured, 383 homes and 200 other buildings lost and 160,000 hectares burnt.[43]

Again the fire-front was along the urban–bush frontier. Norton Summit is near the top of the range east of Adelaide. In 1859 the Rev. Thomas Playford built a house there, 'Drysdale'. The Playford family still owned the house in 1983. Two Thomas Playfords, both premiers of South Australia, lived there. Another family member, Dr John Playford, was planning a family biography. But, as the *Advertiser* reported on 19 February, 'Dr Playford's collection of irreplaceable personal papers, unpublished letters, rare photographs and books, oil paintings and 27 cabinet drawers of documents were lost when the fire destroyed "Drysdale" at Norton Summit'. 'Drysdale' was one of several heritage houses lost that day. Some had survived bushfires for 130 years.

But modern homes were also destroyed. Yarrabee Road is only 3 kilometres from Mount Lofty summit and high above Adelaide's eastern suburbs. It is a residential loop that runs parallel to Greenhill Road which is a winding and in places quite narrow descent from the top ridge of Mount Lofty down to the Adelaide suburbs. Both roads run along a narrow ridge with thick, forested bush on either side. The fire-storm roared up the northern ridge from the valley and struck Yarrabee Road. What happened was made famous by broadcaster Murray Nichol, who went 'live to air' in a Walkley Award winning coverage describing the destruction of his own home. Twenty years later Nichol said:

> It was a pig of a day, right from the very early hours of the morning. We're talking 40-plus heat and 60- or 70-mile-an-hour hot, northerly winds . . . When we got a weather change in the afternoon, that whole flank became a front which swept through where I lived. I knew it was going to happen. I came here with a walkie-talkie radio, also to check on my family, because . . . I knew this was going to be the worst place in the Adelaide Hills.[44]

Going live-to-air that afternoon Nichol said: 'We can hardly breathe. The air is white with heat. There's smoke and it's red and there are women crying and there are children here and we are in trouble.' After ten minutes he ran along Yarrabee Road to his own home.

> At the moment I'm watching my house burn down. I'm sitting out on the road in front of my own house where I've lived for 13 or 14 years and it's going down in front on me. And the flames are in the roof and—Oh God damn it! It's just beyond belief. My own house! And everything around it is black. The fires are burning all around me. And the front section of my house is blazing. The roof has fallen in. My water tanks are useless. There is absolutely nothing I can do about it.

Within 100 metres of the Nichol home five people died. Some houses burnt to the ground, others were spared. The *Advertiser* of 18 February documented the tragedy. 'A mother lay dead in her home; her young daughter was burnt alive in a car outside; a friendly neighbour, alight, ran in agony from his blazing home and slumped to the ground, by then almost dead, as his wife screamed hysterically for help; and an elderly woman who some said died because she refused to leave her home.'

Greenhill Road was also the scene of several dramas. A bus full of 12–17-year-old school students was accidentally diverted there from

Mount Barker Road. They had to drive through choking smoke and a 10-metre wall of flame. Many of the children became hysterical, but were calmed by a girl student, Kerry Stone, who told them to sit down, stop panicking and shut up so the driver could concentrate on getting them through.

This narrow road proved a trap for cars. Union secretary Norm Rennoldson, 45, was trying to join his CFS unit. He told the *Advertiser* on 19 February: 'The impulse was great to get out . . . But if I hadn't stayed in my car I would have been dead. I knew I couldn't go any further. I switched off the engine, lay down on the floor and screamed my head off. My windows started exploding. I have never been so frightened in my life. It was only a matter of minutes, as the blaze passed over, but it felt like hours'. He stayed until the petrol tank caught alight. He then rushed to the car behind him where Gillian Kotz, 28, was sheltering under a blanket. But it soon filled with smoke and they had to make a run for it down Greenhill Road. They were eventually picked up and taken to hospital. Sadly, the car in front of Norm's was driven by seventeen-year-old Russell Perry. He had been trying to get to his parents' home at Mount Barker to help them, but had also been diverted along Greenhill Road, when he was killed. He was driving with the windows open and when he hit the wall of fire the interior caught fire.

Meanwhile at Mount Osmond, Geoffrey Maitland lost seven years' work when a manuscript was destroyed when his house was gutted. A physiotherapist, his previous books had been translated into five languages. Next door, Swiss-born Fred Remmele lost a $300,000 house and an art collection worth half a million dollars. And in the village of Peachtown, south of Hahndorf, ten houses were destroyed, among them three 140-year-old cottages built by German settlers. Twelve houses were destroyed at Bridgewater, and fire surrounded Hahndorf, Crafers, Stirling, Aldgate, Mylor and Mount Barker. One RFS volunteer was killed and another seriously injured when five experienced firefighters were trapped on a steep, narrow fire trail in the Mount Bonython area. They were dealing with a spot fire when overwhelmed by a massive fireball. Two crew sheltered in the driver's cabin of their truck. The three dismounted crew were unable to climb back on board and ran back up the track seeking safety. One was picked up by a passing motorist and the others sheltered in a clearing where they were both burnt, one fatally, the other seriously.[45]

But the fires were not confined to the hills. The first fire reported to the CFS was at McLaren Flat, 35 kilometres south of the city. This fire destroyed the 3000-hectare Kuitpo Forest before being contained near Meadows. The report of a fire in the Clare Valley wine growing district came in 27 minutes later. Much of the valley was burnt and the fire came right to the outskirts of Clare. Sevenhill and Mintaro were also threatened.

The fires with the greatest loss of life were in South Australia's south-east where fourteen people died.[46] While predominately a farming area, much of the district was devoted to plantation forestry, about 25 per cent of which was destroyed in the fires. On the day of the fires the south-east was bone dry, Mount Gambier's temperature was 44°C, and humidity was low. Pam and Brian O'Connor's *Out of the Ashes* describes the day: 'From early morning the wind was a dominant feature of the weather. It was incredibly strong and as the day went on it continued to increase in strength.'[47] It was turbulent and volatile and much of the damage to the forests was done by the wind. In Mount Gambier the fire danger index was in excess of extreme for more than six hours.[48]

In essence, a series of fires broke out sequentially right across the south-east in an area bordered by the coast and the Victorian border. The Coroner found it difficult to ascertain the causes of the fires, but the *Border Watch* newspaper of 18 February 1983 was more forthcoming. Having discussed how people quickly forget what to do on 'blow-up' days, its editorial said: 'Few fires occur, most are lit, carelessly or deliberately'. The country through which the fires burnt was flat, farming land dotted in many places with magnificent red gums, many of them hundreds of years old. There are also swamps in the area, but most were dry due to the long drought. Once the flames penetrated the pine forests crown fires developed and extraordinarily high flame heights were recorded. Pam and Brian O'Connor say that, 'One column of fire photographed . . . was estimated . . . to be 192 metres . . . Although this is believed to be the highest recorded in Australia, it was not the highest observed in the area at the time.'[49] The closest fire to Mount Gambier was near Glencoe, just 23 kilometres north-west of the town. It was in twilight at 3.30 pm with the sun 'a dim, blood-red ball', according to the *Border Watch* of 7 February.

The timber town of Mount Burr was saved when the wind changed direction slightly and the fire passed on either side of it. But the Mount Burr Forest Reserve was largely destroyed. The south-westerly front came through at about 3.45 pm when the head of the fire had reached Dismal Swamp. The ferociously strong winds shifted around from north-west to west-south-west with gusts up to 100 km/h. The direction of the fires shifted to the north-east towards the towns of Kalangadoo, where several houses were destroyed and two men died in a house just outside the town, and Tarpeena, where at least thirteen houses were destroyed and 1200 people sheltered on the football oval with their cars, pets and belongings. The fire burnt right up to the fence. Eventually most of the fires joined up to create one large conflagration which burnt through the Penola Forest Reserve and eventually died down near Nangwarry and the Victorian border. Fortunately, as the O'Connors note, 'as it became dark the winds gradually reduced in speed and swung south-south west'. But that did not mean the fires went out. They 'burned on all that night and for the next two days in forest areas causing substantially more damage'.[50]

Fourteen people died. The cause was the sheer speed and intensity of the outbreaks, especially after the wind direction changed. People quickly became trapped. Typical was Andrew Lemke, 24, a grader driver. He was trying to clear vegetation so a break could be burnt on the Robe–Penola road. Inexperienced with fires, he drove too close to the burning bush. Trapped, he abandoned the grader and two firefighters, Brian Nosworthy, 52, and Paul O'Leary, 25, died when they left their vehicle to try to save him. A young woman, Stephanie Prance from Millicent, died when her car crashed into a tree and she was trapped near Furner.

The worst tragedy occurred in the Kalangadoo area. Margaret Williams, 32, called her neighbours, the Rogers family, saying she was driving her four children to their house for safety. When she did not arrive, Gavin Rogers went out to look for them. He was later found dead close to Margaret's car where she and her children, aged seven to two had died trapped in the vehicle. Her husband, John Williams, lost their home and two-thirds of their sheep. His father and brother were injured as they tried to fight fires on family properties.[51] Gavin Rogers' uncle suffered a fatal heart attack not far from his home which was destroyed. Gavin subsequently received a number of posthumous awards for bravery.[52]

All up, the damage bill for farming properties was $25 million and for the pine forests $60 million. 'The area burnt out was estimated at eight per cent of the South-East, with 300,000 sheep lost and about 10,000 cattle. Between 350 and 400 actual farming properties were fully or partially burnt', noted the *Border Watch* on 21 February 1983.

A month later southern and central Australia experienced floods following above-average rainfall across much of the country. The drought had broken.

Ash Wednesday was followed by the inevitable investigations, including a major inquiry by the House of Representatives Standing Committee on the Environment and Conservation, which produced a voluminous report.[53] A National Bushfire Research Unit was established within CSIRO, led by Phil Chaney. After a promising beginning it fell victim to cuts and government stinginess.

v

By the 1980s it was noticeable that the weather pattern in Australia was changing. With the exception of New South Wales, and to a lesser extent Queensland, the years 1986–2000 were comparatively bushfire-free. This is particularly striking in Victoria where there were only two significant and several minor fires in this period. Tasmania, South Australia, Western Australia and the Northern Territory were almost free of serious fires between 1986 and 2000. The pivot of fire moved to New South Wales which had severe fires in 1987–8, 1990, 1991, 1993–4 and 1997–8, and to southern Queensland with fires in 1992–3, 1994, 1995 and 2000.

Part of the explanation is that firefighting had improved. Far better equipped and trained, large contingents of volunteer firefighters could now be mobilised quickly and fires that might have become dangerous previously were dealt with speedily. Community attitudes also changed.

The message had got through: people no longer lit fires in the open, and pyromania had became limited to anti-social misfits whose purposes were malicious. There were fewer people living in the bush to light fires. Also there was a move away from a preventative strategy based on the theories of McArthur and Luke of broad-scale fuel reduction burning. Many fuel-reduction burns had gone astray, a few developing into wildfires. Part of the problem was the enthusiastic amateurism of some burns where the old pyromaniacal tendencies were subsumed by the adrenalin-run of lighting fires and then controlling them.

The shift was toward a more ecologically sustainable understanding of biologically diverse landscapes and the different fire intensities they could sustain. Trusting folk wisdom was not enough. The new approach was 'mosaic burning'. This posited the fact that specific environments required different approaches. At the same time the university-based professional ecologist was taking over. Forestry was being 'ecologised'.

Certainly the McArthur–Luke theories were not abandoned by everyone. Since the late 1970s bushfire discussion has been characterised by low-level warfare between fire management strategies based either on ecological considerations and science, or convictions about fuel reduction and dependence on local knowledge.

The basic problem is the complexity of the issues surrounding bushfire and its behaviour. Certainly fuel management prevents fires becoming intense, but the key issue is the relationship between weather and fire. It is temperature and humidity that create the conditions for wildfire, and it is the wind that determines the speed and rate of spread. Weather is one element over which we have no control. What we can control is where people place their assets, such as their home. If they insist on living on a ridge-line with bush coming up on both sides to their houses, or building houses abutting or surrounded by inflammable forests, then they have no one but themselves and the local council that granted the building entitlement to blame. Development on the urban-bush boundary remains a serious problem.

In late November 1990 houses were threatened and nude sunbathers and picnickers in the Sydney Harbour National Park at Balgowlah Heights had to be rescued when a north-easterly wind drove a fire through dense bushland on the harbour foreshore. Fortunately the fire was stopped before it reached Reef Beach where the nudists congregated. At the same time an arsonist was active in the Blue Mountains. Then two days before Christmas Sydney was ringed by up to 370 fires, many deliberately lit. The National Parks and Wildlife Service reported that more than half of the fires in national parks close to settled areas that year were the product of arsonists. Up to 45 fires were still burning around New South Wales and in the Sydney area as the New Year approached. There were bad fires along the coast between Sydney and Newcastle, and south of Sydney around the Appin–Campbelltown and Holsworthy areas.

In mid-October 1991 on a day the *Sydney Morning Herald* called 'Our black Wednesday' there were fires along the New South Wales coast from the Hunter to the Shoalhaven and down to the Cooma district. Many were deliberately lit. A mother and a daughter died at Kenthurst in Sydney's outer north-western suburbs in a fire that came out of a gully. A witness said, 'the sound of exploding trees was "like a shed full of bullets going off. The gully was 30 metres deep but the flames were up above us . . . If the wind had changed it would have come back on us"' (17/10/91). The Kenthurst fire was part of a bigger conflagration in the Baulkham Hills Shire that was probably started by a cigarette thrown from a car.

The summer of 1992 was almost fire-free, but the 1993–4 fire season was the worst that New South Wales had experienced for twenty years. All along the coast weather conditions were extreme, with high temperatures and hot winds. Fires began to break out from 27 December onwards and reached their climax in the first week of January 1994 when there were five days of extreme fire danger, usually peaking between midday and evening. The outbreaks were not brought under control until after 12 January 1994. There were more than 800 fires and they extended right down the seaboard east of the Great Dividing Range from north of Grafton to Bega, a distance of 825 kilometres. They were particularly severe around outer Sydney, and it was there that most damage was done as fires penetrated further into urban areas than had ever been experienced before. Many were lit by arsonists; some speculated that up to 90 per cent of those in national parks and three-quarters of all other fires were deliberately lit.

Almost half of the housing losses were in the suburbs of Como and Jannali to the south of the Georges River where 101 homes were destroyed and 94 damaged, even though the area burnt was less than 400 hectares. The other badly hit suburbs were in the north around the Lane Cove and Ku-ring-gai national parks. At the height of the fires on 7 January there were 204 significant wildfires burning in the Sydney area, the Hunter and the Blue Mountains. Four people died and 800,000 hectares were burnt, including 250,000 hectares of national park. Almost 90 per cent of the Royal National Park was burnt; it was estimated it would take 5–10 years for its ecosystems to recover. Almost 20,000 firefighters were deployed. The smoke was so bad at times that planes had to be diverted from Sydney International Airport. At one stage all roads between Sydney and Newcastle were closed, including the F3 Freeway, which was closed for two days. The Woy Woy area was cut off for 24 hours and the main northern railway line was cut for 48 hours.

By Sunday 9 January the focus of the fires had moved to the Blue Mountains. Driven by a 60 km/h wind a fire began in an inaccessible part of the Grose Valley and raced towards Hawkesbury Heights and Winmalee. There were plans to evacuate 16,000 people from the lower mountains and 350 older people were actually moved from two retirement complexes in Springwood. The next day there were fires all over the mountains: along Bells Line of Road from Bilpin and Mount Wilson to Kurrajong, right through the Grose Valley, at Blackheath, and just to the north of Springwood and Winmalee and along the western side of the Hawkesbury River Road at Yarramundi. Wind shifts made it impossible to tell which way the fires would move, although firefighters had a strong defence line between the northern edge of the Grose Valley and the Great Western Highway. It was not until Wednesday 12 January that things began to ease.

What is particularly striking is the number of claims by authorities that arsonists began these fires. Few of the blazes were of natural origin; it was generally accepted that from 75 to 90 per cent of them were deliberately lit. Typical was the Lake Macquarie area where all the major fires were deliberately lit. The local deputy fire control officer told the *Sydney Morning Herald* of 31 December 1993 that, 'It was very rare for a fire to start from natural causes . . . A lot of fires started from people burning off stolen cars and there was an increase in bushfires when the school holidays started.' Throughout the decade there seemed to be an increase of malicious fire-lighting. The government promised stiffer penalties and a $100,000 reward was offered. But arsonists had to be caught in the act of lighting the fire before a successful prosecution.

Perhaps typical of the attitudes of many young men was a report in the *Sydney Morning Herald*'s 'Column 8' of 8 January 1994: a woman doctor driving to work at Helensburg 'saw two men in a white convertible ashing cigarettes on to the road. The air was thick with smoke from bushfires in the Royal National Park, but that did not stop them from tossing butts out of the car. When she yelled at them not to do so, they told her to do something very unladylike. She wishes she had taken down their registration number.' Eventually 72 people were charged under the Bushfires Act and various other statutes.

The Coroner's inquest into the deaths that resulted from the fires highlighted serious organisational deficiencies in the fire services and a number of recommendations were made. These included a clearer line of command, an improved and integrated radio system, and a lessening of the role of local councils. In response the Rural Fires Act of 1 September 1997 created a state-wide chain of command 'from the Commissioner to the firefighter, and placed an emphasis on ecologically sustainable development'.[54] The aim was to co-ordinate firefighting throughout the state. Funding would be more centralised.

A useful article by Helen Trinca appeared in the *Sydney Morning Herald* on 5 December 1997, asking why people lived next to the bush, sometimes in streets along ridge-lines with bush on either side. Trinca's interlocutors said they wanted to live among the trees and agreed this was a risk. She interviewed Dr Tim Flannery and he conceded it is understandable that people want to live in a natural setting; he lost his own house and possessions when fires swept through Jannali in 1994. He said he was opposed to inappropriate development along ridge-lines because the fires race up hills.

> But Flannery doesn't overstate the dangers of the bush . . . 'Thousands more people die from Christmas dinner and being too fat or driving than will die from bushfires'. He says we

can never change the nature of Australia, but we can change the nature of our houses. And there are things that we can do like planting trees that are less combustible than eucalypts and keeping fuel levels down around our properties.

The 2001–2 season began on 30 October and, especially in New South Wales, was to develop into a summer of fire. The worst-hit areas were the Hunter and, from 3 December, the Blue Mountains. Many of these fires were begun by lightning strikes. But this was merely the overture. The most intense and destructive period of the fires was between Christmas and New Year. On Christmas Day fires began across the state. There were extreme weather conditions and unusual fire behaviour due to the variability of the wind and dryness of the vegetation. Comparisons were made with 1994. The summer of 2001–2 was to be the longest continuous bushfire emergency ever in New South Wales, only surpassed by the 2002–3 season. By Boxing Day much of the area surrounding the greater Sydney region was ablaze, there were outbreaks from Grafton to Kempsey, and at Narromine, Mudgee, Cessnock, Oberon, the Blue Mountains, Hawkesbury, Warragamba, Appin, Helensburg, the Shoalhaven, Jervis Bay (where 5000 residents and holiday-makers had to flee from around Sussex Inlet), and Canberra. A fire that began at Warragamba actually reached Stanwell Tops on the coast north of Wollongong in six hours.

While no one was killed, 50 people were injured, 121 homes and fifteen business premises were destroyed and 36 houses seriously damaged. Around 733,000 hectares were burnt.[55] Up to 100 large, out-of-control fires lasted for three weeks and were driven by hot, dry, north-westerly winds. According to the *Sydney Morning Herald* on Christmas Day a firestorm from the Blue Mountains 'was so vast and powerful it created its own cumulonimbus thunder clouds, stretching up more than 4,500 metres. Some witnesses reported lightning similar to the electricity that can be generated by a volcanic eruption'. Firefighters were deployed from New South Wales, the other Australian states, and New Zealand. These were the first fires in which three Erickson Air Cranes were used extensively in New South Wales, especially around the greater Sydney area. They had already been used in Victoria.

Eighty-two other helicopters were used, as well as 24 fixed-wing aircraft. Many national parks were badly affected, with large numbers of native animals dying. Seventy per cent of the Royal National Park, which had scarcely recovered from the 1994 fires, was again burnt, although national parks management were confident vegetation would recover quickly; they were less confident about threatened species. Sydney experienced its worst air pollution in seven years.

Once again, many of the fires were lit by arsonists. Those caught were males and ranged in age from young children to a man in his forties. There was increasing and understandable public anger over the damage caused by arson. This boiled over when police only 'cautioned' and then sent home three fifteen-year-olds who lit fires in a park on the south coast of New South Wales. Premier Bob Carr announced a tougher approach to young arsonists. 'The juvenile offenders will be forced to visit burns victims and doctors, meet victims of fires . . . take part in community service to clean up affected areas and possibly have to pay compensation. Sometimes they will be accompanied by their families.' Carr wanted to 'rub their noses in the ashes they've caused by making them clean up the mess', according to the *Sydney Morning Herald* of 2 January 2002. Eleven children ranging in age from nine to fourteen and two sixteen-year-olds were arrested. Despite the Premier's comments, police merely cautioned most of them. Six men, aged 20, 49, 24, 22, 18 and 21, were arrested and charged.

The estimated total cost of the fires for New South Wales government agencies was $106 million, and the insurance costs were about $75 million. Amidst the chaos and destruction some retained a sense of humour. One couple were evacuating their home as the fire approached on Christmas Day when the wife called out to her husband: 'Oh Ian, we have left the big ham in the fridge'. He replied: 'Well, it will be well done when we get back'!

Later that year, for four days in early December 2002, Sydney was again ringed by 60 bushfires, many of them begun by arsonists. Strong winds, high temperatures and low humidity created the perfect conditions for fires. In November Sydney normally averaged 83 mm of rain; in November 2002 it received 35.7 mm.

Two men died in the fires, one of whom was burnt to death in his caravan. Forty-one houses were destroyed and many major roads, including rail and road links between Sydney and Gosford, were closed. More than 3000 firefighters were on duty, with CFA crews flying in from Melbourne. The worst fires were at Glenorie on the Old Northern or Putty Road to the north of Windsor, around Berowra and Hornsby Heights, in the Blue Mountains, and south-west of Sydney near Holsworthy, Picnic Point and Menai. An eighteen-year-old male from Glenfield was arrested for starting this fire near Holsworthy Army Barracks.

Fires on the urban–bush frontier are probably going to get worse in the next decade as more 'baby boomers' build dream homes in the bush and along the coast.[56] This is happening all around Australia, but especially in the fire-prone south-east. Not only is it destructive of the coastal environment, it exposes many people to serious risk from bushfires. No significant preventative measures will be achievable until governments, particularly state and local, begin to place limits on this type of development.

vi

Usually the sparse vegetation of inland Australia is insufficient to fuel widespread fires, but there have been exceptional seasons, such as in 1920–1 when an area as large as New South Wales was burnt out in the Northern Territory.[57] Between 1968 and 2003 there were a number of significant outback fires across Australia that should be recorded here. These vast periodic fires have generally occurred after good seasons and heavy rain when the combustible material dries out. But because there are so few people in the outback, newspapers tend to give little coverage to such events.

One such fire in January 1969 covered an area of about 900,000 hectares in the arid far north of South Australia 'where fuel is normally scarce and fire rare'.[58] A major fire in the Northern Territory during 1968–9, provided us with the first detailed records of Top End fires.[59] Known as the Killarney–Top Spring fire, it originated from two separate outbreaks: one on 29 August 1969 near Gorrie Station in very isolated country about 130 kilometres south-south-east of Katherine, and the second 150 kilometres further south on the Stuart Highway at Dunmarra. These fires eventually joined up on a broad front. West-south-west of these fires was another on Wave Hill Station which moved south-easterly into the Tanami Desert. There was also a fourth fire in the Newcastle Waters area on the Stuart Highway. More than 5.5 million hectares of the Northern Territory were burnt, although Luke and McArthur say 'the extent of [the fire's] spread into semi-desert country was not determined'.[60]

The focus of the fires moved west and south in the 1969–70 season, and most of the outbreaks were around Victoria River Crossing on the Victoria Highway that runs from Katherine to Kununurra in Western Australia, and around Alice Springs. All up, about 4.7 million hectares were destroyed in 27 fires. McArthur and Luke estimate that 30 per cent of the Northern Territory was burnt in these two seasons.

New South Wales also experienced outback fires. One of these was the Roto fire which began in November 1969 and burnt through December and possibly lasted until 21 January 1970, burning out more than 1 million hectares.[61] Roto, which had seen fires in 1940 and again in 1956–7 and 1957–8, is a tiny settlement on the Sydney–Broken Hill railway line north of Hillston, and halfway between Condobolin and Ivanhoe. Fires occur in this region when 'grass and shrubs are abundant after a good growing season'.[62] There were further serious fires there in December 1974.

During the mid-1970s outback fires broke out again. Above average rainfalls in January–February 1973 led to the end of the drought and two of the wettest years ever in Australia, particularly in 1974–5 when very heavy rain in the inland encouraged the growth of dense vegetation. For instance, Alice Springs received 875 mm from September 1973 to August 1974 in contrast to the town's usual average of 246 mm. As a result of the good rains the remote fire season was probably a one-in-50-year event.[63] As summer began the grass and vegetation quickly cured and fires began around late June 1974 on the Barkly Tableland in the central Northern Territory to the east of Tennant Creek. There were more fires further north near Newcastle Waters, in the north-west along the Victoria Highway and Victoria River, and in central Australia around Alice Springs. The total area burnt in the Territory in 1974–5 has been estimated at 45 million hectares.

In Western Australia it is estimated that 29 million hectares were burnt in fires to the north and east of Kalgoorlie, across the Nullarbor Plain and northwards along the border with the Northern Territory, while in South Australia most of the fires were in the central and north-western part of the state across both pastoral and wilderness land including the Great Victoria Desert and what is now Pitjantjatjara and Maralinga Land. About 16 million hectares were burnt.

Victoria was lucky that season as only about 100,000 hectares were burnt, mainly in the Mallee. In Queensland most of the fires were in the central-west and south-west of the state, with the area to the north-west of Boulia towards Urandangi on the Northern Territory border being badly affected. Some of the fires finally burnt themselves out in the Simpson Desert. A lot of 'protective' fires were lit around the Gulf of Carpentaria and Cape York, burning through 15 million hectares of grazing land. It is estimated that all up the Queensland fires covered about 22 million hectares.

The 1974–5 outback fires we know most about were those in New South Wales, mainly because they received some media coverage.[64] The vegetation growth that season was extraordinary with grass up to 2 metres high in some areas. By Christmas two massive fires began moving in an easterly direction, one on a 160-kilometre front to the south-west of Cobar and east of the Darling River, and another even larger fire south of Ivanhoe and north of Hay. Two smaller fires were also burning, one along the Darling between Tilpa and Louth, south-west of Bourke, and the other north-east of Tibooburra. Right through Christmas of 1974 these fires burnt out of control on vast fronts. Firefighters faced strange ironies. The *Sun-Herald* of 22 December reported: 'Already rabbits fleeing with their fur alight have spread the flames across roads and fire-breaks. As firefighters worked yesterday millions of butterflies blown before the blaze fluttered around their faces.' On Christmas Eve the Tilpa–Louth fire joined up with the fire to the south-west of Cobar across the Barrier Highway to create what was eventually an 800-kilometre front. These fires were brought under control just before the New Year. About 3.75 million hectares of remote New South Wales was burnt and three people lost their lives.[65]

The total area burnt in 1974–5 across the Northern Territory, Western Australia, South Australia, Queensland and New South Wales was about 117 million hectares. This equals 15.2 per cent of the continent's land surface. Only about 20 per cent of these fires were deliberately lit.

The next big fire season in the outback began around Christmas 1984 when uncontrolled fires broke out in the Ngarkat Conservation Park on the South Australia–Victoria border to the north of Keith and Bordertown and in the Danggali Conservation Park on the South Australia–New South Wales border, 85 kilometres due north of Renmark. These fires were driven by strong winds and were still posing problems in the middle of January 1985.

In late December and throughout January 1985 the Western Division of New South Wales had to deal with a number of really big fires after several good seasons of rain and abundant grass growth which had dried out over the early summer. These fires were started by dry lightning storms and driven by relentless westerly winds. Originally there were four fires: a huge one to the north of Cobar, two to the south-south-east around Nymagee, and another in the Yathong Nature Reserve south of Cobar. 'Fire-fighters predicted the worst bushfire season in ten years as six fires burnt out of control, the largest over an area of 40 kilometres long and 15 kilometres wide', reported the *Sydney Morning Herald* on 28 December 1984. A state of emergency was declared in Cobar and there were numerous outbreaks in the nearby Wilcannia district. 'Cobar' is the local Aboriginal word for 'burnt earth', and that was certainly the case in January–February 1985 as more than half a million hectares were burnt and upwards of 40,000 livestock were killed.

By the middle of January 1985, and with no let-up in the westerly winds, fires spread right across New South Wales. The Cobar fires broke out again, there were fast-moving outbreaks near Yass and in Harden Shire, in the Bega Valley, around Captain's Flat, and further south at Albury and Holbrook, in the Kosciuszko National Park, near Mudgee, in the ACT near Cotter Dam and at Mount Gingera, at Moree where Aboriginal children were lighting grassfires, and around Carrathool between Griffith and Hay where 230,000 hectares were burnt. There were also outbreaks in the national parks along the Great Dividing

Range to the north of Sydney and Newcastle, most of which were due to dry lightning strikes. About 20,000 firefighters were mobilised and 800,000 hectares burnt. Sydney itself was largely spared, but there was a particularly nasty series of arsonist attacks in the Kurrajong area around the junction of Bells Line and the Putty Road which caused many problems for firefighters. By the end of this fire season in February three firefighters and one other person were dead.[66]

Nineteen ninety-eight was the warmest year since 1910, when reliable temperature records began. While there were no major fires close to cities, there were a number of significant outbreaks across the outback of Australia, particularly in the Northern Territory in late August, especially around Batchelor and further south near Katherine and the Kimberley where 600,000 hectares were burnt in late August–early September. In October 1998 it was eastern South Australia's turn, with fires south of the Flinders Ranges and north of the Murray River. A fire in the Mallee north-east of Ouyen burnt throughout December 1998, destroying 7000 hectares. Large fires in the south-east of Western Australia burnt 23,000 hectares of pastures and crops as well as 8000 sheep around Condingup, 65 kilometres east of Esperance. In January 1999 there were bushfires in Victoria in the Big Desert Wilderness Park area where 100,000 hectares were burnt. Outbreaks also occurred in South Australia just north of Millicent in the south-east in early January, and later in the month in the Ngarkat and nearby Scorpion Springs Conservation parks, and across the border in Victoria's Big Desert Wilderness Park. All up, 100,000 hectares were burnt. In mid and late May there were more fires in the Northern Territory.

A similar pattern of fire occurred in central Australia in early October 2001, but on a much smaller scale, when outbreaks near Alice Springs burnt about 80,000 hectares of scrub and grazing land, and 50,000 hectares of the Purnululu National Park, near the Western Australia–Northern Territory border, were also burnt.

While 'fires of enormous extent have apparently always occurred in the arid regions in periods following heavy rainfalls,'[67] the outback has otherwise been largely fire free. The one exception is in the Northern Territory, where controlled burning is being increasingly undertaken near the end of the dry season to prepare for regrowth of better feed in the following wet season, thus imitating Aboriginal fire practices.

Until recently, the 1939 bushfires were seen as the archetypical fires of European Australia, with the 1851 fires and Ash Wednesday running close behind. But the new century was to bring an even greater fire threat to south-eastern Australia than Black Friday 1939—the alpine fires of 2003.

PART 3

THE GREAT FIRES OF 2003

7

'STINKING HOT AND WINDY', THE SNOWY MOUNTAINS AND VICTORIA, 2003

i

For many years I shared that dream of the urban middle class: to get a bush block with a shack where I could be close to the environment. However, reality quickly impinged: on 30 January 2003, two and a half months after I bought a remote block south of Jindabyne in the Snowy River country of southern New South Wales, it was completely burnt out in the largest conflagration in south-eastern Australia since European settlement.

In late December 2002 the fires seemed remote from the block. A dry lightning storm had ignited vegetation on the southern flank of the Snowy Mountains in the Kosciuszko National Park (KNP) on 17 December 2002. The fire was just 9 kilometres west-north-west of Mount Kosciuszko. These fires were quickly contained by remote-area fire crews and aerial water bombing. An ignition in the Purgatory Hill–Tuross Creek area about 9 kilometres south-west of Thredbo and close to Dead Horse Gap on the Alpine Way proved much more difficult to contain.

Then, on 20 December 2002 an intense, dry lightning storm started several fires in the KNP on the south side of the Snowy River in the Byadbo Wilderness, just below Milligans Mountain.[1] This is where the Snowy turns from heading west and briefly loops north. After this the river turns south again heading towards Victoria. The Byadbo Wilderness comprises 63,000 hectares of the far eastern side of the KNP. This is difficult country, even for the remote-area firefighting units that were deployed. The Coroner later commented that '[these] fires were identified as having the capability to quickly exceed the capacity of any containment lines'.[2] The outbreaks were intense, the terrain was hazardous and the weather was dry and hot. Other fires were ignited from them. They were never extinguished and in late January joined the massive complex across the whole alpine area. Experienced locals with long family connections to the Paupong district immediately to the north felt that they could have controlled these fires with better support from the National Parks and Wildlife Service (NPWS). Even in the past this area was described by Klaus Hueneke, historian of the high-country huts, as 'the real outback of the mountains'.[3] Nowadays, very few people other than locals go near the area. George Seddon in his book *Searching for the Snowy* (1994) says that, 'There is still a strong feeling against the National Parks Service in this part of the world . . . There is much nostalgia for the "good old days" . . . The stockmen knew it intimately. There was . . . probably better management of fire which could often be put out before it really took hold. On the other hand the cattle compacted the soil, spread weeds, and initiated soil erosion on many steep slopes.'[4] However, given the multiplicity of ignitions, the weather conditions and the pressure on resources, it is doubtful if much really could have been done about these outbreaks. What was significant to me was that this area is just 10 kilometres south of my block. The fire quickly became a reality for me.

It needs to be emphasised that this fire—like almost all of the 2002–3 fires—was 'natural' in that it was not lit by people. The real change that has occurred since 1939 is that everybody in the bush is now much more careful about fire, although the folk wisdom of preventative burning still influences some. Deliberately lit fires are now usually confined to anti-social misfits who largely inhabit the urban–bush fringe. Following police and Rural Fire Service (RFS) investigations of the fires, the New South Wales Deputy-Coroner, Michael Milovanovich, who investigated the Snowy and Brindabella fires, says unequivocally that 'it is clear from the evidence that the fires did not originate from human intervention—but rather the evidence strongly suggests that every fire had its origin from a natural cause, in most cases lightning strikes and subsequent spotting'.[5] In light of the history of fire in Australia that is a significant statement. These fires resulted from a rare concatenation of natural circumstances accompanied by dry lightning strikes in extremely remote country. There have probably not been fires like these in the mountains of eastern Australia since the mid to late eighteenth century. The 2002–3 fires impacted on an even larger area than those of 1939, especially in New South Wales, making them probably the biggest complex of fires since European settlement. They were made up of a series of interconnected conflagrations across the whole alpine areas of Victoria, New South Wales and the ACT.

The 2002–3 fires have been thoroughly analysed. In New South Wales there was a coronial inquiry. Evidence tendered to the New South Wales Coroner included a submission from the Department of Environment and Conservation (DEC) outlining the Kosciuszko fires. There is also a detailed 'Coronial Brief' prepared by the New South Wales Police Service for the Coroner.[6] The Victorian fires have also been studied in detail, especially in the 'Esplin Report' (*Report of the Inquiry into the 2002–2003 Victorian Bushfires*). The Department of Sustainability and Environment (DSE) also prepared a thorough report, *The Victorian Alpine Fires January–March 2003*, including a helpful 'daily narrative'. There was also the 'Nairn Report' (*A Nation Charred*) prepared by a House of Representatives Select Committee chaired by Garry Nairn, member for Eden–Monaro.[7] Perhaps this report is best characterised by the additional comments of the Labor members of the committee: 'Regrettably many Parliamentary Inquiries are established in a highly charged political atmosphere following national disasters, where the media is seeking the sensational story, the community is demanding answers, and the politicians are seeking to apportion blame. These are hardly conducive circumstances for the rational evaluation of evidence, the setting-aside of long-held prejudices and the development of practical recommendations.'[8] State authorities refused to co-operate with the Nairn Committee. Finally there was a Council of Australian Governments inquiry.

The situations faced by the Victorian, New South Wales and ACT authorities were almost identical. There had been a severe six-year drought across eastern Australia. The ten months prior to the fires were the third driest on record. Water levels were low in streams and dams; evaporation rates were high. Some towns, such as Goulburn, were facing the possibility of running out of water. Canberra had seriously low storage, as did Sydney. The Bureau of Meteorology Summary for 2002 said that, 'Widespread dry conditions during 2002 resulted in one of Australia's driest years on record. It was also one of the warmest. The all-Australian average maximum temperature was the highest on record. Preliminary data indicate that, for Australia as a whole, 2002 was the fourth driest year since 1900. The total annual rainfall averaged over Australia for 2002 was 339mm, well below the long-standing average of 472mm.'[9] Temperatures were also high. 'November 2002 was of particular note, with day maximum temperatures nearly 5 degrees above normal. A consequence . . . was to exacerbate the drought by increasing evaporation and thereby further reducing soil, surface fuel and live fuel moisture contents.'[10] Woodlands and forests were drying out. As moisture stress increased the trees dropped more litter. 'When conditions are wetter often only the top layers of the litter bed . . . will carry fire; however in times of drought the full extent of the litter bed becomes available for the carriage of fire.'[11] Preventative burning would not have lessened the impact of these fires. The Bureau of Meteorology warned in late 2002 that conditions were similar to those before the 1983 Ash Wednesday and 1939 Black Friday fires. There were many days of total fire ban.

By New Year the December outbreaks in Victoria and New South Wales were more or less under control.[12] However, on 8 January 2003 the fires returned with renewed intensity. The situation deteriorated when a cool change which was made up of many thunderstorms and dry lightning strikes moved across south-eastern Victoria, southern New South Wales and the ACT on the evening of 7 January and early the next morning. Lightning strikes ignited 72 fires in New South Wales, at least three in the ACT, and 87 in Victoria.[13] These ignitions occurred across a wide area in Victoria, but in New South Wales were largely confined to the mountains and alpine areas of the KNP and along the Brindabella Ranges to the west and south-west of Canberra. Dave Darlington, NPWS Snowy Mountains regional manager, said that 'a massive storm cell passed over the KNP. As many as 60 lightning strikes were recorded hitting the mountain range along the full length of this 690,000 ha. Park. The weather was extremely hot with strong north-westerly winds. Because of the hot, dry, windy conditions the fires took hold across a wide area almost immediately making containment in the first 24 hours virtually impossible.'[14] It is important to remember this. While some criticism might be justified, the authorities could not have foreseen the ultimate consequences of all these outbreaks.

This was just the beginning. For a month the weather remained consistently and relentlessly hot and dry (the temperature was often over 40°C), with west-north-westerly winds up to 110 km/h that constantly drove the established fires onwards and regularly caused them to break containment lines. In other words, things remained 'stinking hot and windy', as Elly Spark from the Sydney Weather Bureau so succinctly put it.

ii

As a result of the dry lightning storms about twenty fires were ignited on the western side of the Snowy Mountains, with other strikes in the northern section of the KNP. By 11 January there was a complex of interconnected fires in the Jugungal Wilderness Area about 20 kilometres south of Khancoban on the Alpine Way. On 9 January this road was closed from Tom Groggin to Murray One Power Station. Later investigation by the police and RFS showed that there had been 'in excess of 700 lightning strikes which resulted in at least 38 confirmed ignitions' in the KNP.[15] Meanwhile the Byadbo fires continued throughout Christmas and New Year and gradually came together to create one large conflagration covering much of the wilderness area. It remained isolated until late January.

To the west of the Snowy Mountains lightning strikes also caused a series of fires on 8 January in the Tumbarumba–Batlow–Tumut area.

A reconnaissance pilot, Matthew Pope, 'observed and identified no less than 19 separate fires' around midday on 8 January in a triangle from west of Batlow to Tumbarumba to Yarrangobilly north-east of the Snowy Mountains Highway. 'There was no evidence to suggest human involvement in [these] fires.'[16] They were not threatening homes or property.

It was difficult for New South Wales authorities to respond to these multiple ignitions.[17] Firstly, they had to be assessed as to their seriousness and then resources had to be found to deal with them. The RFS, the NPWS and State Forests were already committed to fighting fires in other areas (for example, a fire south of Braidwood and the Wakefield fire on the Victoria–New South Wales border). The unstable air, smoke and storm activity made it difficult for helicopters to insert remote-area firefighters into many isolated fires. A planning meeting was held on 8 January in Jindabyne, attended by the RFS Commissioner, Phil Koperberg, as well as police and NPWS representatives. A 'Section 44' had already been declared for the fires in the Byadbo Wilderness on 21 December 2002. Section 44 refers to the Rural Fires Act which has provision for the RFS Commissioner to take charge of firefighting operations. He usually delegates his authority to an Incident Management Controller (IMC) who assembles an Incident Management Team (IMT), comprised of RFS officers, land management agencies and volunteers. Two more Section 44s were declared for the Tumut–Tumbarumba–Gundagai area and the Yass area. Command posts were set up at Jindabyne and Tumut. The Yass fire was intimately connected with the fire that swept into Canberra and will be dealt with in the next chapter.

The DEC submission to the Coroner outlines the problems facing firefighters from the beginning of the campaign. 'Initial direct attack was a limited option on the vast majority of these fires due to the rate of spread combined with the terrain, continuing instability in the afternoon [of 8 January], and remote locations.' The fires were progressively grouped into complexes and the main aim was 'to keep these complexes within secure lines thereby minimizing the likelihood of one complex joining up with an adjacent complex'.[18] By 10 January roads through the park were closed except the Snowy Mountains Highway, which remained open during daylight hours until 13 January. The fear was that the fires in the KNP to the south-south-east of Khancoban would join up with the Pinnibar fires, a large complex just 12 kilometres over the border in Victoria.

By Monday 13 January there were three main fires: at Hannels Ridge, just below the Kosciuszko main range on the Alpine Way, at Scammells Ridge, east-south-east of Khancoban, and at Tooma to the west of the KNP. There was also a large fire in the remote northern section of the park at Yarrangobilly, just north of the Snowy Mountains Highway. This fire spotted 30 kilometres away into the Tumut River Gorge south of Cabramurra, the highest town in Australia (1400 metres). Alpine ash (*Eucalyptus delegatensis*) stands on slopes in excess of 40 degrees occur in this area and there was immediate concern for the safety of Cabramurra residents.[19]

For three days from 16 to 18 January the 150 people in Cabramurra fought a battle against surrounding fires. Most residents stayed to defend their homes. There were four smaller fires, some threatening private assets, especially an outbreak close to the valuable softwood timber plantations to the north of the park and east of Tumut.[20] From 13 January onwards Jindabyne was regularly covered in smoke. Often heavy smoke prevented aerial firefighting, the insertion of remote-area firefighters, and surveillance of outbreaks.

There was extreme fire weather on Friday 17 January with high temperatures and westerly winds up to 75 km/h. Driven by a large convection column close to Cabramurra, the Tumut River Gorge fire was spotting up to 15 kilometres in an easterly direction. This led to ignitions almost 20 kilometres away at the northern end of Lake Eucumbene, close to Adaminaby township. Outlying settlements were placed on high alert.

That day there was 'an all-out effort to keep fire perimeters within forested areas of KNP and to minimize impact on private property. A further underlying objective was to minimize threats to assets within KNP including ski resorts and heritage sites.'[21] Guthega, Charlotte Pass and Smiggin Holes were evacuated. Later that evening and during the next morning 1700 visitors at Thredbo were removed, leaving only permanent residents and essential staff. The Tenth Thredbo Blues

Festival was cancelled. The Alpine Way was closed and did not reopen fully until 6 February.

The next day (Saturday 18 January) the Snowy Mountains Highway and the Kosciuszko Road were both closed. Throughout the whole fire period Thredbo remained threatened either by a direct bushfire front entering the village, or by the effects of spot fires. However, the assessment was that the New South Wales Fire Brigade and experienced resort staff and locals 'proved to be particularly effective' in defending the village, and no assets were lost.[22]

Saturday 18 January was the day the fires hit Canberra. The destruction wrought in Canberra helped to focus national attention on the fires, although it did tend to grab headlines and push the KNP fires out of public consciousness; at this stage 130,000 hectares of the park had been burnt. Worsening weather conditions on 18 January saw the Kosciuszko fires breaking out of containment lines in many areas and continuing to expand. A Section 44 was declared for the Cooma–Monaro shire as the Tantangara fire moved eastwards towards the farming districts of Yaouk and Shannons Flat, 35 kilometres northeast of Cooma. This fire was also close to the southern edge of the outbreaks in the ACT's Namadgi National Park. The consciousness was now growing that the fires in New South Wales and Victoria could easily join up to form a massive conflagration. Dire predictions were being made that wildfires would not stop until they reached the coast.

As well as being part of the two complexes of fires that joined and burnt right into the Canberra suburbs on 18 January, the Namadgi fire also moved eastwards towards the Murrumbidgee River and the Monaro Highway which heads south from Canberra to Cooma. Nick Goldie, a local firefighter whose house near the Murrumbidgee River was right on the front line, described this fire:

> There's a wall of flame coming down the hillside across the river . . . it's a huge fire—all the little ones have joined up . . . I spent the day in Colinton One, a new fire truck we are very proud of (big, ugly, slow), fighting a fire on the hillside above Michelago, on the river side west of the village. Flames, crashing branches, panicked cattle (and alpacas!). And anxious

landowners with garden hoses. We were told don't fight the fire, it's too big, just protect houses . . . here's a wall of flame across the [Murrumbidgee] River, but hell, it's the same wall of flame that was there last night and the night before. I've spent the last two days cruisin' in one or other fire truck, up and down the smoke darkened [Monaro] Highway, and up all the little roads off into the bush to find remote homes and see if there is anyone there . . . This has convinced me that my own house is the worst designed, worst situated house in the district. Prettiness is not enough, as Ms Nightingale said . . . Of course I haven't been able to go shopping . . . Food isn't a problem during the day: we get Salvation Army sandwiches or hamburgers, and tonight at the fire station where there were many visiting fireys and police-persons and a helicopter and a grader with driver, they produced something called by all 'road-kill stew'. It was very good, served on a plastic plate with a spoon and with rice cooked as your grandmother cooked rice. I enjoyed it, with a beer, and so did Sam the dog.[23]

One of the other problems facing Nick and his colleagues was protecting the electrical power lines to Cooma, Jindabyne and parts of the New South Wales south coast. The continuing threat to Canberra and the fires along the Monaro Highway also had serious resource consequences for the KNP. It meant that many interstate firefighters who would have headed to the park were now deployed elsewhere.

Feeding the firefighters was a massive logistical task. For instance the Rocky Plains Brigade in the Shannons Flat district, north-west of Cooma, organised 70 volunteers who prepared over 5000 meals over six weeks. A similar process occurred at Numbla Vale and Dalgety.[24] Breakfast consisted of bacon, eggs and sausages, lunches were packed with sandwiches and cool drinks, and an evening meal was provided. This gave local women not involved in the firefighting a sense of participation and helped them deal with the endless waiting at home worrying about their husbands and partners out on the fire line.

Sunday 19 January was a day of extreme fire danger. To protect Thredbo 28,000 litres of fire retardant was dropped around the village.

The danger of the Tumut, Kosciuszko and the Victorian Pinnibar fire complexes linking up continued to cause grave concern. There were also worries about a fire on Brindle Bull Hill above Thredbo spotting eastwards across the park 13 kilometres to the farming and grazing lands along Barry Way at Moonbah and Ingebyra, south of Jindabyne. A population of about 700 lived in this area. Further east, 10 kilometres along the next ridge, was my block. It was now literally in the firing line!

On Tuesday 21 January the Perisher fire burnt down the steep 600-metre escarpment towards the Thredbo River and the park boundary, and threatened a number of private properties. At the same time the Moonbah–Ingebyra area along Barry Way was under threat, and Eucumbene Cove was still affected by the fires to the west of Adaminaby. The fire in the ACT's Namadgi National Park now moved south along the western side of the Monaro Highway threatening localities around Bolaro, Shannons Flat and Yaouk. Firefighters faced 'extreme weather conditions'.[25] With threats increasing to private property outside the park there was a growing number of complaints by local landholders that road closures were preventing them getting back to their properties to protect them.

Over the previous days a number of controlled back-burns had been carried out, mainly by the NPWS. There was much criticism of these attempts. For example, there was considerable concern over back-burns in the area around Tom Groggin Station where the Alpine Way runs parallel to the upper reaches of the Murray River. The station is actually in Victoria and was just 12 kilometres from the Victorian Pinnibar fires. On the afternoon of 17 January, Trevor and Lynda Davis, who ran Tom Groggin, became concerned when the NPWS lit a back-burn at Murray Gates on the river below the station where it enters a gorge often used for white-water rafting. In evidence to the Coroner, Trevor said he considered this a 'huge risk' and told the NPWS that, 'This is not a good time to light a fire and there [is] a total fire ban in place'. His experience had taught him that weather conditions at Tom Groggin were very different from those at Jindabyne.

That afternoon 'all hell broke loose. The winds picked up and the fire jumped the River.' It took the Davis family, their staff, and the NPWS four hours to get things back under control. Trevor is adamant that this back-burn invaded Victoria and burnt toward the Pinnibar fires along two creeks to form the feared cross-border conflagration.[26] It was these kinds of incidents that led to criticism of the NPWS and the RFS for ignoring local knowledge. There is substance to these complaints.

Back-burning continued on 23 and 24 January around Lake Eucumbene, at Adaminaby, along the Thredbo River, as well as along Barry Way, and residents were on high alert. Snowy Plains was evacuated. During the next two days firefighters tried to consolidate containment lines and, with extreme weather predicted for 26 January, attempts were made to complete back-burns and mopping-up operations. Exhaustion was setting in and emergency personnel were beginning to suffer fatigue-related injuries. The fires had burnt more than 250,000 hectares of the park and more than 1600 emergency services personnel were engaged using ten fixed-wing aircraft and 25 helicopters.[27]

Sunday 26 January, Australia Day, was a day of extreme fire danger with temperatures in the forties and north-west winds gusting to 60 km/h. All of the fires in the KNP took on new life and there was intense spotting over a wide area. The Telstra optic fibre line from Thredbo to Jindabyne was burnt out. Jindabyne itself was under threat from spotting and many landholders along the Thredbo River and Alpine Way felt let down that fire services were withdrawn from their properties to defend the town against a fire that never came. Many remained with their properties and were well prepared. However, there was considerable disquiet among residents about attempted back-burns with incendiary devices dropped from helicopters to try to protect Jindabyne; one property owner described this as 'poorly exercised and not properly controlled'.[28]

The next three days were spent trying to consolidate containment lines and strengthen property protection. There were still serious threats along Barry Way and nearby Gullies Road. There was a prediction of extreme fire weather for Thursday 30 January. The fire danger rating was predicted to be 'above extreme'.

The forecasts were right. The last Thursday of January was the worst day. There was a powerful north-west wind gusting 60 to 70 km/h, and very high temperatures. Phil Koperberg, quoted in the *Sydney Morning Herald* of 31 January, described the conditions as 'about the worst imaginable . . . almost unprecedented in ferocity'. The fire chief said that fire behaviour

> was absolutely horrific. That there was no life lost and property loss was kept to a minimum . . . was a credit to all those involved. Almost 100,000 hectares of forest and grazing land [were] burnt in extreme conditions in about 12 hours in the southern Kosciuszko fires alone . . . Significant resources were deployed to Jindabyne to ensure spot-fires did not take hold within the town . . . A large expanse of rural land in the Paupong-Numbla Vale area was also heavily impacted by a fire front.[29]

Winds spread hundreds of spot-fires. There was thick smoke and embers rained on Jindabyne and around the whole alpine area. Previously burnt areas re-ignited.[30] The fire spotted across Barry Way and burnt eastwards straight through my block. It was not until 6 February that the loss of the shack was officially reported.[31] It was the only one destroyed in the Snowy River Shire. This fire was finally stopped in cleared country 40 kilometres away around Numbla Vale.

iii

The greatest fear around Jindabyne that day was that two men had been killed. In fact they hadn't, due largely to their own skill and bravery. Kerry Wellsmore and his brother-in-law, Mark Thoha, were trapped in Reedy Creek just below my block. The Wellsmores have been farming in the nearby Paupong Valley since the 1840s, and locals like Kerry know the country thoroughly. Mark, from Sydney, had gone with Kerry on horseback across two ridges to the western Ingebyra side to assist a bulldozer driver who was trying to cut containment lines between the Paupong Nature Reserve and the KNP. He ran into trouble as he worked around spot fires and was forced to retreat. This is dry, eroded country that has lost a lot of topsoil from grazing and overburning. The creek beds are usually dry, but there are waterholes, a few of them quite deep.

The situation deteriorated rapidly for Kerry and Mark.[32] Kerry describes their predicament:

A spot fire broke out on the right side and we really only had one option and that was the middle of the creek. And the harder we rode, the harder these two fires pulled across the valley [towards] each other. We could see we were going to get cut off, which we did. We rode right to the base of the fire and we had to make some pretty quick decisions. We thought of riding through it; we couldn't have been that far from burnt country [behind the fire]. But it was coming over the top of the tress, pretty fierce. You could see back underneath the bush for about 300 metres. The fire was coming up off the ground, so our only option was to turn around and ride back until the fire got us.

We got out of the creek temporarily and went around the side of the hill. The fire started to gain on us . . . it kept pushing us farther to the right [back down toward the creek]. I knew we could die in the creek. We managed to get to the fire trail. Mark started to hesitate about the situation and felt that our best options were to follow the trail back to Ingebyra. But I said the fire we just rode away from was going to cut us off. Our only option was to get to water. To get there we had to cut across this ridge. It was all bush. There was no track. Every decision had to be immediate. There was just the roar of the fire, the shimmer of the heat, the ash falling on us, burning, the choking smoke, kangaroos running into you, running past you, they were frantic trying to get out of the fire, just like we were. We kept riding. By this stage our horses were pretty well buggered. The biggest fear was whether we could get far enough on the horses before we had to start on foot.

We got to the last fence and our horses couldn't get over the fence and we had to leave them. By this stage the fire was gaining pretty much on us. We had no options. We unsaddled the horses, everything, pulled the bridles off, and decided to run. I said to Mark: 'Just stay focused, watch where you run, don't look back'. I thought to myself what happens if one of us falls over and breaks a leg or ankle? Do we stay and help the other, or do we just go it alone? That sort of played on my mind a bit.

What Kerry wanted to do was to get to a waterhole in Reedy Creek with an overhanging ledge for protection. But they had to get down the hill to the creek. 'From the time the fires cut us off and we abandoned our horses', he continued, 'everywhere we seemed to go there [was] fire, from behind us or from the side; it seemed to be everywhere. The last two or three hundred metres [down the hill] we were just sliding on our bums, rolling and hitting trees, it was that steep.' Eventually they got to the creek.

> We had a drink. We were pretty exhausted by that stage, but the adrenalin kept us going to look for other options. If the fire got to where we were, what was our next option? We just tried to find the best spot to hide under a ledge to get out of the heat. What we feared was that the air would have the oxygen taken out of it if the fire came down through the gully. I thought maybe we are going to die. I just said to Mark they survived the 1939 bushfires at Paupong in the sheep-dip, so we can survive this.

Kerry's father, Don Wellsmore, had often recounted the story that the family and neighbours took refuge in a sheep-dip on Black Friday. The embers fell on them, their house, 'Honeyvale', was burnt down and for the first time Don saw his grandmother cry. Kerry said that 'those sorts of images stay in your mind and you try to relate your situation to those times. You know, how they coped with it. So I was pretty confident to a certain point that things were going to be right once we got to that waterhole. Same as what they were in '39 in the sheep-dip'. They got to the waterhole. 'There were a couple of brown snakes there, five or six lyrebirds, wombats, and big long lizards. They knew something was wrong. It was just the noise of the fire, the roar of it, you'll never forget it. The birds were totally quiet, the lyrebirds were just quiet. It was a relief. You felt at home a bit.'

After the fire passed Kerry scrambled up through the smoke to the hill and his satellite phone worked. He rang Barry Aitchison, the deputy fire controller, but due to the smoke they could not get a helicopter up immediately, although one did try that night, unsuccessfully, to locate them. Kerry also called his wife Cecilia back in the Paupong Valley where there was also thick smoke and a serious fire threat. The fire that trapped Kerry and Mark had now roared up the mountainside and over the ridges and down towards Paupong.

The two men spent the night in a cave. They were freezing 'wishing to God we were sitting up to one of the fires'! They rolled over back-to-back, 'anything at all to keep warm'. Just before daybreak the smoke cleared and the wind dropped. 'The day was so perfect and still. There was no sign of any wind. It was a beautiful, clear morning.' The rescue helicopter got to them about 6.30 am. As they were winched into the chopper Mark said he was 'just relieved we did not have to walk out of there'. The two men then spent the next seven days with locals and firefighters from Dalgety, Dubbo and Wagga Wagga defending their properties in the Paupong Valley.

By the beginning of February 337,000 hectares of the park had been burnt, as well as 11,500 hectares of private land. But things were easing. A north–south containment line was secured and maintained over the next few weeks 'with thorough mop-up and patrol despite the constraints which a perimeter of many hundreds of kilometres brought with it'.[33] From Saturday 1 February onwards the main focus of operations moved eastward with firefighting concentrated around Paupong, Numbla Vale and Bombala, although there were flare-ups elsewhere. The fear was still abroad that the fire could reach the coast. Over the next few days 'fire-fighting operations concentrated on strengthening containment lines on all fires burning within and east of the KNP'.[34] There were still fires burning around Avonside, Paupong, and Tingaringy on the state border to the south-east of the park. By 9 February favourable weather helped contain all active fires. The Section 44 declarations were gradually revoked. It had been an extraordinarily difficult period.

Between October 2002 and March 2003 1,465,000 hectares of land had been burnt in New South Wales with a fire perimeter of 10,340 kilometres. The Snowy Mountains complex accounted for more than half the total area destroyed. There was a total of 915 ignitions.[35] Without doubt it was the largest fire in the European history of New South Wales.

iv

Victoria fared even worse. The Victorian 'Daily Fire Narrative' of the Department of Sustainability and Environment (DSE) runs for 59 days, from 8 January to 7 March. It is a detailed historical account of the Victorian fires. The Esplin report on the fires begins: 'Late on the Tuesday, January 7 and early the next day, dramatic, dry thunderstorms brought widespread lightning strikes that ignited over 80 fires in Victoria's North-East and Gippsland. Many were in mountainous, forested areas of the Alpine National Park' (ANP).[36] So together with 42 ignitions in New South Wales, the south-eastern corner of Australia faced 122 fires breaking out in a less than 24-hour period. 'The remote country in which the ignitions started, as well as the weather conditions, compounded this.'[37] As a result not all fires were detected on the first day. The aim was to deal with ignitions immediately, and the DSE had 280 staff working on the fires from day one, with a further 265 quickly brought in from elsewhere.[38]

'As new outbreaks were detected, fire managers were faced with a number of complex and difficult decisions. They needed to set priorities for the allocation of firefighting resources and needed to redirect resources already deployed to existing fires.'[39] Many of these fires were in remote, rugged terrain close to the New South Wales border. Much of the criticism directed at authorities was that they did not act quickly enough. 'It was suggested that weather conditions for the first ten days following the thunderstorm should have permitted a more aggressive initial attack than was initiated.'[40] But, as Bruce Esplin, chair of the Inquiry into 2002–3 Victorian bushfires, points out, it was easier said than done to mount an 'aggressive attack'.[41]

These were not Victoria's first fires of the season. Initial outbreaks had occurred as early as September 2002. On 17 December dry lightning began two fires in the Big Desert Wilderness Park in Victoria's northwest, abutting the South Australian border. This fire quickly spread into the neighbouring Wyperfeld National Park. It was the largest fire in Victoria in twenty years and was very difficult to control because of the undulating sand dunes and the dense mallee scrub. Altogether 181,400 hectares were burnt.

Two days before 8 January a fire with the potential to cause considerable damage had broken out in East Gippsland at Yambulla Creek, a tributary of the Genoa River in the Coopracambra National Park on the Victoria–New South Wales border. This fire absorbed considerable resources, including 180 DSE firefighters, bulldozers and fire-bombing aircraft. The outbreak was contained on 11 January, having burnt 3000 hectares. There were also outbreaks in South Gippsland in the area between Yarram, Traralgon and Sale, particularly in and around the Holey Plains State Park, with other fires in North Gippsland near Heyfield. These south and north Gippsland fires were brought under control by Monday 13 January. But as a result of these outbreaks considerable resources were already tied up when the lightning strikes occurred on 7–8 January.

The situation late that Monday afternoon was that there were nine active fires which had already burnt 66,000 hectares of north-east Victoria. There were essentially five fire complexes. These were ultimately to become one massive conflagration that reached its apogee seventeen days later on 30 January. The main problem was that most of the outbreaks were in steep, forested country and were difficult to access. In order to locate them and develop strategies to deal with them, the DSE used infra-red, aircraft-borne cameras which scanned the state's forests and national parks. The results were used to place fire crews and control lines. When necessary one of the 10,000-litre, water-carrying Erickson S–64 Air-Cranes (one of which was christened 'Elvis') were activated, although they came at a cost: they used 2000 litres of fuel per hour and employed ten people.[42]

To deal with them efficiently the fires were grouped into regions. In the Upper Murray region between Tallangatta and Corryong and south to Dartmouth Dam, there were four fires burning, including one at Mount Mittamatite between Corryong and the Murray, a group in the Mount Pinnibar area in rugged country west of the Murray and close to Tom Groggin Station on the New South Wales border, another at Mystery Lane just to the west of the Pinnibar complex, and a fourth at Cravensville, a tiny settlement at the end of the Tallangatta Valley. To the south of the Upper Murray region was the Swifts Creek region. Here the main outbreak was the Razorback fire, close to the Omeo Highway and north-west of Benambra. It had already burnt 4500 hectares and had a perimeter of 29 kilometres. West of Swifts Creek and south-west of the Upper Murray was the Ovens Valley region centred on Porepunkah between Myrtleford and Bright on the Great Alpine Road. There was a series of six fires burning in this rugged region. The worst was at Andersons Peak on the Mount Buffalo massif; it was about 2800 hectares in extent and burning in inaccessible terrain. Another fire was burning very close to the peak of Mount Bogong in the ANP; close by was the much smaller Bald Hill fire. Away to the south-east near Harrietville were the smaller Mount Feathertop and Cavalier Spur fires with flames reaching two to three times tree height. To the north of Mount Bogong was the Mountain Creek fire. There were strong winds overnight on 13–14 January, and the fire area in north-eastern Victoria increased.

v

On Saturday 18 January the Canberra fire disaster occurred, focusing media attention away from the alpine fires. That day there was a major expansion of the ANP fires. The DSE Daily Narrative says that by the beginning of Day Twelve (Sunday 19 January):

> around 1,400 personnel from Victorian government agencies, around 500 CFA personnel (supporting local volunteer brigades), 24 aircraft, over 200 specialised vehicles and 60 bulldozers are working to contain the fires. The firefighting effort is assisted by the arrival of 13 New Zealand alpine fire specialists who are dispatched to Bairnsdale. About 80 Defence personnel arrive.[43]

On 18 January temperatures were in the high thirties with wind gusts up to 35 km/h. A number of the fires had joined. The Pinnibar and Mystery Lane complexes had joined up with adjacent outbreaks in the KNP to form one conflagration. Fires around Mount Bogong and Mount Feathertop had now linked up, and the fire on Razorback in the Swifts Creek region had joined the Bogong fires. These were now managed as the 'Bogong Fire Complex'.[44] The Mount Buffalo firefighting effort was now focused on asset protection, particularly around the small towns, such as Harrietville and alpine resort areas at Mount Hotham, Dinner Plain and Falls Creek, where the effort to save the ski resort went on for almost eleven days from Friday 17 January. The small town of Mitta Mitta on the Omeo Highway north of the ANP was also under threat.

The previous day (17 January) lightning had ignited a fire in rough country in the northern part of the Snowy River National Park, just south of the isolated village of Tubbut in East Gippsland. The next day another fire broke out west of Gelantipy on the Snowy River Road.[45] These fires were contained but they 'placed further demand on DSE and CFA at a time when resources were described as "very stretched" '.[46] To the west-north-west the situation at Benambra, 21 kilometres north of Omeo, shifted from fire suppression to protection of assets after a house and outbuildings were burnt. This was the south-eastern edge of the Bogong fire complex. Benambra (population 150) was highly organised. John Cook, divisional commander of the CFA, says that:

> local landholders with UHF radios spent days and nights at specified high points as spotters and radioed in to the control-centre every half hour to report. All grader and dozer operators had spotters and locals with them to direct them . . . Nearly all households had a UHF radio to listen in on so as the women knew where their menfolk were, and what was happening. This stopped any panic in the community. The women set up a catering roster to feed the strike teams and locals. The air support helicopter and spotter plane from Benambra air base kept the ICC and local CFA leaders informed . . . Community meetings were held . . . to keep the public informed and refugee areas were pointed out, as well as safety procedures to follow.[47]

The heavy smoke, which covered most of the state, made it difficult to get firefighting aircraft into the air. There was thick haze over Melbourne with the worst air pollution since the 1983 Ash Wednesday fires. And, as if this was not enough, six new, deliberately lit fires started near Beechworth on Tuesday 21 January, a day of very low humidity, 50 km/h winds and temperatures of 36°C.[48] Two of the fires became problematic: one was west of Eldorado close to the Chiltern–Mount Pilot National Park, and the other was threatening the outskirts of the village of Stanley, south-east of Beechworth. After what firefighters called a 'ferocious run', the outbreaks were eventually contained on 27 January after assistance from three water bombers and the air-crane. These fires absorbed considerable resources; more than 500 firefighters, 97 tankers and nine bulldozers from an already overcommitted firefighting force.

On Day Fifteen, Wednesday 22 January, things became acute in the resort area of the ANP. As a result of the prevailing south-westerly winds 'residents of Bright, Wandiligong, Freeburgh and Porepunkah were warned the Mount Buffalo fire was travelling in their direction. Fire was within 500 metres of Harrietville, and the town of Dartmouth was under threat.'[49] In Bright residents were told for the second time in 24 hours, according to the *Age* of 23 January, 'to leave their homes immediately or stay and be prepared to deal with spot fires and flying embers. By late [Wednesday] night many of the town's 3000 residents had fled, taking their possessions packed in cars. Most of Bright's shops were closed and the summer tourists were gone.'

Many complained that the worst part was waiting for fires to come. The *Age* commented that residents had been

> on tenterhooks since Monday when the Mount Buffalo fire first threatened . . . Their nerves are now completely frazzled . . . The thick white smoke that is smothering Bright makes everything worse. A film of ash lies on cars. Visibility is down to 50 metres . . . Suppression efforts . . . were focused on asset protection in and around the townships and private property. The Weather Bureau predicted dreadful fire weather for the coming weekend . . . a number of communities went on heightened alert.

The DSE Daily Narrative comments matter-of-factly: 'The Bogong fire is now expected to join the Pinnibar fire in the coming few days'.[50]

Saturday 25 January was the worst day so far. There was a total fire ban for the whole state and Melbourne's temperature reached 44.5°C, the second highest on record since Black Friday. Across Victoria hot to very hot conditions obtained with moderate winds and low humidity.

Sunday 26 January was also a day of total fire ban with extremely high fire danger. That day's DSE Daily Narrative reported that:

> the major fires continue to be the Pinnibar complex, the Bogong complex and the Mount Buffalo fire, which have collectively burnt an estimated 316,000 ha. ...The two largest fires (Pinnibar and Bogong) are now close to merging... Spot fires around Mount Hotham and Dinner Plain continue to cause problems. The fires move south and south-east towards Benambra and Omeo where private property is under threat... Extreme and erratic fire behaviour is experienced across northern and southern sections.[51]

Bogong Village lost three houses and a twenty-bed lodge, and six houses were destroyed at Mount Hotham village.

Close by is the Cobungra View Estate, a rural subdivision nestling among trees on a low hill overlooking the historic Cobungra Station on the Great Alpine Road about 6 kilometres from Mount Hotham Airport and 21 kilometres west of Omeo. This is classic 'high country' and is attractive because of its close proximity to the snowfields. On Monday 27 January most people in the estate were well prepared for the fire, and Strike Team 1106 was sent from the Airport to help protect the residents as fire approached.[52] Three large tankers and two 'Pigs' (4X4 utes with water-tanks) went to the estate.

About 11.30 am the Great Alpine Road was closed behind them and at 11.45 spotting blocked the only other escape route. Mark Reeves from Dinner Plain CFA said it was becoming 'increasingly hot and [a] strengthening wind was blowing. We realized we would need to establish a sheltering position and establish a "self-protecting" mode.' They quickly arranged the 15-tonne Isuzu trucks behind a tank and a shed. Mark Reeves describes what happened next.

> From our apparently secure perch in the cabin, with woollen blankets over us, we had an amazingly good view of the approaching fire... Over the following fifteen minutes or so, we were witness to some of the most incredible spectacles

I have experienced. The flames ripped around us. Tornados of fire looped and danced. Several times our truck was engulfed in absolute blackness, darker than a moonless night. It was soot, ash, embers and earth, instantaneously baked dry and hard and blown away. Our truck shuddered and shook in the hurricane . . . I watched what I thought was bare earth flame under us, the adjacent tankers and truck . . . I recall watching Jan Sully in the command vehicle next to me and . . . it was like watching muffins cook in the oven! The glass became too hot to touch. The heat haze shimmered . . . At various times the tyres of the adjacent tanker began to smoulder and burn, and a quick radio call for them to move forward or back snuffed it out . . . It was a bizarre dance of hot trucks in a firestorm . . . After 20 minutes the main flame front passed, and we were left in a 'post-nuclear' smoking and flaming landscape.[53]

One of their biggest problems was breathing, even with respirators. The air was hot, foul, acrid and smoky. They wanted to help the residents but had to wait another ten minutes before they could move because of the intense heat. The winds remained ferocious. Mark Reeves says that well after the front had passed it was 'an incredible scene, still flaming and scorched bare to the extent the earth seemed sterilized and vaporized'.

They then worked with the well-prepared locals to save as many houses as possible. Despite fires in all of them, ten houses were saved. However, the Cobungra View flats and six houses were destroyed.

Back in Benambra the locals had been fighting fires for days on end. Fifteen-year-old Georgina Williams said that, 'The waiting was the worst, knowing it was on its way, knowing we were going to have to fight . . . Days passed as it edged towards us . . . People were exhausted just waiting. The waiting was slowly getting to people, the gas-bagging pros had run out of topics, two weeks was too long.' The fires returned on Australia Day. Georgina says that about lunchtime:

all hell broke loose. The fire was coming from all directions and the wind was relentless. The fire was spotting anything up to ten kilometres in front of the main line. People had brought their dogs, cats, poddies and chooks into the fire station and tied them up to the fences around the school. People were frightened. The sky was black and the trees were moving viciously in the wind. The fire was so close . . . generating the wind we were getting and the dense black smoke made breathing too hard. About 6 pm things started to calm down, people started going home to see what was left.[54]

Three houses were burnt, but after a strenuous defence and a change of wind direction, the town was saved and the fire headed toward Omeo, covering the 21 kilometres in less than two hours.

In Omeo already exhausted fire crews had to face the fire-storm. The Omeo District Hospital evacuated its eighteen patients to Orbost and Bairnsdale. The hospital had been burnt down in the 1939 fires. Highly organised locals fought the fire with flames 20 metres in the air; it was as dark as midnight. The fire went through from midday to about 2.00 pm, when Omeo was saved by a southerly change that forced the blaze back upon itself.

The town of Swifts Creek, 27 kilometres south of Omeo, was threatened by an intense ember attack. Meanwhile road access and power had been cut off to Benambra and aerial bombing was hampered by thick smoke.

When the Pinnibar and Bogong complexes joined on 28–29 January there was continuous fire across the mountains from Canberra in the north to the Bogong High Plains and Mount Hotham in the south-west. There were now only three fires in Victoria: the Mount Buffalo National Park fire (35,000 hectares burnt), the Bogong complex of fires (270,000 burnt), and the Pinnibar complex (74,000 burnt). By the next day 'a further 20,000 hectares of national park and state forest [had] been burnt by [the] fires'.[55] The Mount Buffalo fire joined the massive complex near Bright on 1 February. There was now a continuous, gigantic conflagration stretching right across the Victorian and New South Wales alpine country. Altogether 465,000 hectares of Victoria had now been burnt. This had increased to 700,000 hectares by Sunday 2 February.[56] Much of the increase had occurred on Thursday 30 January.

At this stage over 3700 firefighters were deployed, including 1770 from government agencies, 1500 from the CFA, 160 from the ADF, 313 interstate firefighters, 33 New Zealand Alpine specialists and 22 American remote-area firefighters ('smoke jumpers') who arrived on 29 January. Five Erickson air-cranes were operative in New South Wales and Victoria.[57]

vi

Weather forecasts for Thursday 30 January, in both Victoria and New South Wales, predicted hot to very hot conditions in excess of 40 degrees, freshening north to north-westerly winds and a fire danger beyond extreme. A total fire ban was called for the whole state. Right across the mountains people went on high alert again, but as a local policeman, John Kissane, commented 'people had gone past the panic stage' because many had already experienced up to three earlier bushfire alerts. The predictions for Thursday proved to be accurate, and Esplin says that 30 January 'was probably the single worst day of the fire season with intense and erratic fire behaviour'.[58]

The Daily Narrative recounted what happened:

> Severe fire weather predicted yesterday saw winds in the higher altitudes of the alpine areas exceeding 90 kph a little after 0300 hours. Wind speeds across the fire area strengthen during the morning and temperatures rise. By early afternoon numerous communities in the area are under significant threat. They include Mount Hotham, Omeo and Benambra in the south, and Mitta Mitta, Eskdale and Harrietville in the north. The work of fire crews and local residents sees property damage kept to a minimum ... Significant stock losses are also sustained. Fire behaviour is at times extreme and erratic, and under the influence of strong and gusty north-west winds the fire spreads rapidly in a south-easterly direction across the Tambo and Buchan River Valleys to Wulgulmerang and Gelantipy.

Nevertheless there was considerable damage to private property.[59]

One of the extraordinary phenomena of large bushfires like those of 30 January is the ability to produce their own thunderstorms and lightning. There seem to have been several thunderstorms that afternoon east of Gelantipy in the Snowy River National Park. The report on the Victorian alpine fires of January to March 2003 describes the process:

> Large bushfires ... generate heat, smoke particles and paradoxically water vapour. The heat from a fire can generate an updraft which carries the water vapour and smoke particles up into the atmosphere, spawning pyrocumulus clouds. In some cases these pyrocumulus clouds can become fire generated thunderstorms, with the likelihood of thunderstorm generated phenomena; lightning, erratic winds and rain.[60]

This fire also created a fierce 'fireball' captured by satellite image which showed 'a dramatic, intense area of fire approximately ten kilometres wide and two deep in the Dargo River Catchment near the Great Alpine Road'.[61]

That Thursday afternoon they had to fight again to save Omeo. This was the third time in less than three weeks. Thick smoke and fire, driven by 80 km/h winds descended on the town at 1.30 pm on Thursday and soon spot fires were breaking out everywhere, but strike teams were quick off the mark and damage was eventually limited to four homes. One of the fears was that if a pine plantation on the edge of town caught fire it could destroy half of Omeo. A spot fire actually started in the plantation at 2.23 pm but it was quickly put out. The heat was intense. Residents reported seeing large fireballs flying around the town. Even trees at the edge of the football oval, which had become a refuge, caught fire.

Among those fighting the fires in Omeo was Stan Warren, 72, who saved his weatherboard home. The *Age* of 31 January reported his victory: 'It was the second time fires had threatened the 100-year-old house. "The fires came through here in 1939 and she survived that, and she's survived again. But only just", he said ... "I heard the roar and rushed out there with a hose and buckets. It was coming up from the front fence and then she came in at the back of me. A neighbour tried to put that one out on his own".'

Stan had already lost stock and property in the Bingo Munjie area, north of Omeo. Spot fires from around Omeo reached as far as Swifts Creek, 27 kilometres away. Rain and a cooler change eventually came later in the afternoon across many parts of Gippsland and up into the mountains. The fires were dampened down.

One of the major runs of the 30 January fire was in an east-south-easterly direction from the Bogong South fire toward the Snowy River Road and the Snowy River National Park around Gelantipy, Seldom Seen, Wulgulmerang and across to the village of Tubbut. Another fire made a run from south-east of Mount Hope on the state border towards the Snowy River. These fires were particularly bad, and hard to deal with because the isolation of the area meant that a small group of people were left almost entirely to their own resources. They had waited for two weeks knowing that eventually they would be on the front line, and the intensity of the fire that went through their district was the most severe that occurred across Victoria. Many felt let down by the authorities. There were fire trucks in the area but nothing was done to assist isolated residents and families.

Beth Allen was the nurse at the Gelantipy Bush Nursing Centre on the Snowy River Road. The fire did not quite reach this far south due to a wind change, but Beth reported that she:

> walked outside and heard the most incredible noise—like a great waterfall or of huge trucks on a freeway. It was the roaring of the fire about two or three kilometres away . . . The sky to the north was a deep red-black colour, but the wind had dropped and apart from the roar of the fire there was little else to hear . . .
>
> The next day the road was cleared. I was able to go up to see how people fared and to witness the devastation caused by the fire. The fire spotted to within a couple of cooees of the Bush Nursing Centre, and hit for real a couple of kilometres up the road. A narrow track had been cut through the hundreds of trees that had fallen across the road; luckily we had a chainsaw

because more could have fallen at any time. As we travelled slowly up the road, we passed houses burnt down, paddocks burnt, fences burnt. Driving through the bush near Seldom Seen was quite disorienting—it was like a moonscape. Like a bomb had been dropped. All that remained was a few burnt sticks where before it had been a bush land of snow gums. The service station at Seldom Seen had been burnt down... As we came out of the bush into the cleared area at Wulgulmerang, [we saw] the sickening sight of many head of cattle that were fleeing the fire and had been burnt to death by the side of the road. I can't describe the sight—it was heartbreaking... The air rang with the sound of gunshots as farmers shot their injured and dying stock. Dead sheep lay in black piles waiting for excavators to come and bury them.[62]

It was some time before a clear picture emerged of what happened. According to the DSE report, 'The fires were started by a combination of spotting and lightning strikes... [which] most probably originated from a build up of pyrocumulus clouds due to the alpine fire on the previous day'.[63]

Multiple fires were still burning strongly the next day in the Gelantipy district, the Snowy River National Park, and the ANP to the north. The *Age* of 22 February noted:

The wonder is that no one died. David Woodburn abandoned the battle to save his home and the Seldom Seen Roadhouse to shelter in his dam with his dog, nine geese and a passing kangaroo. Bill Livingstone ran from his burning house, injured his knee jumping from the veranda and, pulling a blanket over his head hid behind a grader until the fire passed. One of his dogs reappeared from God knows where the next morning, waiting in its charred kennel. 'We can laugh about it now', said Helen Bowman at the gallows humour rippling around neighbours at the uninsured wreckage of the roadhouse. 'But at the time it was desperate. It was the most horrific day I've ever had in my life'.

Twenty kilometres north at Suggan Buggan, perhaps the most isolated spot in Victoria:

> Clive Richardson watched 20 years of work go up in smoke when the fire swept through his uninsured organic orchard. His pistachio trees were promising their first decent crop. He lost kilometres of irrigation piping and thousands of trees. 'I'm just waiting to see what sprouts back', he said. 'I haven't got to the point of giving up but it is looking more like I will have to find some income from other sources to rebuild it—back to the rat race. Yuk.' (*Age*, 22/2/03)

In fact, as Clive realised, 'the main fire never got here [Suggan Buggan] and this side of the range was burnt by ember attack'.[64] The bush was only burnt in patches in the deep valley. To the east of McKillops Bridge across the Snowy, as well as to the south around Buchan, the threat from fires in the national park dragged on for weeks. 'A long campaign was waged to secure control lines to stop impact on private grazing land.'[65]

There was a sense in which everything built up to a crescendo on 30 January, and after that there was a gradual winding down. The fires continued and there were new ignitions on 31 January from lightning, but the early February weather gradually became milder with reduced fire activity. The first week of February was spent strengthening containment lines, back-burning, fighting spot fires, mopping up and beginning rehabilitation. Widespread rain in Gippsland in mid and late February took the pressure off fire crews even though the rain caused bad erosion and wash-aways in some areas and lightning ignited more fires. By 1 March the fires were 95 per cent contained and the crisis was declared officially over on 7 March. The DSE Daily Narrative says that this was 'the largest forest fire to occur in Victoria since 1939 . . . The total area burnt by all of the eastern Victorian fires ignited on 8 January 2003 was around 1.12 million hectares . . . It was probably the longest campaign in the recorded history of firefighting in Victoria.'[66] If the area burnt in New South Wales is added to the Victorian total, 2.58 million hectares of land were affected, the largest fire ever in the European history of south-eastern Australia.

vii

The length of the season had serious implications for those who confronted the fires. Most outbreaks in the past were over in a day, or a week at most. But the 2003 fires lasted from late December, intensifying on 8 January, and not finishing until early March. Many communities faced an extended period of firefighting. Towns like Jindabyne and Thredbo are not only dependent on the winter snow for their economic viability, but summer brings many bushwalkers and visitors to the high country, and these contribute about 20 per cent 'of the 5500 jobs and the $608 million generated by tourism in the region each year'.[67] Yet some had no cash flow for six weeks or more while Jindabyne was shut down and the entire energy of the whole area was devoted to firefighting. Many of the Victorian alpine towns like Harrietville, Falls Creek, Bright and Mount Hotham, are also dependent on visitor numbers. But for six weeks residents of these towns lived on a knife-edge and were constantly confronted with high levels of smoke inhalation with visibility reduced at times to less than 30 metres. Bright (population 3000) was closed to tourists until the emergency was over.[68] The Victorian alpine area is estimated have lost $60 million in tourist income.

Many lost assets—up to $30 million in damaged infrastructure in Victoria alone. Losses were also high in New South Wales. Snowy Hydro faced replacement costs of $6.5 million, and Telstra over $4 million to replace optic fibre cables.[69] Smaller businesses were also affected. For instance Christine Schatzle and her husband own the property 'Pender Lea' on the Alpine Way between Jindabyne and Thredbo. Not only did they have to fight the fire unassisted, they also suffered considerable income loss. She told the coronial inquiry, 'As a result of the fire we have suffered substantial financial loss including damage to property and fencing. Our business has suffered great losses from cancelled bookings at the time. I estimate we have lost $150,000 which is not claimable through our insurance company, GIO Insurance'.[70]

Rural residents faced greater loss. Some experienced long periods of constant firefighting which took them away from their farms and normal employment. In Victoria many dairy farmers away fighting fires had to employ temporary help to keep their farms viable. The experience of Craig Allen from 'Spring Creek', a property bordering the KNP, illustrates what some rural people faced. 'We were directly affected by the fires that burned around our property from 20 December 2002 until 24 February 2003. As a member (and president) of the Ingebyra Bush Fire Brigade, I spent 14 days in active fire fighting and patrols in KNP and around Ingebyra. When the fires reached our property I spent a further 14 days actively fighting fires on my own property.' He also highlights the consequences of fatigue. 'Criticism is often made of the NPWS and Forestry requirements that crews be stood down after a certain period of fire involvement. The failure of local RFS officers (professional and volunteer) to do likewise resulted in overtired and possibly dangerous crews and certainly impaired the quality of decisions being made.'[71]

Volunteer firefighters also ran the risk of emotional burnout. The adrenalin-charged high of fighting fires can be followed by insomnia, anxiety, heavy drinking, flashbacks, difficulty adjusting to reality and even relationship collapse. Most people move on quickly but, as CFA counsellor Jim Unkles warns, confronting a bushfire can be a terrifying experience: 'It's overwhelming. It can cause hysteria, panic. People talk about the noise—like roaring thunder, like a locomotive.' Unkles says this affects even senior, experienced fire officers.[72] Psychologists worked on-site with Victorian departmental staff throughout the period of the fires. They reported that:

> The main issue faced by . . . staff as the duration of the fire campaign continued was fatigue amongst workers on the ground. This was counteracted by high morale to begin with, and an observed spirit of mutual co-operation and friendship. As the fire campaign went on, however, there was an identifiable and understandable decline in morale generally. Frustration was evident . . . This translated at times into a sense of hopelessness and helplessness amongst staff. There were also many people who experienced close encounters with dangerous situations, and for these individuals there was a sense of bewilderment, shock and disbelief at what had transpired. Fear was also experienced by some staff, depending on their proximity to the fire front.[73]

General practice doctors in fire-affected areas, like Mark Robinson at Mount Beauty, reported that:

> people forgot their coughs and colds and minor ailments. As the fires continued the pattern of presentations to doctors changed, becoming mainly trauma and occasional acute medical presentations... Despite incessant, unrelenting smoke cover, presentations with respiratory illness and asthma did not increase. Many people suffering chronic airway illness either left the area or stayed inside their homes. The wearing of facial masks in public became common practice... Recovery for many in the community will be slow, having suffered significant emotional and psychological trauma. Others have suffered financial loss, including volunteers who were absent from their workplaces, sometimes for a six week period. Some elderly and disabled were stressed by the displacement and the threat they experienced when moved to evacuation centres.[74]

One of the key elements in the 2002–3 conflagrations was the widespread, effective use of aerial firefighting. What is often not appreciated is how inherently dangerous these operations are. Brad van Wely, writing in *Flight Safety Australia*, says, 'There is an element of risk in all aerial operations. But when an aircraft is carrying 9000 litres of water above a scorching cauldron of fire through a dense smoke haze, the stakes are even higher... There has been no major research on aerial firefighting accidents and incidents in Australia.'[75] US studies have found that smoky, turbulent, gusty conditions in mountainous terrain were dangerous, with a key issue being low flying in high-density altitude, that is in circumstances in which air pressure is low due to the heat of the fires. This reduces lift and engine performance when power is needed to carry heavy loads of water. 'While air-cranes like Elvis grab the headlines, the Bell 212 and Squirrel are the real workhorses in the firefighting effort.'[76] These helicopters can carry water either in belly tanks or in external bambi buckets hanging below the aircraft.

Victorian fire authorities have also developed four specialist groups of firefighters who rappel from helicopters into remote or inaccessible areas. 'Only a few helicopter pilots are qualified to drop the rappel teams into hazardous areas. The teams carry a huge amount of equipment, including chainsaws, rakes, ropes and water. They are self-sufficient for up to 16 hours in remote locations . . . Heat and smoke, and operating in uncharted terrain make the assignment extremely difficult for even the most experienced pilot.'[77] Fixed-wing aircraft were also used, although there was much criticism that these were not used effectively in New South Wales.[78] As a result, the Nairn report recommended that the Commonwealth should provide funding for aerial firefighting.[79] In response the Howard Government announced funding of $16.5 million over three years in the 2004–05 budget for the National Aerial Fire-Fighting Centre.[80]

viii

One of the most contentious issues that arose after the fires was that of hazard reduction burning. This became a touchstone that focused a range of other issues simmering away in the community. It was clear throughout the period of the fires and afterwards that those who favoured regular burning, such as the grazing lobby, used the occasion to push their point of view and to blame everything on the political influence of 'powerful city greenies' who were accused of blocking regular, prescribed burning on ideological grounds. Here we should not underestimate the role of the psychology of blame. Much of the attack was an expression of the usual need to find scapegoats and deal with a genuine sense of loss. How governments respond to this is important. Governments in Victoria and the ACT initially offered help, but once their bureaucrats got hold of the process people had to go through complex hoops to get minimal assistance. While both states produced comprehensive reports, these simply fuelled community anger. They failed to deal adequately with the emotional response of people to seeing their livelihoods and sometimes their homes literally go up in smoke, and their landscape destroyed.

The New South Wales Premier, Bob Carr, was far more intelligent. Aid was both promised and delivered without a great deal of red tape. A project was funded whereby people could tell their stories of the fires.

This is important as a kind of public acknowledgment of what were often terrible experiences. The entire process was controlled through the Premier's Department, thus preventing minor bureaucrats erecting an obstacle course for victims. Other governments were not as helpful.

The annoyance felt by people with inept government responses was widespread. This kind of displaced anger is particularly difficult to manage and bureaucracies rarely deal with it adequately. Many people were still feeling its raw effects and ideologically driven groups attempted to channel these feelings. The conservative Institute of Public Affairs held a conference on the bushfires on 11 March 2003. Foresters from the hazard reduction school spoke. While they 'declined to speculate' on why prescribed burning had not been carried out, they did mention 'that there were elements of "community opposition" to such burning'. Predictably, participants who had been affected by the fires were more bluntly forthcoming. They 'were unhesitating in naming the influence of vocal but ill-informed green groups as the likely culprit' for the intensity of the bushfires. Participants were also concerned about property rights, particularly 'the expectations of private landholders that adjoining public lands should be properly managed', and they explored 'the legal redress they have when that management fails'.[81] The IPA even brought in the Melbourne *Herald-Sun* journalist and right-wing commentator, Andrew Bolt, to discourse on 'Green Religion: the triumph of a set of mystical values over science'.

Other dissatisfied groups of farmers and foresters emerged, such as the 'Stretton Group' and the 'Bush Users Group'. Arguing that the Victorian Esplin inquiry was a whitewash, the Stretton Group even obtained an opinion from Collins Street Senior Counsel, Allan Myers, who was critical of the inquiry's findings about 'fuel reduction burning'.[82] Myers had 'grave misgiving about the manner in which the report is written'. He thought it was 'highly generalized and qualified' and contained 'a great deal of peripheral and irrelevant material'. He said this was particularly applicable in the 'sensitive area' of fuel reduction burning. He cited critically a chapter in which 'members were unable to reach any conclusion about Aboriginal burning practices'; surely justifiable given we know so little about these practices. He claimed that the chapter on high country grazing ended with 'an inconclusive observation', yet what Esplin clearly concludes is, 'There is currently no scientific support for the view that "grazing prevents blazing" in the high country'.[83] Myers claimed that, 'The next chapter appeared to focus on the breeding habits of the mallee fowl'. This is a cheap shot given that the chapter is devoted to fires in the Mallee where 183,000 hectares were burnt.

In order to push their view on the need for preventative burning, two members of the Stretton Group used a burnt stand of native cypress pine (*Callitris endlicheri*) abutting the Snowy River Road near Willis on the Victoria–New South Wales border for a media 'exposé' of a Parks Victoria policy that had led to 'environmental disaster'. Two *Age* journalists, who visited the area with Stretton Group members, described the burnt stand as 'a landscape forever altered by a bushfire so intense that there are no seeds left from which saplings can spring'(3/8/04). How the journalists were able to make this judgement after one brief visit is not clear. It is simply wrong to say that the landscape was 'forever altered' by the 2003 bushfire.

Firstly, the Snowy River Valley had already been decisively altered and degraded by cattle grazing and constant burning by settlers for 130 years prior to the 2003 blaze. Grazing, burning and the arrival of the rabbit around 1900 led to a massive depletion of native undergrowth and subsequently to the loss of 15 to 30 centimetres of topsoil in the area. Cattle were withdrawn between 1944 and 1967 when the KNP was established. So the real, long-term destruction had already been wrought *long before* the 2003 fire.

Secondly, *Callitris* pines are ancient trees whose ancestors evolved in Gondwanaland.[84] However, once fire became endemic in Australia, fire-sensitive species like *Callitris* were confined to less combustible areas where they had a chance to recover between major fires. Because of this they have developed survival skills. Otherwise they would have been eliminated long ago. They are seeder species that maintain their seeds in cones. When fire comes through the cones open and the seeds fall out and a stand can gradually re-establish itself if conditions are right. If these particular trees do not re-seed the reason is that there is no topsoil in which to re-seed as a result of erosion, rather than the intensity of the 2003 fire.

Thirdly, what the article does not tell the reader is that while there are some burnt-out areas of cypress pine, these are surrounded by unburnt, perfectly healthy stands. Less than 10 per cent of the cypress pines on the Victorian side of the border have been burnt. There are still vast stands of these trees surrounding the much smaller areas that have been burnt.

ix

There is no doubt that native animals suffered terribly and many were lost in the fires. Nevertheless, Australian fauna is well adapted to fire. Again, they would not have survived if they were not. In their natural state populations can bounce back as the fire-adapted ecology of areas recovers. But the problem is that the habitat of many native species has been so modified or destroyed by our activities and the invasion of predators, especially in sensitive alpine areas, that animals are restricted to increasingly isolated pockets. Breaking up their habitat makes it difficult for them to move from pocket to pocket and they are unable to interbreed. This is especially true of rare and endangered species like the mountain pygmy possum and the corroboree frog, which are confined to specific alpine locations. The real issue here is not the intensity or extent of fire, but the activity of humans, which causes extinctions.

One of the fears of conservationists during the East Gippsland fires was that a tiny remnant group of the Victorian sub-species of the brush-tailed rock wallaby (*Petrogale penicillata*) in the steep and inaccessible Little River Gorge in the remote Snowy River National Park would be wiped out. Altogether there are only about 40 individuals left; most are in captive breeding programs. Less than a third are in the wild. The Victorian brush-tails are a sub-species that are genetically different to those in New South Wales and Queensland, where there are more secure colonies.[85] During the fires Raz Martin, a strong defender of brush-tails in the wild, was scathing about the failure of Parks Victoria to do reduction burning in primary brush-tail habitat in the gorge, which in places is 300–400 metres deep with sheer sides. He told the *Age* on 25 January 2003 in the middle of the crisis: '"The gorge hasn't been burnt since 1952, and the amount of fuel there is terrifying" . . . "The litter is a metre deep . . . If it's over 32 degrees we can't go [into the gorge] because one spark from a falling rock sliding down the hill and it could go up . . . I've begged the department to do reduction burns or let me do it. But they don't believe anybody understands fires but them".' He accused Parks Victoria of being dominated by an ideology which defied common sense. He believes that, like the practices of Aboriginal people, the only way to manage the landscape is through broad-scale, low-intensity regular burning.

Careful study of the animals and their environment followed the 2003 fire. The conclusion:

> It is likely that the fires have had a net positive effect on the rock-wallabies . . . two years after the fire. Observations of animals caught since the fire suggest that they are in unusually good condition and breeding has occurred post-fire. It remains to be seen whether better condition of the mothers, increased forage availability and quality for at-heel young, and reduced predation, will combine to allow the recruitment of breeding individuals into the population for the first time in many years.[86]

If not hunted to near extinction or placed under other destructive pressures, Australian animal species are very well adapted to fire. The very first animal I saw on 'the block' a month after the fire was a female lyrebird. Three years later there are lyrebirds, emus, wallabies, wombats and kangaroos. A family of wedge-tailed eagles has been present since early 2003. For them to survive there must be food on the ground.

Dr Andrew Claridge of the NPWS, who had been involved in a study of quolls in the Byadbo Wilderness, found that rather than being wiped out, a considerable number had survived and that seven juveniles had been added to his study population after the fires.[87] During the fire the animals apparently retreat to dens beneath granite boulders and then adapt their eating habits to what is available. Wombats also survive. They retreat to their burrows and as long as excessive smoke does not smother them, they can easily outlast the starvation conditions that inevitably follow. Even the corroboree frog and the mountain pygmy possum have been found in surveys since 2003.

During the fires Dr Ken Green of the NPWS spent a day defending a prime swamp habitat of the black-and-yellow-striped corroboree frog. There are only about 250 of these critically endangered frogs left in isolated bogs in the KNP. With fire on a 200-metre front threatening the bog, Ken beat back the flames from about 8.00 am to 5.00 pm, armed only with his fire-resistant jacket. His only help was a single water-bombing helicopter. He admits that the smoke was dreadful and that he spent much of the time coughing and spluttering. But both Ken and the frogs survived.[88] He points out in the *Age* of 2 June 2003 that Australian alpine animals are well adapted to the occasional fires which are an integral part of the ecosystem. 'Every species we expected to be out there was out there somewhere . . . There are pockets where the population has been untouched and they're ready to move back into the burnt areas. Even in the totally bare places two months after the fires we were still catching all the species. I was amazed because I didn't expect to catch anything.' In other words fires, even massive conflagrations, are simply not the ecologically destructive events that some contend.

x

An important connected question is whether the alpine fires were a 'one-in-200-year-event'. 'No' says the DSE's *Victorian Alpine Fires* book. It argues that ' the probability of similar events occurring the next decade or two is reasonably high . . . Periodic summer droughts, multiple ignitions from lightning strikes, and days of extreme fire danger will continue to be part of the Victorian environment and may even be exacerbated by the effects of global warming.'[89]

This statement is partly true. The historical evidence is not well defined, but it can probably be safely asserted that major fires in the Australian Alps naturally occur about every 60 years or so, with a massive fire about every 200 years. Part of the problem is that the 'burn, burn, burn' regime distorted the natural cycle of fire. Certainly, control of deliberately lit fire now means that we might slowly be able to recover the natural rhythms of burning by fires resulting from lightning. However, the unknown factor in the future is the influence of global warming. The snow cover might disappear, and with it the flora and fauna dependent on it. But whether the mountains will dry out, or whether rainfall will increase across the high country creating an almost cool temperate rainforest situation, is unknown. While global warming is well established, its effects in specific areas are yet to be discerned.

What is clear is that the answer to the south-eastern Australian cycle of fire is certainly not to be found in deliberate, broad-scale, reduction burning of forests, even if this were possible. Nor is it letting the logging industry or cattle into national parks to 'thin' the forests and undergrowth out. There are strong scientific and historical arguments to show that such interference with the forests and their natural rhythms is precisely what makes them vulnerable to fire.

But the worst danger confronting the environment is the modern infatuation with 'management' and the belief that linked with technology, humans can improve on nature. This kind of arrogance ignores the fact that the forests and mountains have been 'managing' their own affairs reasonably well for long before humans actually descended from the trees themselves. Nature is not there to be manipulated, nor does it exist merely for our convenience or exploitation. We need to withdraw and practise humility and allow nature to look after itself. It is only when we do this that we are able to marvel at its complexity, study it with integrity and understand it with respect.

So the answer is not regular controlled burning of forests, which would be an impossible task given that there are 7.7 million hectares of forest in Victoria alone. As Dr Ian Lunt, a vegetation ecologist at Charles Sturt University, says, 'If every forest [in Victoria] was burnt every five years, then over one and a half million hectares would have to be burnt every year. This is much more than the total area burnt in the 2002–3 fires. Single fires rarely cause widespread ecological damage, but repeated fires at short intervals endanger many plant species and damage ecosystems.' He says that there needs to be good hazard reduction programs around assets which need to 'be strategic and carefully targeted'.[90] As Mike Leonard, fire manager for the DSE, pointed out in the *Age* on 23 January 2003 during the height of the fires, 'Even if the state could afford it, the impact of frequent—every five years or so—fuel reduction burns would be devastating for some animals and forests. Scientific studies have shown that because regular low-intensity burns reduce the shrub layer the forest's structure is weakened and ground-dwelling birds and mammals that depend on shrubs for protection often die from predators and habitat loss.' Studies by CSIRO showed that 25 species (thirteen native, such as eastern grey kangaroos and twelve introduced, such as foxes and feral cats) benefited from regular burning, and 26 species (25 of them native) were disadvantaged by reduction burns. 'A bushfire will wipe out many animals, but once the forest regenerates the pre-fire animal population will increase tenfold', according to Dr Peter Catling of CSIRO Sustainable Ecosystems, quoted in the *Age* of 23 January 2003.

The most problematic report on the 2002–3 fires was the House of Representatives Select Committee Inquiry, *A Nation Charred*. Perhaps the best that can be said for it is that it is the history of everything that went wrong in the firefighting. The fundamental problem was the political context in which it was established. The state and territory governments did not participate. They were concerned that the inquiry was biased against them and the land management policies they had followed, and they were uneasy with federal interference. Unsurprisingly, one of the recommendations of the inquiry was that the federal government play a much greater role in the control of bushfires. In a summary statement the Nairn Inquiry quotes favourably from the submission of the Institute of Foresters of Australia describing a kind of divide between the city and the country:

> We see the community divided over fire management and the divide ... deepening ... On one side of the divide are some influential environmentalists and academics, supported by inner-city residents not threatened by bushfires, and not responsible for bushfire management. These people in general advocate a hands-off approach to land-management, where 'natural' events like bushfires are allowed to run free. On the other side are rural people, fire fighters, foresters and land managers who are responsible for values threatened by bushfires. The latter tend to advocate an interventionist approach, where steps are taken to minimize risks before fires start, as well as having in place a well-equipped rapid-response fire fighting force.[91]

This statement caricatures 'urban greens' who live in the unreality of inner suburbia, versus the real, 'practical' people who live in the bush. It takes no account of ecological studies, of the fact that environmentalists spent most of their lives in the bush, and it presents all country people as supporting the ideology of regular burning. This was the essential problem with the Nairn Inquiry Report: it reinforces divides rather than building bridges. It gives voice to untested opinions which may be genuinely and sincerely held. But just because 507 people gave evidence to the committee does not guarantee that their interpretation is definitive. The evidence presented does not give the complete picture. It is a biased report; it couldn't be anything else.

The fires of 2003 were not confined to New South Wales and Victoria. There were also fires in Western Australia, the Northern Territory, South Australia, Queensland and Tasmania. The Northern Territory experienced 'the most significant burning in desert country for 25 years' and altogether 38.4 million hectares were burned between August and November 2002. It was also a severe season for Queensland, especially along the rural–urban frontier. While South Australia was at risk of severe fires, they did not eventuate and the outbreaks experienced were relatively minor. In Western Australia a total of 656 wildfires burnt 2.11 million hectares of land in the south of the state. They were particularly bad around Manjimup in the far south-east. In Tasmania 58,000 hectares of forest and grassland were lost in the most severe and difficult fire season since 1967.[92]

However, the most spectacularly destructive of the 2003 fires were in the Canberra fire-storm of 17 January.

A PERFECT FIRE DAY, CANBERRA, 17 JANUARY 2003

i

The aircraft, an 87-seat, high-winged, four-engine BAe 146 jet, in Qantas livery, was descending downwind into Canberra from the north about 2.40 pm on Saturday 18 January 2003 during the most destructive bushfires ever experienced in the ACT. It was flying along the edge of a dense, expanding smoke cloud driven by erratic west-north-west winds gusting up to 70 km/h, in a temperature of 37°C, all combining to create severe low-level turbulence. The atmospheric instability was increased by 'a 14 kilometre high cumuliform plume of dry, unstable air above the fire'.[1] The plane was flying southwards past the western side of the airport to make a U-turn and then head back northwards towards Runway 35. But the dense smoke cloud was forcing the aircraft more and more towards the east across the extended centre line of Runway 35, the main north–south runway. It was a nightmarish approach. In order to position the aircraft for a final approach toward the north, the pilot was forced to make a tight, low-level, left turn somewhere above

Harman Naval Station near Queanbeyan about 6 kilometres from the runway threshold. Showing the crew's complete professionalism the plane made the difficult turn from south to north smoothly to line up for a final approach and quick landing on Runway 35, no doubt to the great relief of all aboard.[2]

But if their problems were over, Des Fooks' worries were just beginning. Des still lives in Monkman Street, Chapman, a south-western Canberra suburb. By 4.00 pm that terrible Saturday afternoon Des's house was a smouldering ruin.[3] He says that:

> About midday, from listening to local [ABC] radio, I became conscious that there was an extreme fire danger in the local area. It was an eerie day for indefinable reasons. The sky was a funny colour—the trees had an odd hue to them—and everything was very still. I live adjacent to the bush and there are birds galore and a constant clatter of bird calls. That day there was not a single bird to be seen and no noise at all ... The gum trees in particular had somehow picked up tinges of blue and pink which made them startlingly beautiful.

Around midday Des's daughter, son and two friends arrived. They all walked to the top of Cooleman Ridge. Its highest point, Mount Arawang (765 metres) is one of the best vantage points around Canberra. Des says:

> The wind, which was almost still lower down, was more than gale force at the top. It was barely possibly to stand erect without clutching on to something for support. With a 360 degree vista for kilometres all we could see was some smoke way in the distance to the north-west. There was no obvious cause for concern ... We continued to listen to ABC Radio, which on that day and since, has provided an absolutely brilliant communication link for the community ...
>
> Reports suggested that things were worsening at an alarming rate but, from what we could see high up on Cooleman Ridge, there were no signs of imminent danger ... A little

after 3.00 pm . . . a massive fire that had not been visible a half-hour earlier could be seen racing through the pine forests above Eucumbene Drive on the border of the suburb of Duffy . . . The fire was moving at an unbelievable pace . . . It was across a wide front and possibly fifteen to twenty metres high . . . Helicopters were buzzing everywhere . . . By this time I had a greater sense of the seriousness of the problem but—maybe in some stupid sense of denial—I still felt reasonably safe. Even so I went into the house and grabbed some documents and some photos and put them in a box, thinking all the time what a nuisance it was because I would soon be unpacking them.

The whole feel of the atmosphere changed just before the arrival of the fire.

Within the space of minutes things changed rapidly. Even though it was only 3.30 in the afternoon it had suddenly become very dark, not like night time, but an evil-looking blackness. Colours had become even stranger. Nothing was natural. The air temperature was unbearably hot. There was a constant roar like the sound of a 747's engines at take-off. We decided to reposition our cars so that they pointed down the driveway . . . Things were suddenly grim . . . [and] became even more eerie. For a minute or two the roaring sound ebbed—the wind seemed to abate—and the air temperature dropped from searing heat to just plain hot. We stood there with hoses at the ready and then, in an instant, it was like somebody had opened the door of a blast furnace. The heat and the wind were buffeting us and day turned to darker than night. There was a sheet of flame in front of us that stood fifteen or more metres high. Large pieces of burning objects were in the air. We ran for our lives.

Fortunately they had the cars ready for a quick escape. Yet even in the familiar surroundings of his neighbourhood, Des found it hard to keep to the road. 'The headlights penetrated barely beyond the bonnet of the car . . . and at one brief stage I found myself driving across someone else's front lawn.' The family had decided to meet at the Southern Cross Club in Woden if they became separated. Des says that when he got there, 'It was a surreal experience. I walked into air-conditioned comfort, bright lights, TV monitors showing the tennis and horse racing, poker machines tinkling away and patrons sipping drinks—all this only two kilometres from a holocaust. Bizarre!'

ii

In common with the rest of eastern Australia, Canberra was in the grip of a drought in early January 2003.[4] One good effect was that regional wine-makers had a high quality, bumper crop which promised some of the best reds from the area for 30 years. But ACT dam levels were down to 54 per cent of capacity. There was certainly an awareness of the danger of fire, with ACT Chief Fire Control Officer, Peter Lucas-Smith, warning Canberra residents to prepare their homes for the worst fire conditions for twenty years. The bush was bone dry, there were high winds, low humidity and above-average temperatures.

During the afternoon of Wednesday 8 January the ACT was hit by the dry electrical storms which also affected much of Victoria and south-eastern New South Wales.[5] Around 2.30 pm there was a number of lightning strikes along the ridge of the Brindabella Range, on either side of the Mount Franklin road. As a result three fires were ignited in the ACT's Namadgi National Park, and in New South Wales' Brindabella National Park at McIntyres Hut on the Goodradigbee River, together with two fires in the KNP, south-west of the NSW–ACT border. There were fire-spotters on duty in all four fire observation towers in the ACT.[6] Soon after the lightning strikes, the spotter in the Mount Coree tower (1421 metres) reported smoke coming from behind Webbs Ridge Road at McIntyres Hut in New South Wales about 6 kilometres west-north-west. 'Over the next hour numerous smoke sightings were reported.'[7]

The ACT Emergency Services Bureau (ESB) immediately sent reconnaissance aircraft to check the strikes. The RFS and NPWS in Queanbeyan were informed about the New South Wales fires. Two ACT ground crews were dispatched along the Mount Franklin road, reaching the two nearest fires at Bendora Dam and Stockyard Spur about 6.00 pm. It was eventually decided to withdraw these crews because of the inaccessibility and the danger of direct firefighting at night in this remote location. The Ron McLeod inquiry set up by the ACT Government commented: 'ACT Bushfire Service management . . . was initially confident that the fire could be extinguished—either by suppression or self-extinguishment—in the first 48 hours and that this confidence was based on experience. They acknowledged the severe climatic conditions, but their initial view was that the fire could nevertheless be swiftly put out.'[8] Next day the fires were still burning and indirect ground attack proved unsuccessful, so helicopters were brought in for water bombing. Actually, little was done on 9 January to deal with these fires. 'The objective was to keep the fires contained to their smallest possible size with aerial support. Heavy plant was being organised to assist. As weather conditions worsened in the afternoon the fires expanded fairly quickly.'[9] ACT ESB recognised that 'some unusual fire patterns [were] occurring'.[10] Specialist remote-area firefighting teams arrived.

The McIntyres Hut fire in New South Wales was about 200 hectares in extent in steep country making it almost impossible to attack directly. The danger of the fire spreading to the ACT pine plantations to the east of Mount Coree—and remotely to the city itself—was recognised and a local bushfire emergency was declared. Next day the NPWS and RFS decided to attack the fire indirectly and try to contain it 'due to the steep terrain, difficult access and unpredictable fire behaviour'.[11] Containment lines were decided upon and crews retreated to these in an attempt to hold the fire. Late on Wednesday afternoon the smoke from these fires spread eastwards across Canberra. The *Canberra Times* reported the next day that, 'While the flames posed no risk to people or property, the choking smoke ruined washing on clothes lines, sent holidaying school children scuttling indoors and reduced visibility for drivers'.

iii

Much of the subsequent criticism of the bushfire authorities in the ACT and New South Wales centred around their failure to tackle these ignitions within the first two days. Peter Clack in *Firestorm* and Chris Ulhman's *Stateline* report on ABC TV were both very critical of authorities for their failure to extinguish these fires immediately, and for the lack of long-term preventative burning.[12] Ulhman commented that, 'Many experienced fire-fighters believe that it was in those crucial early days that every possible resource should have been thrown at the fires'. Enthusiastic volunteer firefighters wanted to tackle the outbreaks head-on, but fire controllers were hesitant to commit personnel in remote areas where people could easily become trapped. The Brindabella fires required firefighters to walk several kilometres to the fire-ground through difficult country, and they had to create containment lines using hoes and picks. Nevertheless, the McLeod Inquiry tended to agree with the critical analyses:

> ACT authorities should have been aware that [these] bush-fires . . . would be very difficult to extinguish once they gained a hold and that on unfavourable days the risk of spotting would be considerable where fires were in an area with high fuel loads. These circumstances ought to have alerted the authorities to the absolute importance of trying to put out any fires as quickly as possible, when they were small . . . The responses to all the fires in the first few days present a picture of a measured approach to a threat that was growing on a daily basis—as opposed to an all-out attempt to beat the fires from the outset, using every resource at the ACT's disposal . . . In my opinion the tendency to view the situation from a 'best case scenario' perspective had the effect of understating the risks.[13]

However, the New South Wales Deputy State Coroner, Carl Milovanovich, formed a different view. While not addressing the ACT situation directly, he is quite clear that he didn't think that the New South Wales authorities were remiss and tardy in their response to the McIntyres Hut fire.

> I am of the view that to have sent fire-fighters into the Brindabella Range on the afternoon of the 8th of January,

2003, would be contrary to all the basic fire-fighting knowledge and would have placed professional and volunteer fire-fighters in potentially grave danger. The decisions that were made on the afternoon of the 8th of January must be examined in the light of what was known then . . . The decision to contain the McIntyre's Hut fire was in my view, having regard to all the circumstances the correct decision at that time. The fact that the fire was contained for the best part of 10 days would suggest that the strategy was working.[14]

Milovanovich is right. Everything is easier in hindsight. The McIntyres Hut fire only blew up on Friday 17 January when 'perfect' fire conditions developed. Certainly the ACT ESB could have been more proactive in the first few days of the fires, but that is a post-factum judgement. The fact is these fires would have been ultimately controllable if extreme, 'perfect' fire weather had not occurred on 17–18 January. The real reason why ordinary fires become massive wildfires is not because of so-called 'fuel loads', but because of the weather, over which we have no control.

iv

From 10–12 January the ACT authorities worked to contain the fires in the Brindabellas, while the RFS and NPWS worked on the McIntyres Hut fire. It had tripled in size by Friday, but was then contained for a week. The ACT fires slowly continued to grow in size. On Friday 10 January, 324 hectares of the Namadgi National Park, which takes in much of the Brindabella Range, had been burnt. This increased to 580 on Saturday, 1440 on Sunday, 2250 on Monday, and 2850 on Tuesday.[15] The number of firefighters had risen from 110 to 140.

The Brindabella fires suddenly hit the headlines when a firefighting helicopter on lease to the ACT ditched into Bendora Dam after engine failure. The unconscious pilot, who was trapped under water in the upturned cockpit, was rescued by Euan McKenzie, who dived from the Southcare rescue helicopter. He was quickly assisted by ACT Chief Minister, the dour Jon Stanhope, and bushfire chief, Peter Lucas-Smith, who happened to be on an inspection tour nearby. Pictures of McKenzie, assisted by Stanhope and Lucas-Smith, stripped to their underpants, swimming to the shore of the Bendora Dam with the unconscious pilot put the fires back in the headlines. Lucas-Smith acknowledged that 'we will still be fighting these fires, without rain, in the next two or three weeks'. Canberra was covered in a smoke haze.

While the fires only increased marginally in size on 14–15 January, it was becoming clear that things were going to get worse before they got better. 'Extreme fire weather conditions' were predicted for the coming weekend, although all fires seemed to be holding within containment lines. On Thursday 16 January with the temperature at 33°C and low humidity, concerns were growing. The ACT Cabinet was briefed. 'The Cotter [Dam] catchment, ACT pine plantations and Tidbinbilla Nature Reserve . . . and [Canberra Deep Space Communications] Tracking Station were all listed as potentially under threat.'[16] Also listed was the 'urban edge'. Phil Koperberg, RFS Commissioner, said clearly that, 'The current weather forecast and the fact that the vegetation in . . . southern NSW and the ACT is extremely dry means that the potential for fire to impact on increasingly more populated areas is very high'.[17] The Bendora fire had now spread into the Brindabella Valley where about 30 properties were under threat. A total fire ban was declared for the ACT for an unprecedented five days from 17 to 21 January.

The situation became threatening on Friday 17 January. The temperature reached 36°C, humidity fell to 15 per cent, and a north-west wind was gusting up to 35 km/h. The fires began moving south-east and east with spotting occurring beyond containment lines. The closest fire was now 25 kilometres from the edge of the Canberra suburbs.

It had been known for many years in forestry circles that the real danger to Canberra was from the north-west, from the McIntyres Hut direction. Graham Franklin-Browne, who lived at Stromlo Forestry Settlement said that in the days leading up to the fire, 'I felt a little paranoid about this . . . We all had the uneasy feeling that these fires were somehow different from the dangers we had experienced in previous years.'[18]

With the fires growing in size the staging areas that supported the firefighters were removed from Bulls Head and Orroral Valley in the mountains to playing fields near the ESB headquarters in suburban Curtin. Already 7193 hectares of the Namadgi National Park had been burnt. It was predicted that some sensitive areas would take 40 years to recover. Lucas-Smith told the *Canberra Times* on 17 January that he didn't think 'there is any threat to the urban edge [of Canberra]. We'll certainly see a lot more smoke than we've seen in the last few days.' Fire experts were talking about six weeks of effort ahead and the need for rain to put out the fires completely.

On late Friday afternoon with winds gusting up to 65 km/h and the temperature reaching the forties, things started to get out of control. Ferocious conditions developed with fires escaping containment lines in the evening and during the night. Burnt leaves were falling in the suburbs. ESB director, Mike Castles, conceded in the *Canberra Times* of 18 January, 'It's the worst conditions we've had . . . This looks like continuing for three or four days.' Properties in the Brindabella Valley were threatened, the Tidbinbilla Nature Reserve was in immediate danger with firefighters struggling to protect rangers' homes, and a spot fire was burning less than 2 kilometres from the Uriarra pine plantation. The nearest fires were now 12 kilometres from the edge of the Canberra suburbs.

Life went on normally in the city. People were aware of the fires, but somehow lulled into a false sense of security. The authorities had tried to convey calm, and it was only as people began listening to local ABC radio in the blistering heat of early Saturday afternoon that they realised that something disastrous was happening. But no one had a clear picture of events as communications quickly collapsed. In fact an unmitigated disaster was overwhelming the city.

v

At 9.30 am on Saturday 18 January the ACT ESB was told by the Weather Bureau that the day would be one 'of extreme fire danger'. Humidity was low, 'Winds were blowing from the north-west to the west, and numerous sub-weather patterns were occurring around the fire, partly as a result of the convection column that was being generated up to 14 kilometres above the fire'.[19] Wind patterns became increasingly unpredictable, and by 2.30 pm it was gusting to 52 km/h, and up to 78 km/h at 3.20 pm. The temperature and humidity made for 'perfect' fire weather: at 12.42 pm the temperature was 37.4°C and it was still 33.6°C at 7.00 pm. Relative humidity was 8 at 2.50 pm and fell to 4 at 4.30 pm.[20] Given these conditions it was inevitable that by midday all the fires had reached massive proportions with fronts in excess of 40 kilometres. Eventually they merged to form a gigantic conflagration on Canberra's western edge. There was absolutely nothing now that the authorities could do to stop the fire.

The exact route of the fires towards Canberra is not clear and even the precise time they hit the suburbs is confused because of the ember storm that preceded the main front.[21] But what seems to have happened is that by Friday evening, as 'perfect' fire conditions developed, the outbreaks were about 15–20 kilometres from the western edge of the city in the mountains to the west of the Paddys River Road and south of Tharwa on the Naas Road. By 8.50 pm spotting began to occur over containment lines. There were serious threats to the 42 rural landholders in the Tidbinbilla Valley south-west of the city as the fire had now reached the Tidbinbilla Ranges. To the north, the McIntyres Hut fire was spotting close to the ACT border near the Blue Range, and the southern portion of this fire had linked up around Piccadilly Circus with the Brindabella fire. The situation was getting increasingly out of control as the night passed.

At Tidbinbilla Nature Reserve, National Parks staff, including Geoff Underwood, had been fighting the fires for 30 hours, with spot fires invading the reserve throughout the night. It is in a 'U'-shaped valley with mountains surrounding it on three sides. The reserve was home to endangered species, including brush-tailed rock wallabies, freckled ducks and regent honeyeaters. It was also home to 26 koalas, as well as Geoff Underwood. He says that around midday Saturday, after only half an hour's sleep:

> I was awoken by the phone and told I had 15 minutes to get out of the house! . . . The winds had shifted and strengthened and the Reserve was now blanketed in smoke. Lots of the grassland and bush . . . was burning . . . At approximately 1.15 pm . . . we

saw a fire tornado which was sucking a column of flame at least 100 metres into the air . . . At around 2.00 pm the gates of hell blew open! . . . Winds in excess of 120 kph were recorded at the Visitors' Centre. A Navy chopper pilot water-bombing an adjoining property told me he was recording gusts of up to 160 kilometres per hour. This was no ordinary running-flame-type fire, but can best be compared to a massive blast of burning gas, a firestorm of an intensity never before experienced by my colleagues, some with more than 40 years of firefighting experience. The firestorm raged for over an hour . . . By 4.00 pm, things had quietened down enough to go into the reserve to assess the damage. Nothing could have prepared me for the death and destruction we were about to find . . . Of the 5,500 hectares that make up Tidbinbilla Nature Reserve, 99.9% were not just burnt, but incinerated. Dead and dying animals littered the valley floor . . . It was like a nuclear holocaust that had occurred in the beautiful, tranquil valley that I had been lucky enough to call my home and workplace for many years.[22]

All staff residences were destroyed, including Geoff Underwood's, and only five of the brush-tailed rock wallabies, a few ducks, some red and grey kangaroos, and one extraordinarily tough female koala, who was only discovered six days after the fires, survived. The devastation of Tidbinbilla Nature Reserve is one of the truly serious, long-term tragedies of the fire. Across the road at Miowera two houses were burnt.[23]

Meanwhile south of Tharwa on their farm in the narrow Naas Valley, Sarah Martin and her parents were surrounded by fire and fighting to save their animals and house. Because a fire on Mount Tennant to the north had burnt across the Naas Road, the local Tharwa brigade was temporarily unable to get through to the Martins and their neighbours. The Mount Tennant fire was particularly intense on the steep banks of the Naas River just before its confluence with the Murrumbidgee.

When Sarah rang 000 she was told by the male operator 'Just do what you can to protect yourself'. She says he sounded 'so frightened himself that, if the hills weren't burning around me, I might have spent the next five minutes counselling him'![24] The Martins took refuge in their dam, but made frequent forays out to extinguish spot fires and care for their animals. Sarah says:

> So it was that a city girl without a clue went out with her Dad, with a wet towel and a fire hose that wasn't working properly, to put out what was to be the first spot fire of many that day . . . I was especially amazed at how combustible the cow shit was. It burnt like fire starters . . . It's an amazing sound, the sound of a bushfire. I can only compare it to the volcano I had climbed in Guatemala a year earlier. The roar of the wind and fire coming out of the bowels of the earth. As I looked out at the hills, which were now all alight, I heard the same roar and the crackling sound of trees toppling over. But all I smelt was smoke . . . As Dad and I fought the smaller fires, we swore at each other like troopers, me calling him a few things that I'm not allowed to write down, and he calling me a lot of things back . . . Luckily, just as the fire spread to within a metre of the house, the fire trucks from Tharwa arrived.

The devastation of ACT rural properties was one of the underreported aspects of the Canberra bushfires. All up, 87 rural houses and properties were destroyed and 14 rural houses damaged.[25]

By midday Saturday the fire was one massive conflagration on a 50-kilometre front. The entire western flank of Canberra was exposed from the northern suburbs of Dunlop and Macgregor, to Tharwa in the south, although back-burning there the night before had made it safe. The south-central portion of the fire swept down from the wooded foothills of the Brindabellas with incredible intensity. Passing the Tidbinbilla Nature Reserve on a roughly north-easterly path it entered the Pierces Creek Pine Forest. Among the highly flammable radiata pines it gained tremendous momentum. Racing through them it headed towards Cotter Dam. It was a fire-storm, a series of cyclones of fire, with winds in excess of 150 km/h.

Tim Borough, Fire Safety Officer at the ANU and a volunteer firefighter, was at Tidbinbilla Nature Reserve. He says: 'The storm, instead of hail or rain, was actually fire embers that were hitting us.

Everything was just black and red... At the same time there were also fireballs. There were just big balls of hot gas flying along the ground... probably about 100 feet across... I took a photo just before [one] hit a tree... the tree exploded! It's quite an amazing phenomenon. I've never seen anything like it in all of my life.' Borough reports that when the fire emerged from the nature reserve it headed off across wide, open spaces. 'These fireballs were just zooming around as if everything was very flammable. Whereas the flammable material out there is very low, its just a little bit of stuff on the ground, dried grass on the ground.'[26]

If the Tidbinbilla Nature Reserve was destroyed between 1.15 and 2.30 pm, then the fire moved extraordinarily quickly. It raced across the open farmland towards the Canberra Deep Space Communications Complex, which went off-line and was successfully defended by its own staff and RFS volunteer firefighters. It then headed across the open bush and grassland towards the Murrumbidgee, which was seen by many as a natural barrier. On the way towards the suburbs, the fire passed through the Cotter River Reserve where the pub, shop, road bridge and forestry houses were destroyed. The Cotter Pub had a reputation for fine dining and its owners, Ludwig and Leila Misner, almost lost their lives trying to save their dog.

Just under 4 kilometres away from Cotter was the Uriarra Forestry Settlement on the northern edge of the Uriarra pine forest. This had been used during the previous week as a base for the RFS as they fought the fires across the New South Wales border. Some Uriarra residents claimed that their water was used up in a useless fight, and they were abandoned when the fire burned through the forest that surrounded the settlement almost completely. All twenty wooden homes were destroyed. One resident, Kevin Cotter, told the *Sydney Morning Herald* on 21 January that, 'I had to drive out of there through a wall of fire. Everything I have has been destroyed... and I don't have any insurance... They [emergency services] left us in the lurch.' Meanwhile the Murrumbidgee proved a more psychological than real barrier when the fire easily crossed the river and headed towards the southern suburbs and into the Canberra Nature Park on the edge of Kambah.

vi

On the north-west side of the city the McIntyres Hut fire was also wreaking havoc. It had built to enormous proportions. By midday it had crossed the Mountain Creek Road and was heading towards the sewage treatment works on the lower Molonglo River. Here Canberra's waste water is treated and eventually returned to the river system. There were few trees in the area and, in the prevailing drought, very little that was combustible, but around 2.30 pm the fires hit the plant in several waves.

According to Peter Clack, a former *Canberra Times* journalist and author of *Firestorm*, a couple of regular ACT fire brigade units were present at what he describes as an extremely confusing scene.[27] He claims that between 5000 and 10,000 litres of chlorine was stored on site for water purification 'next to two storage tanks of diesel fuels and a bank of liquid petroleum gas cylinders inside a locked yard'. He says that if the chlorine ignited it would have exploded and spread in a poisonous, toxic cloud across the Belconnen area that would have done more damage to human and animal life than any bushfire. He maintains that disaster was only prevented by two crews of Canberra firefighters and RFS units and a district officer who risked their lives in a desperate battle to make sure the fires did not reach the chlorine tanks. Clack says: 'The life and death struggle to prevent a second disaster taking place was not publicised, its potential for death and tragedy swallowed in the aftermath of the firestorm'.[28]

These claims were strongly disputed by ActewAGL Corporation which runs the plant.[29] They claim that *Firestorm* is 'factually incorrect'. Chief Executive Officer, John Mackay, says: 'The tanks described by Clack did not contain chlorine but ferrous chloride which is not particularly dangerous in the context of a fire'. There was chlorine on the site but it was 'several hundred metres away from the location [described by Clack] in a specially constructed fire proof building' which had been checked that morning. The corporation does not deny that the fires caused 'serious damage' to the plant 'which was struck [by fire] at least three times during the afternoon of 18 January'. In the subsequent inquiry report, McLeod's comments seem non-committal: 'The [fire] crews at Lower Molonglo were faced with particularly adverse conditions in an isolated environment. Their actions did limit the fire's impact on the facility, despite two Fire Brigade pumpers becoming inoperative because they caught fire.'[30]

About a kilometre away, firefighter Matt Dutkiewicz was dealing with another crisis. He was deputy captain of the Rivers Volunteer Brigade based on the Cotter Road at Weston Creek. Early in the morning of Saturday his crews were patrolling the Cotter Road, the Kambah Pool Road and the Uriarra Road. Sections of these roads ran through the Stromlo Forest. By midday the fire was close and Matt found himself on the front line.

> At 12.30 pm we were advised by the communications centre that the fire had jumped the [Murrumbidgee] River on the [heritage-listed] 'Huntly' property on the Uriarra Road. The tanker I was on was asked to help contain it. On the way to 'Huntly' the centre radioed us with a priority message that NSW fire crews had been overrun and safety was the no. 1 priority. We were told to do property protection only and steer the fire around the houses. (*Canberra Times*, 25/1/03)

After arranging his tankers and firefighters to protect the main house, he was told there was a smaller timber cottage down towards the Uriarra Road. They decided to check. 'We arrived at the cottage to find the owner herding his cattle toward the house and the woman who lived in the weather-board cottage washing it down. The fire was some distance away in the valley.' Matt felt safe because the drought meant there was no vegetation left to burn. All that remained was a wind-row of trees about 50 metres away. He decided to place the truck side-on to the approaching fire thus allowing it to be driven back and forth spraying and protecting the house. Suddenly the fire appeared.

> The smoke hit [Matt's crew] first, followed by a gust of wind that broke most of the wind-row in half. I remember saying 'Where the hell did that come from?' The split second that followed

gave way to a wall of flame that was three times higher than our tanker. I yelled to the crew to get down behind the heat shields at the back as the flames hit the truck. The heat was so intense that it cracked the mirrors. We activated our protection sprays but most of the water seemed to evaporate before it reached us . . . We were spraying both hoses at the fire. I was so scared. I thought we would all be killed. The flames licked our arms and faces and burned our airways.

The house caught fire. The animals were also trapped. 'The cows the farmer had rounded up into the paddock were screaming like little kids. I will never forget the sound and sight of burning livestock.' Matt yelled to the crew to back the truck up so they could rescue the woman who was inside the house. As the truck was backing up, he jumped off and ran to the house, gasping for breath. Having got the woman out of the house, 'I yelled to my crew to help me because I had become disoriented on the way back to the truck with the woman . . . My eyes were filled with black soot and I couldn't breathe. I didn't think I'd make it back to the truck so I dropped to my knees to try to get some clear air.' After reaching the truck, 'We drank, coughed and spluttered for a few minutes trying to work out exactly what had happened.' In the middle of it all Matt's brother rang about protecting their parents' house in Duffy.

After the fire-front passed they extinguished the cottage fire. There was little damage. They then returned to 'Huntly' which was saved, although the gardens were destroyed. Steve Rutter, another volunteer from the Rivers brigade, says he had never seen 'a fire move so fast. It was frightening. The ground had barely any grass to burn, but it was pre-igniting before the flames got to it.'[31] While the Rivers volunteers were fighting the fire at 'Huntly', fourteen of their cars were destroyed at their base on the Cotter Road. Among them was Matt's. He was eventually helicoptered out to Canberra Hospital, having trouble breathing. 'I remember looking back over Woden/Weston Creek from the helicopter pad on top of the hospital. It was a sight I will never forget—it could have been midnight but it was only 5.00 pm. It looked like the whole of Canberra was going to burn down.' While he was fighting the fires at 'Huntly', Matt's parents lost their home in Burrendong Street, Duffy.

vii

By now much of the western edge of Canberra was burning. The fire that Matt and Steve and their colleagues faced out at 'Huntly' had picked up an even more massive velocity as it burnt through the Stromlo Forest and raced toward the Australian National University's 80-year-old Mount Stromlo Observatory. Mount Stromlo stands 782 metres high in the middle of what was a pine forest. First established in 1924, the observatory has become a world-famous research and teaching facility. It has one access road which climbed up from the Cotter Road through the pines to the summit where the telescopes, offices, technical buildings and staff residences were. As the fire swept eastwards into Stromlo Forest, those still on the mountain were given twenty minutes' notice by police to evacuate just before 2.30 pm.[32] By now the fire had such velocity and power that nothing could have stopped it.

Three people remained on the mountain. One was left behind accidentally. He was an overseas visiting fellow, Oliver Schnurr, who was asleep in the bachelors' quarters. The others were Mark Bacon, who lived in residence 8—a volunteer firefighter who decided to stay, assisted by his friend Andrew Thompson, also a firefighter. They had met through their mutual interest in medieval re-enactment. Mark and Andrew decided to stay inside the brick house. 'You wait inside the house', Mark said, 'which has been secured as best you can, you have all the protection that you can wear, you pull down all the roller shades and the curtains and close them up because the glass tends to heat up and crack and then blow in or out depending on the pressure variation'.[33]

Andrew says the fire-front came up Mount Stromlo out of the north-west, just before 3.00 pm. As it approached the smoke became heavier, the heat increased and there were a lot of embers.

> It became progressively darker and the fire-wall got progressively taller as it got closer . . . 40 metres high maybe at that stage . . . It was like . . . an orange wall hidden behind the smoke . . . and it was getting progressively hotter and the wind was starting to increase quite dramatically. As the front got closer the wind was

very intense . . . I hadn't expected [it] to be that bad because, I guess, most of the experience of bushfires I'd seen were quite small by comparison . . . I put on an extra pair of overalls, had a hard hat on, had the goggles on.

At this stage the electricity went off and water pressure collapsed. Andrew and Mark were standing in the kitchen and 'through the window we could see the approaching fire', Andrew said,

> dust, smoke and flame. And then the noise started. Whereas before you could hear . . . the helicopters going around [outside] . . . the noise just picked up. We couldn't hear anything once the flames started to arrive. It was quite surprising. I mean you hear all this anecdotal evidence about what to expect and you just imagine that they're all blown out of proportion because of the circumstances and because it's such a very exciting, I suppose exhilarating, experience . . . You just hear this funnelling of the flame coming towards you and, at about say five and ten metres in front of the actual flames themselves, anything that was combustible would suddenly erupt. The actual radiant heat was so great that the flames did not have to come in contact with it because it was already that hot as it approached . . .
>
> I turned to Mark as we were looking out the window and I said to him 'Is this the appropriate time to tell you just how frightened I am?'

Whirlwinds circled inside the fire as the leaf and ground materials ignited. Andrew says he lost all sense of time for a period. The garage on the western side of the house caught fire as did the car parked in it. The fire-front seemed to have passed through in about twenty minutes, but the house had to be checked, especially after a window on the northern side popped. 'That's when we had to start doing our tour. We'd each go around the house, check to see which windows had blown in and try to secure [them] as best we could, put out any embers and then we constantly went around inside the house to make sure that nothing was igniting.' Eventually they went outside to put out spot fires.

I stepped outside and everything was in flame, so I looked at everything because you're naturally drawn to it. It's a fascinating circumstance to be in . . . Its incredibly warm . . . its almost like a nurturing cocoon of warmth in your overalls because you feel protected. I had three layers on . . . It was like an incredibly hot summer day and yet on the odd occasion when the wind actually blew up underneath the guard of my helmet I'd really appreciate just how hot it was outside . . . You can't rush because if you rush you take in too much hot air and smoke.

Meanwhile Oliver Schnurr was asleep in the bachelors' quarters. He woke up to find fire all around the building. He had been up the previous night observing and had gone to bed about 11.00 am. He hadn't heard the police sirens and shouts to evacuate. He had previously read a pamphlet about bushfires and followed it to the letter. It told him to stay inside while there were flames around and to cover himself, stay low near the floor and use a wet cloth for breathing. As soon as the front passed the pamphlet told him to leave the building and take refuge in an already burnt-out area. Covered in a wet blanket he went to the childrens' playground and stayed there until he saw Mark and Andrew. He then joined them in looking after Mark's house.

By 5.00 pm the men could explore the whole site. The destruction was horrendous. Nothing that anyone could have done would have saved the observatory from the massive fire that came out of the pine forest. The devastation of Mount Stromlo was the most significant intellectual, educational and research loss, not only in the Canberra fires, but probably in all Australian bushfires. Five telescopes: the 74-inch, the 50-inch and the 30-inch reflectors, and the 26- and 9-inch refractors suffered irreparable damage. The historical 1920s administration building, housing the library, was gutted, although the walls remain standing. The valuable workshop complex was completely destroyed. The Near-Infrared Integral Field Spectrograph, which had taken four years to build and was ready for delivery to an observatory in Hawaii, was ruined. Seven of the twelve residence houses were gone, although Andrew and Mark's efforts had not been in vain; residence 8 is still there. The more modern buildings survived, as did the computer centre. The damage bill was considerably more than $50 million. It was impossible to work out why some buildings were destroyed and others survived. The dynamics of the fire were bizarre.

What is even more bizarre is that such a valuable intellectual and cultural asset was completely surrounded by an inflammable pine forest. Not to have realised the danger to both the observatory and nearby suburbs betrays a negligence by successive federal and ACT governments that is almost criminal.

viii

While Oliver was asleep and Mark and Andrew were watching the dynamics of the fire-storm, down near the Cotter Road other disasters were unfolding. The fire was racing through the pines towards Stromlo Forestry Settlement and ACT Forestry Headquarters. There were twenty houses in the settlement occupied by older people who had long-term connections with ACT forestry, and a number of others placed there by Housing ACT. Ione Kitson who was there with her brother, Derek Hamilton, told the *Canberra Times* of 20 January that she looked 'to see my car was on fire—all the cars in the street were on fire, and the power poles were burning. There was a huge wall of flame and it was so hot it was white in parts. We had no way to escape because it was on all sides so we ran into the house.' Derek and Ione made the right decisions. 'There were fireballs outside and it was raining cinders on us even though all the windows were closed. The kitchen started burning and we were hosing the walls just to slow it down and give us some time for the main fire to pass through.' They retreated to the hallway along the floor to get under the smoke and get some clear air. Even though they were only a metre apart, they could not see each other and had to scream to communicate. They quickly soaked themselves in the bath and ran out just before the entire timber house exploded in flame. Outside they found a small pocket of fresh air and a fire hose and 'used it to protect themselves briefly'. They were saved by a fire truck which suddenly appeared. It was at the settlement that the fire claimed its first victim: Dorothy 'Dot' McGrath, 76, died from burns fighting to save her home. All but three of the twenty houses in the village were burnt to the ground.

About 400 metres beyond the settlement through the pines were the suburbs of Duffy and Holder and, behind them, Rivett. Most Canberra suburbs are small. Duffy comprises about 35 streets and cul-de-sacs. Holder is a little larger. The front-line streets are Warragamba Avenue running along the northern edge of Duffy and, at right angles, Eucumbene Drive running along the western edge. The pine forest came right up to these streets, literally no more than 15 metres from the houses across the road.

About 150 metres from the intersection of Eucumbene Drive and Warragamba Avenue, as the fire raced through the pine forest, an ACT Fire Brigade tanker was drenching homes and trees. Paul Fixter was one of the firefighters. He says they could see a gigantic 100-metre high fire-front on Mount Stromlo about 2 kilometres away. The fire was heading straight for them.

> Within the next few minutes, the sky turned to midnight, the wind became unbelievably strong and the air became unbearably hot. When the fireball hit the trees only metres from us we were engulfed in embers and flames and . . . struggling to breathe. The three of us hid beside the truck struggling for every breath of air and drenching ourselves with two fire hoses just to stop burning. The flames were coming over the top of us, as well as underneath [the truck] along the bitumen. We made the decision to try to get into the cabin and escape that area, but it took at least two attempts to all pile in the vehicle from the passenger side. Once in we drove south along Eucumbene Drive and stopped again to try to make a stand or do something, but when we stopped we were hit again with the firestorm. We drove another few hundred metres and saw a lone volunteer firefighter with a hose. We told him to come with us, as we knew what was about to hit us again, and drove down Renmark Street, where we were confronted with so many fires it was overwhelming. We . . . started putting out house and spot fires for another seven hours.[34]

The fire hit Warragamba Avenue between 3.15 and 3.20 pm. Anthony Coles from Sydney was not far from Paul Fixter's fire unit.

He was in Jemalong Street, a block back from Eucumbene Drive, helping protect his parents' home.

> Being naturally curious and not wanting to miss out on any action I decided to walk up the street towards the pine forest to see if I could observe the fire. I was within 200 metres of the forest, looking up towards the conifer-lined ridge. It was now completely dark, but I could see the outline of the pines silhouetted against the red glow of the large fire burning behind them on the reverse side of the slope. I stood there for another few minutes and, as I watched, huge tongues of red flame appeared suddenly along the length of the ridge, leaping high into the air. At the same moment I was blasted by a wall of superheated wind coming down from the ridge. The wind at the time was approaching gale force, and was tearing branches from the nearby trees. For a few seconds I stood there completely dumbstruck . . . What seemed like a safe distance moments before, now seemed like suicidal lunacy. When my capacity for coherent thought returned, I realised my parents' house would shortly be in serious trouble.[35]

The police reported that the first house in Duffy caught fire about 3.22 pm.

After hours of struggle Anthony's parents' house was saved by the family's efforts and the brief arrival of a fire unit—probably that of Paul Fixter—to hose the house down.

Anthony tried to help save other houses in the street. 'Here and there, the odd intrepid resident managed to keep the flames from taking hold, assisted . . . by neighbours, but in most cases it was hopeless. As more and more houses began to catch people returned to protect their own homes, and it became a matter of every man for himself.' Sixteen houses were destroyed in Jemalong Street. Anthony comments that, 'The smoke was incredibly thick and I spent as much time gagging and vomiting as I did fighting the fire . . . The smoke also stung the eyes and, as more grit was blown into them, it started to feel as if they were filled with sand.'

One street that suffered horrendous losses given its length, was Tanjil Loop. For anyone aware of Black Friday in Victoria the name 'Tanjil' is a stark reminder of the terrible wildfires around the tiny village of Tanjil Bren in Victoria. Even though it was five blocks back from the main fire-front, eighteen semi-detached townhouses were destroyed in Tanjil Loop. The fires spread from the BP service station about 20 metres away on Burrinjuck Crescent, which had caught fire from embers. It exploded, with flames shooting high in the air and cinders sprayed everywhere. The townhouses in Tanjil Loop were quickly engulfed.

Duffy lost 228 houses. Nearby Holder lost 34, and Rivett sixteen. Three people died: Douglas Fraser, 60; Alison Tener, 38, the mother of three boys; and Peter Brooke, 74, who most probably had a heart attack.[36]

Just to the south of Duffy and Rivett along the hills that comprise the Cooleman Ridge section of the Canberra Nature Park is the suburb of Chapman, where Des Fooks was preparing for the worst. Just a block away from him in Chauvel Circle, Eda and Tony McGloughlin were in a similar situation.[37] Their house in Chauvel Circle was on the lower side, to some extent protected by the low rise around which the circle ran and it was some distance back from the nearby nature park. This was not the type of suburb in which you would expect a massive bushfire. Nevertheless the McGloughlins prepared for the fire by hosing the house down. Friends from Rivett came to pick up their animals. The cat ran away, but they were able to take the dog. The cat survived by hiding in a drain and reappeared the next day. The friends took the car they had packed. 'Then we just worked for the next several hours', Eda said, 'Tony at the front, our daughter down the side, and I ended up at the back of the house. So I actually couldn't see houses exploding and catching fire, which was probably a good thing! But we could hear the explosions. There were fires all around us, there were spot fires everywhere . . . The noise was like bombs going off, trees were exploding, it was pitch black. But we had water pressure, so we just kept working and working and working.' At one stage they abandoned the house and drove to the friends in Rivett, but soon returned. The side fence and the garage of the house next door had caught fire. Eda said:

The smoke cleared a bit and I saw a spotter plane and it seemed like it was going to go past us. We ran up and down like banshees. It did see us because they were pretty low. Then two water bombers arrived and dropped water on the garage. That put the fire out. We could handle the fences. Then the fire brigade arrived . . . We had a wooden garden room at the back and the fire was just about touching it. I said 'Just give me a hand. I really need help'. 'We'll get to you when we can, lady', one of them said. 'We're still on our own', I thought. So we just kept working, and the wind had changed by then. That's really what saved us.

Tony says they spent the whole night going around the circle putting out fires. Eda says: 'We acted, I think, sensibly, we helped ourselves as much as we could, and we were also very, very lucky. That element of luck cannot be understated.' Five houses out of twenty survived in Chauvel Circle.

Another arm of the fire burnt roughly eastwards through the Stromlo Forest to the north of Duffy and Holder along the northern side of the Cotter Road towards the Tuggeranong Parkway. On the edge of the forest near the junction of the Cotter Road and the Tuggeranong Parkway is the RSPCA Animal Refuge. Most of the 120 animals had been evacuated, but two brave animal lovers rescued the remaining dogs and cats, as well as extinguishing fires in the administration centre, the cattery and kennels. Only a couple of kittens and some birds perished. Less than a kilometre away 25 dogs and cats were killed when the Woden-Weston Animal Hospital in Holder was destroyed. The man who was in the process of buying the hospital told the *Canberra Times* on 21 January that 'hospital staff had been rounding up the animals, praying help would come, when a firebomb suddenly lodged in the roof and they fled for their lives'.

Across the road from the RSPCA is a very modern building, the Australian Defence College. The campus houses the Australian Command and Staff College which prepares middle-ranking officers for appointments in their services, and the Centre for Defence and Strategic Studies, which tailors its courses for senior officers of colonel or equivalent rank who are on the way to senior command. The fire burnt right up to the fence of the college but the buildings were saved when the sprinkler system was activated. About 600 metres across the Cotter Road were a couple of other complexes. The Orana School was saved by parents and staff. Close by is the Weston Police Complex and Training Centre, and the Weston campus of the Canberra Institute of Technology, both of which survived, although there were significant fires in the police complex.

ix

It was at this point that central Canberra had a lucky break. Around 1.30 pm on Christmas Eve 2001 suspicious fires had broken out near 'Huntly' on Uriarra Road, as well as in the pines in Stromlo Forest on the Coppins Crossing Road. The fires joined, crossed the Tuggeranong Parkway, and burnt through the pines to the shore of Lake Burley Griffin near the Scrivener Dam. Arson was suspected. By the time the fires were brought under control 1500 hectares of the north-eastern end of Stromlo Forest had been burnt. In the process the National Zoo and Aquarium were threatened, and a spur had burnt right up to the edge of 'Yarralumla', the residence of the Governor-General.[38]

So on 18 January there was a 3-kilometre fire-break. This was large enough to slow the fire down. Yet, so great was its momentum, it still burnt right to the edge of the Black Mountain Reserve. If the eucalypt forest on Black Mountain had gone up, the northerly front of the fire would have penetrated into the National Botanical Gardens, the CSIRO laboratories, the Australian National University, and eventually the city centre. Another front of the fire probably would have burnt eastwards towards 'Yarralumla'. The Governor-General, Peter Hollingworth, was not at home on 18 January. Like the Prime Minister he lived most of the time in Sydney. The fire did cross the Parkway and the Molonglo River below Scrivener Dam into an open and largely treeless area that housed horse holding paddocks and the National Equestrian Centre, which was destroyed. Miraculously the 120 horses at the centre were saved, but 600 sheep and cattle were killed. The proximity of the fire to the city shows just how vulnerable the pine forests had made Canberra.

Developed along the lines of a 'bush capital', the natural landscape is retained through the Canberra Nature Park. Part of this is a bush corridor that runs along the low ridge that separates the Weston Creek suburbs from those of Woden Valley. Parallel to the ridge is the four-lane Tuggeranong Parkway. The fire burnt in from the northern Cotter Road end of the park, crossed the ridge and threatened houses in Curtin and Lyons which bordered the Woden Town Centre. The ACT ESB headquarters were then just five blocks back from the nature park. The building lost power intermittently for three hours from 4.30 pm onwards and the emergency power only supported the communications and operations room. At one stage there were firefighters on the roof of ESB headquarters putting out spot fires. Four houses were destroyed in Curtin and Lyons. Six houses were lost in Weston on the other side of the parkway.

Further south the nature park widens out to take in the area around the 855-metre Mount Taylor, wedged between the suburbs of Kambah, Fisher, Pearce and Torrens. To the south-west of Kambah is another section of the nature park around the Urambi Hills. The fires reached the Urambi Hills after they crossed the Murrumbidgee. It was inevitable that spot fires would reach Mount Taylor from one or other direction. Just after 4.00 pm Mount Taylor exploded. Some observers said that in the darkness it resembled an erupting volcano with glowing embers. Kambah was now wedged between fires along the Cooleman Ridge in Chapman to the north-west, fires in the Urambi Hills to the south-west, and Mount Taylor to the north-east. Three major roads, the Tuggeranong Parkway, Drakeford Drive and Sulwood Drive, became impassable because they were surrounded by outbreaks on both sides. Thirty-seven houses were lost in Kambah and two in Torrens.

The fire also penetrated across to the back of the suburb of Farrer along the Farrer Ridge which connects with the Mount Taylor section of the nature park. Local resident Ann Ford, whose house backs onto the ridge, described what happened there. 'At about lunchtime the smoke in the sky was growing darker and redder. We walked to the Ridge behind us to gaze with horrified awe at the flames way in the distance. Still, there was no threat to *our* suburb, *our* property.'[39] But the situation quickly got worse.

The ABC began to sound strident warnings and there was news of houses being burnt. The threat was growing . . . The fright increased when we heard that Mount Taylor was burning . . . The sky was black, worse than night, a portent of real terror. In the eerie glow of the street light we saw a dark mound on the driveway. With sad horror we realised that an eagle had fallen from the sky and was lying like a crumpled umbrella in the blackness. If such a creature could be brought down what would happen to the rest of us? The fear grew to fever pitch when a hot gale blew from the north-west. In the blackness it was the worst thing I've ever experienced. I tried and tried to find a number for WIRES—just someone to come and help the eagle. Why were we humans so helpless? Our son yelled the fire was on the ridge . . . My frantic call to 000 was calmly handled. I was able to speak to the fire service. Yes, Farrer Ridge was burning. But we were on our own. 'Dress appropriately, keep hosing, don't panic. Good luck.' This honesty galvanised our sense of purpose.

What happened was that locals gathered to stop the fire, as did many people all over Canberra. It came in three waves but was beaten off each time by a small group of ad hoc firefighters. In the end no houses were lost in Farrer.

x

Throughout the afternoon Tritia Evans, an experienced producer at the local Canberra ABC station, 666, found herself on the front line of communication.[40] That afternoon she produced almost seven hours of live fire coverage. The station had been broadcasting the national ABC sports program, *Grandstand*. At 1.10 pm they broke in with a ten-minute report on the fires after the news from Sydney. At 1.52 there was a further update, and at 2.05 a report that the fire had crossed the Murrumbidgee River.[41] At 2.29 they went local with reports of the speed of the fires which were starting to spot into the Canberra suburbs, and a report from Ginninderra Falls that the flames were four times the height of the trees across the Molonglo River. At 2.32 pm they broadcast the Standard Emergency Warning Signal for the first time with a message that a major deterioration in the fire situation had occurred. Then for nine and a half hours they covered nothing but the fires. This is a fundamental function of the ABC: to act as the major means of communication in a serious emergency. Radio 666 did this with extraordinary integrity. Right at the beginning there was a serious glitch, but it was not the fault of the ABC. A State of Emergency had been declared at 1.45 pm. But this information was not received by ABC radio. McLeod says that 'for reasons never made clear to the Inquiry' the message did not get through for three-quarters of an hour. The fact is that, for whatever reason, there was a complete breakdown of communication from the ESB. 'The delay between the ESB's release of the message and Radio 666 receipt of it was explained to the Inquiry as technical—a fault in the automatic bulk-addressing function on ESB's fax'.[42]

Tritia Evans takes up the story. She was at Canberra Civic Centre at lunchtime and says there was 'a black, heavy sky. I thought the fires must be close . . . but all must be fine or else we'd have been told'. She returned home to find a message on her phone from Julie Derrett, Acting Program Director. '"Can you come in?", Julie asked. "Bring the papers. Everything seems fine, but we'll be there just in case"'. Tritia arrived, newspapers in hand, at the ABC Northbourne Ave studios just before 2.30 pm. 'I was about to make a cup of tea when Julie said "We're on!". Papers put aside and no tea for hours'. The first crisis was finding the Standard Emergency Warning Signal. 'Mad panic. Where's the tape? I think it's in Liz McGrath's office [ABC local manager]. Who has the key? Eventually it's found. It's just a siren, over and over, no words . . . And we have to play this every fifteen minutes!' By 3.00 pm the phone was constantly ringing. 'I was madly trying to confirm the accuracy of information passed on from listeners. It was hard to do this. We eventually went with gut instinct, especially when more than one caller was saying the same thing'. At around 3.50 pm a hysterical woman on the phone said: '"The whole of Duffy's gone. My house has been burnt down". I insist that we have to verify this information. We have NO idea what's actually going on and how bad this is. As it turns out the lady was telling the truth, but it was hours before I believed her'.

The 666 program record shows that they were both accurate and keeping up with extraordinarily fast-moving events on a series of fronts. Following a report from the sewage treatment plant, at 3.02 pm Jon Stanhope was interviewed on the State of Emergency. At 3.13 reporter Michael Turtle was leaving Duffy describing 20 metre high flames, at 3.15 the newsroom reported houses on fire in Eucumbene Drive, and at 3.33 Michael Turtle was back on air reporting people evacuating from Duffy, gardens burning and no firefighters in sight. Leanne Scott was at ESB HQ reporting Peter Lucas-Smith admitting that houses were lost, that the fire was on the urban edges and that the situation was out of control.

Meanwhile, experience kicked in with Tritia co-ordinating a chaos of information, checking its accuracy and getting it to air. 'We're keeping the control room calm. Speaking quietly, slowly. Eye contact with the presenter, telling her "You're doing well". Phones going wild with colleagues in the control room answering them, hovering to help. I try to keep the presenter's screen clear of everything but vital information. Kill the fly buzzing around. A light moment in a dark day. Reporters and other staff phoning in, fires everywhere. We thought it was just Duffy. But there seem to be many fronts. What's going on? Are they being deliberately lit?' Tritia says she put 'listeners to air if they seem credible. Working our way through a list of 'basics': police, fire, emergency services, airport, weather bureau. This is Radio on the Run, adrenalin-powered'. It was 'seven hours of guessing in the dark. Something terrible has happened, but how terrible? Communication with emergency services, *very* ordinary.'

When she left work around 9.30 that night Tritia faced a surreal world.

> Outside the town is hot and thick with smoke. Now it's real! Inside, although seemingly at the centre of things, it was all an abstraction. Outside people are on a knife's edge. At work I was inside a cool, air-conditioned studio, going through the paces but not really understanding what had happened. Full understanding would not really kick in for three weeks, by which time I felt it would not be too voyeuristic to drive through the fire-affected areas.

Tritia is too hard on herself. Not only did 666 win a 2003 ABC Local Radio Award for 'Outstanding Coverage of a Local Story', but they were overwhelmed with thanks from the community. Perhaps the ultimate accolade came from the *Canberra Times* on 25 January 2003. In a wonderful cartoon by Geoff Pryor a visitor listening to the radio in a taxi from the airport hears the presenter say, 'You're on ABC Triple 6 and we now bring you the latest fire update'. The out-of-town passenger says, 'Six, six, six. Isn't that the devil's number?' The driver replies, 'Not round here it isn't!'

But Saturday was not the end of it for 666. Right through the rest of January there were alerts and dangerous fire days. In fact, Thursday 30 January had the worst weather conditions of the whole period. Tritia says that, 'The next few weeks burn everyone out. No programs are planned. We just go to the studio and make it up on the go. More fire alerts, more information on how to protect your home. Thousands, yes, thousands of callers. I've turned into a social worker. Never again!'

xi

Back on that terrible Saturday things became clearer as afternoon became evening. By 8.00 pm the situation was more or less under control. Four lives were lost, there were many injuries, including three people with severe burns which resulted in them spending months in hospital, more than 500 homes destroyed (although it took the authorities a couple of weeks to establish the exact number), and many others badly damaged. Irreplaceable institutions, like the Mount Stromlo Observatory, were lost. Seventy per cent of the ACT's nature parks, forests and pasture were severely burnt. Countless native and domestic animals were killed or injured. Damage was estimated at $300 million.[43]

I visited Chapman the next day and saw for myself the destruction in Lincoln Close and Doyle Terrace. Further north was Monkman Street and Chauvel Circle. The whole area looked like a bomb site. Large gum trees were ripped from the ground, powerlines torn down, large brick houses destroyed, people standing around stunned. Tim Grainger captures the aftermath exquisitely:

People talk about the incredible noise of the firestorm—the roar of the wind, the boom of houses exploding and the incessant screech of sirens—but for me the loudest part was the silence that followed. After the fire had passed, the smoke had cleared and the wind died down, there was this surreal silence. Real, true silence—no dogs, no cars, no kids, no choppers, nothing. No sound but shock. Families walking aimlessly, speechless, unable to comprehend what had happened. Houses reduced to little more than broken bricks and twisted metal. Power lines strewn across streets, shells of burnt out cars and every shade of green now a single shot of black. Ask someone if they were okay and you'd get a teary-eyed smile, but not a word was said. What is there to add to the sight of your house, the next door neighbour's and the place behind you reduced to a pile of burning rubble?[44]

But this silence was soon followed by the viciousness of those who were looking for someone to blame. The blame game follows all bushfires, and in a town where politics dominates you expect everyone to have an 'opinion'. Bushfires render people impotent before the overwhelming force of nature and to regain some sense of control over their lives, individuals often lash out. But 'fire' or 'nature' are rather nebulous targets, so people displace anger onto organisations or others. This is the psychological mechanism of displacement.

The 'greenies' were an early target with a Holder resident yelling out to *Canberra Times* journalist, Megan Doherty, 'Tell the greenies to stick it [the fire] up their proverbial' (*Sunday Times*, 19/1/03). Soon some were blaming the ACT Fire Brigade, the ESB and the National Parks, both ACT and New South Wales. Never one to miss an opportunity for publicity, the then Territories Minister, Wilson Tuckey, accused the ACT of having inadequate hazard reduction policies. Prime Minister John Howard quickly stepped in to 'reinterpret' Tuckey's outburst. But the instant-expert Tuckey was still at it days later, this time criticising New South Wales fire and national parks administrators, leading then New South Wales Environment Minister, Bob Debus, to remark 'I sometimes think he won't be happy until we've asphalted over Mount

Kosciuszko'. Former ACT Chief Minister, Kate Carnell, joined in by saying that logging should occur in national parks in order to reduce fire hazards. After resigning as chief minister when her government developed and enforced the very policies she was now criticising, Carnell had been employed by the National Association of Forest Industries. She said that, 'I think after this weekend all of us have come to realise the dangers that these national parks pose'. Why didn't she do something about it when she was chief minister? She was supported by the Forestry Division of the CFMEU, the union covering loggers.

But the most appalling piece of all was by that doyen of the chattering classes, Padraic P. McGuinness, columnist in the *Sydney Morning Herald*. In an insensitive and ignorant piece on 21 January, McGuinness wrote 'It is too much to hope that the bushfires in Canberra will lead to a rethink of the role of Canberra in our national life'. He opined that people in Canberra lived out 'a middle class fantasy, much of which was financed by taxpayers'. He spoke of:

> the 'toytown' government, totally incompetent and unable to plan for even the most obvious contingencies . . . The trouble in Canberra has always been that upper-level bureaucrats and academics have had too much input into the planning process, and as always, created a dysfunctional socialist utopia . . . The best thing that could be done now, after the victims of the fires are helped to rebuild their lives (though the uninsured should have to bear the cost of their own folly) is to abolish the ACT as it exists, pare it right back to essentials, and let most of its citizens live normal lives in NSW.

It was gutter commentary of the worst sort.

However, there were more rational critics. Commentary began mildly and intelligently enough with CSIRO bushfire specialists, Jim Gould and Phil Chaney, quoted in the *Sydney Morning Herald* of 21 January, agreeing that Friday and Saturday were extraordinary days with 'perfect' bushfire weather. Gould said, 'I think it is very difficult to combat a fire under the very extreme conditions we had on Friday, and particularly Saturday, with the large ember attack . . . What we are trying to figure out is where the large ember attack came from.' Chaney said: 'You could have lined up the whole United States fire service and they couldn't have stopped these fires. Once it gets to that level there is nothing anybody can do.' Nevertheless, Chaney had long been critical of the build-up of fuel in ACT forests which, he said, was the result of what he called misguided conservation-minded policies. He told the *Herald* that:

> firefighters had only one option when it came to beating wildfires. 'From the scientific point of view the one thing you can manage in a bushfire situation is the amount of fuel'. He said controlled burning reduced a fire's fuel supply giving firefighters a better chance of beating it . . . a programme of prescribed burning would also equip the Territory with better expertise, more experience and equipment crucial to dealing with emergency situations.

Chaney claimed on the ABC program *Stateline* that his theories, which were popular in the 1960s and 1970s, had been neglected because of the influence of the environmental movement. '"The tide of public opinion has changed dramatically", he said . . . Mr Chaney accepted that he was out of step with the environmentalists.'[45]

Val Jeffery would agree with much that Chaney said. Jeffery is a man of many parts. Not only does he run the general store in Tharwa, a village of four short streets beside the bridge over the Murrumbidgee River, he is also the Southern Area Volunteer Fire Brigade captain, the postmaster, the unofficial mayor, and the best-known man in town. Usually Tharwa is a quiet village, but on 18 January it was on the front line. The village was threatened by fires that came in a northerly direction from the Namadgi National Park. But Jeffery's brigade had already done back-burning the night before, so the fire was successfully stopped just after it crossed the 1384-metre Mount Tennant. Jeffery and the local Tharwa volunteers had saved the village, as well as homes and properties in the surrounding area, including that of Sarah Martin's parents.

Val Jeffery has been fighting fires for 50 years. He is a firm and very outspoken believer in hazard reduction, and he contends that decisions about actual firefighting should not be left to bureaucrats and land administrators in central offices, but should be made at the fire-front by local-level firefighters. Jeffery had warned before Saturday that if the fires had not been brought under control by Thursday 16 January there would be no hope of ever controlling them. Many Canberra people, including Eda McGloughlin, felt he had been vindicated. 'Nobody listened to Val Jeffery. He was made to look like a fool, to feel irrelevant. But they saved Tharwa and it was shown that he did know what he was doing', she said.[46]

Another who agrees is Peter Clack. His book, *Firestorm*, is critical of what Clack calls politically correct politicians and environmentalists. He argues that there is a kind of alliance between them to prevent the views of more experienced people being heard. 'A farmer would say go ahead and burn. But the Greens provide conflicting advice. A Green alliance is in the majority in academia and a Green dogma permeates the Left of Labor politics.'[47] No evidence is offered for these assertions. Actually, Clack overestimates the influence of environmentalists on the Labor left, many of whom have taken a decidedly anti-green position on a number of issues. By the simple device of capitalising 'Green' he turns what is at most a loose agglomeration of people, many of them scientists with years of study of Australian biota and expertise on continental fire regimes, who share roughly similar views but disagree on details, into a tight-knit politically aware force driven by dogmatic, unexamined theories. Clearly Clack knows little about the diversity of environmentalists. He argues that these (capital 'G') Greens 'dominate' government agencies like the New South Wales and ACT parks service and Environment ACT, and that senior bureaucrats feed these views up the chain to politicians. Certainly there is a consensus among many modern land managers in embracing ecological values, but to turn this into a 'Green' conspiracy is the equivalent of finding Reds under every bed.

Clack also argues that there is a loss of the accumulated experience of volunteer firefighters that leads to bad mistakes being made when serious fires threaten. He cites the example of the ACT ESB acting too slowly in dealing with the fires in the Brindabellas and then compounding their mistake by using inappropriate back-burning which 'created a single line of fires . . . that eventually joined up the McIntyre's Hut and Bendora fires'. His evidence is an [unnamed] volunteer firefighter: 'The

ACT had three very small fires and eight days later it was a solid line of fire joined up by back-burning'.⁴⁸

The essential problem with *Firestorm* is that while there is some cogency in the argument that the authorities were dilatory in tackling the fires in their earliest stages, Clack represents the point of view of those who believe that only regular controlled burning and the reduction of fuel loads can save us from conflagrations like 2003. The book only deals with the arguments of those disgruntled with the 'centralising policies' of the ESB, Environment ACT, the parks service, and unnamed 'conservation-minded land managers'. More considered reflection might have produced a more balanced book.

xii

Four official investigations followed the 2003 ACT fires. Of these the McLeod Inquiry was the most comprehensive. McLeod, a former Inspector-General of Intelligence and Security, and former ACT and Commonwealth Ombudsman, delivered a reasonably balanced if soft report which would have pleased the ACT's Stanhope Government.

McLeod's first recommendation centres around the question of fuel management. He says that the ACT authorities were too slow during the first two days 'when the fires were most amenable to extinguishment'.⁴⁹ While he is quite clear that, 'Increased emphasis should be given to controlled burning as a fuel-reduction strategy', and that the public should be educated 'about the beneficial and protective aspects of fuel-reduction burning', he is far from embracing the 'burn, burn, burn!' approach.⁵⁰ He also recommends the establishment and maintenance of fire trails both inside and outside national parks. He supports the use of aerial operations against bushfires especially in conjunction with the New South Wales RFS. He calls for the establishment of a new operational command and control facility for the ESB and reorganisation of incident command and control. It should be noted, given the unjust lambasting that ESB received from some in the ACT community, that McLeod says: 'Criticism should not be levelled at staff at ESB for the loss of control and confusion that occurred at the height of the fires during the afternoon of 18 January. They were battling against impossible odds and, despite being completely overwhelmed, they struggled on.'[51]

While his recommendations are sensible and moderate, what McLeod completely fails to take into account is the very specific circumstances in the western suburbs of Canberra on 18 January. They were surrounded by pine forests, which had been originally planted in the late 1920s. Lane-Poole, the Commonwealth Inspector General of Forests, had told Stretton in 1939 that, 'The policy is to convert the whole of the useless scrub eucalypt timber . . . into plantations of exotics' in order to pay for ACT fire protection! 'Pine is not so inflammable [as eucalypt]. That is definite. You have got a much better fire break with pine than you have with eucalypt.'[52] While this is clearly wrong, it needs to be remembered that the pines were planted when Canberra was only tiny. There were no suburbs anywhere remotely near the pine forests. It was the expansion of Canberra in the decades from 1960 to 1980 that led to houses being built directly across the road from such the pine forests. The fact that houses were allowed to be built literally across the road from forests beggars the imagination. The real responsibility lies with the development authorities, successive ACT governments, and with those residents who claimed that their quality of life demanded the proximity of these forests. In a reasonable, but unfortunately timed opinion piece in the *Canberra Times* on 14 January 2003, five days before the fires, the economic, recreational and even environmental virtues of the pine forests were outlined in contrast to eliminating them for housing development.

Subsequent to the fires there has been much debate as to what should replace the pine forests. The report *Shaping Our Territory* recommends that east of the Murrumbidgee a series of forest parks and an arboretum be set up, along with urban development. West of the Murrumbidgee, in the lower Cotter water catchment area, the Stanhope Government declared it planned to replant the whole area with industrial scale radiata pine forestry, despite strong opposition from the scientific community and the public. However, after twelve months of doing nothing the ACT Government has come to its senses and at the time of writing it has more or less conceded that the only hope for the area is replanting with native vegetation. Experts like Professor David Lindenmayer told the *Canberra Times* on 13 May 2006 that 'If the Government is visionary and prepared to invest in that option, it'll be a winner'. Who knows if politics and vested interests will get in the way of this vision, but for the moment it looks as though the Stanhope Government has seen the possibility and will end commercial forestry in the area.

In his concluding remarks McLeod asks an extremely important question: were the 2003 fires historically unprecedented? This question is important because, according to McLeod, it colours judgements not only about responses to the fires but, much more critically, it influences the judgements we make about the causes of the 2003 fires.[53] If 2003 was a one-in-50 or 100, or even 200-year event, then we will have a different view of how these fires came about and how we should respond to them, than if we were to have similar fires every twenty years due to a lack of regular burning. A major problem assessing this is that we do not have an adequate historical perspective through which we can compare bushfires.

McLeod's report reflects this. Early on he gives a potted history of European fire in the ACT. At the conclusion of this he says that 'the 2003 fires were not a one-in-one-hundred-year event'.[54] He reaches this conclusion by arguing that 'bushfires are a natural part of the Australian environment' and that 'There appears to be some substance behind the proposition that the longer the period since a major bushfire, the more severe a bushfire is likely to be when it does happen', a proposition that might seem obvious, but is not necessarily correct.[55] Forests are continually recycling forest litter and regular burning interrupts this process and encourages the growth of very combustible shrubs, bushes and undergrowth. McLeod refers to the work of Joan Webster and Luke and McArthur on bushfire cycles, according to which exceptionally bad bushfires occur every 22 years.[56] While there is some evidence to support this, a lot depends on what is meant by 'exceptionally bad bushfires'. No one would argue that there have been serious fires in the ACT on a reasonably regular basis. But none as severe as 2003. McLeod himself provides the maps that demonstrate this. In Appendix E to his report he sets out eight maps of previous fires between the 1919–20 season and 2001–2 season.[57] None of these fires came even close to the extent of the 2003 fires. The nearest were in 1925–6 and 1938–9.

McLeod correctly says that 'the 2003 fires led to a larger footprint than any of the previous major fires in the last eighty years'.[58]

But it is not just a question of 'footprint'. The fact is that it was a much larger conflagration, that it behaved differently, and all of the evidence from trained and experienced observers is that somehow 2003 was 'different'. As McLeod himself points out, even in its early days the 2003 fire behaviour was 'unpredictable'.[59] This fire had unique characteristics: the convection columns above it eventually reached a massive 15 kilometres in height. It was driven and accompanied by a tornadic wind. One observer, Ian Macarthur, in a chopper fighting the 2003 fire says, 'I was seeing flame heights two to three times the height of trees, but what impressed me was the flame depth—between 50 and 100 metres'.[60] Most people were unable to say precisely how it was different, but they were certain that the 2003 fires were unique. The reason is simple: such fires only occur rarely. The only parallel is 1939. Even McLeod himself concedes that the 2003 fires were 'unique'.[61] Yet he still concludes in an amazing non sequitur that 'it is the view of the Inquiry that it would be misleading to regard the event as a one-in-100-year occurrence', although he had said in the previous sentence that this 'was probably the most severe fire experienced in the region in the last 100 years'.[62]

Even allowing for the fact that paucity of sources makes the reconstruction of the history of fires in the ACT region difficult, there is little doubt that 2003 was probably the most severe fire for a couple of hundred years. The ACT ESB are certainly right when they say that 'the scale and impact of the [2003] fires were "well beyond anything seen before in the ACT" '.[63] There is no doubt 2003 was the biggest fire since 1939. Certainly in terms of area burnt 2003 was the most extensive fire since the arrival of Europeans in the ACT region. The dendochronological record in the Brindabellas indicates a considerable increase in fires from about 1860, the period that widespread European pastoral activities began in the mountains. Major studies were completed by L.G. Pryor in 1939 and by J.C.G. Banks in 1974.[64] Banks says that:

> For . . . the ridge top along the Brindabella Range, the study identified that over the past 220 years there had been 100 fire

years . . . Pre-1860 there were only twelve fire years giving a mean fire-free interval of 7.1 years, post 1860 there were some 88 fire years with a mean fire-free interval of only 1.3 years. This represents a 5.5 fold increase in the fire years frequency after 1860.[65]

The conclusion is inescapable: European settlement means more fire. But what Banks' study does not indicate is the extent and intensity of these fires.

Another part of the problem is the political reality of separate jurisdictional regions and inquiries. To separate the Brindabellas and the ACT from the context of southern New South Wales, Victoria and south-eastern Australia is to introduce a totally artificial division that is not there in nature. The ACT is geographically part of a much larger region and the ACT bushfires cannot be separated as though they were somehow unique. Seen in a bigger context the Canberra fire was really the northern edge of the massive fire in New South Wales and Victoria discussed in the previous chapter.

Finally, a brief note on the ACT Coronial Inquest into the 2003 fires conducted by ACT Coroner, Maria Doogan. This investigation has been beset with problems and its objectivity has survived a Supreme Court challenge. Its conclusions were not available at the time of writing, and the legal entanglements in which it has been enmeshed do not augur well for any information emerging from it which has not already been covered in this book.

PART 4

THE GREAT FIRE DEBATES

TO BURN OR NOT TO BURN?

i

The Aboriginal fire debate takes us into very contentious territory. The essential problem is that it intersects with two much-debated issues: the first is the treatment of Aborigines since 1788, and the recognition of their rights and position in contemporary Australian society. The second is the relationship between modern Australians and the landscape they inhabit. The question of how one views traditional Aboriginal attitudes to fire intersects closely with these issues.

Fire was a central element in Aboriginal life simply because it has been an essential component of the Australian landscape for aeons. Dean Yibarbuk, a Maningrida man from Arnhem Land, says that traditional people 'burn to hunt, to promote new grass which attracts game, to make country easier to travel through, to clear the country of spiritual pollution after death, to create firebreaks for later in the dry season and a variety of other reasons which "bring the country alive again"'.[1] In the Northern Territory Aboriginal people are being encouraged to advise and participate in attempted reconstructions of traditional burning patterns. 'Aboriginal fire management' is the new catch-phrase.

Before examining the interconnections between Aborigines and fire, we have to go further back into the history of the landscape to see the intimate relationship between fire and the omnipresent *Eucalyptus* and the other dominant species, *Acacia*, the wattles. As Australia separated and drifted north over the last 40 million years from the original super-continent Gondwana, it very slowly began to dry out. At the same time the eucalypts began their long march to dominance, probably originally as weeds on the edge of widespread rainforests. The cool temperate *Nothofagus* beech forests, characteristic of Gondwana, maintained their dominance until major climate changes about 5 million years ago. The atmosphere cooled down and dried out, and there is evidence that there was an increase in wildfires between 5.1 million to 2 million years ago. During the recurrent ice ages the *Nothofagus* rainforests gradually disappeared from most areas on the mainland, although they survived in Tasmania and Victoria's Gippsland. The cool temperate rainforests expanded or contracted according to wet or dry cycles. However, in southern Western Australia these rainforests disappeared completely, probably because they were too far from remnant rainforests elsewhere to be re-seeded when the right conditions returned.

The eucalypts and acacias gradually assumed dominance across the landscape. They were well adapted to a dry climate and intermittent drought, with extensive root systems that are efficient in finding water and moisture in the soil. Not only could they live in dry conditions, they developed a special relationship with fire, which depended on lightning for ignition. An intimate interrelationship between weather, fire and a eucalypt-dominated biota determined the future of the Australian landscape. Many eucalypts have thick bark which insulates the living structure of the tree from the heat of fire. Some eucalypt species only release seed after a fire and the seeds germinate in the remnant ash-bed. They grow upward quickly, competing with each other, but without competition from other plants. So for several million years before the arrival of humans fire was already a central element in a plant biota dominated by eucalypts.

By the time Aborigines arrived some time between 40,000–60,000 bp the eucalypts and acacias monopolised most of the forest landscape, except for remnant *Nothofagus* rainforest along the Great Dividing Range and Tasmania. There is debate about the exact arrival time of humans on the continent. Archaeologist John Mulvaney considers that the evidence is not there to support arrival dates of 50,000–60,000 bp as some argue. Writing in the 1999 edition of *The Prehistory of Australia* he says, 'An early landfall in Australia or New Guinea may be discovered, but it remains hypothetical today'.[2] However, another historian, James Kohen, dates human arrival at between 50,000–60,000 bp, and David Horton, editor of the *Encyclopaedia of Aboriginal Australia*, says it was about 50,000 bp.[3]

ii

The post-1788 settler assumption was that these early arrivals and their descendants were primitive hunter-gatherers who wandered across their tribal lands eking out a basic existence, lacking the technical expertise to manipulate or impact upon their environment. It was not until the late 1960s that this view came under serious challenge, parallel with Aboriginal people themselves demanding greater justice from white society. Politics and history intersected. Norman Tindale was one of the first to question the dominant settler view of Aboriginal primitiveness. In 1974 he argued that the first humans on the continent were not passive but used fire as a farming tool to create and open up grassland. Tindale held that none of the original vegetation remained because of Aboriginal activities. As a by-product, many large marsupial species became extinct.[4] Tindale's theory was developed by Rhys Jones who argued that Aborigines used the fire-stick as a skilful 'farming tool', modifying the landscape as they went. Jones said that they increased the productivity of the land, replacing the mature closed forests with grasslands and open, park-like woodlands. Aborigines also 'farmed' kangaroos through the use of fire, although as David Horton has pointed out, archaeological evidence shows that kangaroos were actually an insignificant element in the Aboriginal diet.[5] Jones argued that there has never been a completely 'natural' environment nor any 'wilderness' (that is, landscape unoccupied and unmodified) since human arrival in Australia. Thus, in Jones' view, the whole of the Australian landscape was an artifact.

Several other scholars have followed this line. Sylvia Hallam's work is based on the records of the early settlers of Western Australia and she argues that they often noted that the Aborigines used fire to farm the land and that it had a benign effect.[6] Eric Rolls went further, claiming, 'It seems impossible to exaggerate the amount of burning in Aboriginal Australia'.[7] He quotes the first New South Wales Surveyor-General, Thomas Mitchell: 'Fire, grass, kangaroos, and human inhabitants, seem all dependent on each other for existence in Australia, for any one of these being wanting, the others could no longer continue'.[8] Mitchell described open forest lands near Sydney which he claimed were created by Aboriginal burning.

The theory of fire-stick farming has become the 'new orthodoxy' and the argument has been adopted by many who want to support Aboriginal people by maintaining that they were successful and thoughtful environmental managers of the landscape. It is also used as an argument by those who maintain that the legal notion of *terra nullius* was a fiction conjured up by the settlers to support occupation. If you 'farmed' the land you obviously owned it. According to Horton, 'The politics of the situation crystallised when anthropologists, on the side of the angels in the fight for Land Rights for Aboriginal people, began to see Australian hunter-gathering as a form of farming, and the Australian landscape as a managed landscape'.[9] This argument has been widely accepted by scientists and lay people.

In a variation on the Rhys Jones theme, Josephine Flood's *The Archaeology of the Dreamtime* (1983) maintained that human arrival in Australia more or less paralleled the decline of the megafauna, the large marsupials that once populated the continent, such as the massive wombat-like *Diprotodon optatum*, the kangaroo-like *Procoptodon goliah*, and the tiger-like marsupial carnivore, *Thylacoleo carnifex*. Simultaneously, several large birds and reptiles also died out. Flood concedes that climate change and the great dry at the end of the Pleistocene, an intense ice age that began about 20,000 bp and lasted until just after 10,000 bp, was probably the last gasp for the megafauna, but she says that the main period of extinction preceded this. Since the only new element in the equation was humankind, she says that Aborigines 'may have caused the extinctions by a combination of hunting and . . . use of fire, which drastically reduced the animals' habitats'.[10]

Elements of the Flood argument have been restated by Tim Flannery, in his widely read book, *The Future Eaters* (1994).[11] His basic contention is that a kind of 'blitzkrieg' occurred when humans first arrived. Like everyone penetrating an alien landscape, the first Australians went through a stage of eating their future until the extinction of their food sources forced them to live more harmoniously with the environment. They were responsible for overkill of many of the megafauna, which for millennia had grazed on the trees and shrubs and fertilised the ground with their droppings. Their demise led to a tremendous build-up of plant fuel in the environment, as well as an impoverishment of the soil. This build-up led to massive fires.

Flannery maintains that the first settlers only gradually worked out how to live in harmony with the environment. Part of the process was that they slowly adopted a fire management strategy that involved a complex process of regular burning that rendered the countryside almost 'park-like' with the wide open woodland described by the first Europeans. In other words, the fire-stick farming hypothesis. He argues that Rhys Jones was mounting 'a serious challenge to the concept of *terra nullius*, for he saw the Aborigines farming the land, albeit through the use of fire'.[12]

Flannery's views have been widely discussed in the media and scientific community. The majority of scientists disagree with him. Most consider that climatic factors rather than 'future eating' were central to the extinction of the megafauna. Flannery returned to the issue recently in the *Quarterly Essay*, 'Beautiful Lies'.[13] He repeats the basic themes of *The Future Eaters* and he paints a picture of an Australian landscape as 'a vast, 47,000-year-old artefact' shaped by hunting and fire. But with white settlement 'the pyrophobic Europeans took control of the fire-stick and the vegetation changed'.[14] As a result, he argues, bushfires became devastating and frequent in south-eastern Australia.[15]

iii

There are a number of assumptions operative in these arguments. The first is that Aboriginal people were identical across the continent and that they followed the same fire regime even in widely different landscape settings. No account is taken of local variations. It is also assumed that Aborigines occupied the entire continent. In addition, to call the settlers 'pyrophobic' is very misleading. The historical evidence over the last 225 years is that Europeans were the opposite: they were pyromanic, especially after they realised that fire was a potent means of clearing forest and bush. Flannery doesn't acknowledge the active role settlers played in deliberately lighting fires in order to 'tame the wilderness' and to force upon an unwilling landscape the agricultural practices of Europe.

The response of the then New South Wales Environment minister, Bob Debus, to the Flannery essay brings us back to the argument about contemporary hazard reduction regimes. Debus argues persuasively that the key issue is weather, and that 'extreme fire events coincide with extreme weather events'. He points out that we now face global warming and that the actual number of days of total fire ban are increasing. 'Add in the fact that south-eastern Australia has been experiencing the worst drought on record and that [in January 2003] there was an unusually high number of dry lightning strikes', and one can begin to comprehend the conditions that led to the 2003 Alpine fires. Debus points out that the Australian Alps 'were one of the areas we know Aborigines specifically didn't *ever* burn extensively or frequently'.[16] Environmental historian William Lines is right when he says that 'Aboriginal occupation only lightly touched the environment and did not fundamentally alter the natural fecundity of the land'.[17] Flannery's theories are perfectly legitimate within the context of scientific discussion. Knowledge advances as theories are developed and critiqued. But the problem is that because of his profile his ideas are prominent in public discussion and the media. Important nuances are lost when one is talking about events buried deep in history and sophisticated discussion quickly degenerates as theory modulates into 'Facts' that are then used to promote particular political and economic arguments.

As a result the fire-stick farming theory is now used by people with a pro-development ethos whose aim is to exploit as much of the landscape as possible for economic gain. The political implication of this is that what was a dispassionate scholarly discussion has now become a minefield. Jones, Flannery and others did not create this problem, but their ideas have been seized upon 'in order to legitimise intentions which have long been [out] there'.[18] What has happened is that a coalition has formed including cotton- and grain-growers, beef producers, loggers and big agribusiness, many of whom favour regular burning and large-scale land clearance of so-called 'regrowth'. This coalition has used the fire-stick farming hypothesis to promote extensive burning as a hazard reduction tool, as well as justifying the 'thinning out'—that is, logging—of old-growth forests on the grounds that these are not part of the 'natural' ecosystem. Their argument is that if we can recover the Aboriginal way of 'managing the landscape' we will find a convergence of modern theories of land-clearing and regular burning with pre-European Aboriginal fire regimes.

This is precisely the argument of the pamphlet *The Australian Landscape—Observations of Explorers and Early Settlers* published by the Murrumbidgee Catchment Management Committee.[19] This pamphlet extracts scattered observations from the writings of early Europeans and argues that many of them mention smoke, fire, or conflagrations of some sort, concluding from this that Aboriginal people were constantly lighting fires. *The Australian Landscape* argues that as a result of Aboriginal fire husbandry the whole feel of forested areas was of an open park, rather than a thick, closed forest. 'To Aborigines fire was seen as necessary to clean up the country. They regarded unburnt forest or grassland as being neglected . . . It was seen as doing their duty by the land.'[20] In order to prove this the pamphlet is made up of a tissue of quotations from disparate historical sources of varying significance, loosely arranged around the themes of fire regime, forests, and regrowth. The essence of the argument is that the extracts and comments from the first Europeans in Australia indicate that the countryside was 'a park like landscape of grasslands and grassed open forests with very few areas of thick forest'. The pamphlet claims this was the result of regular, wide-scale Aboriginal burning. It is Hallam and Rolls' argument applied Australia-wide. *The Australian Landscape* says that when the Aborigines were displaced and their fire regime ceased, there was a massive regrowth of thick forest and scrub understorey that destroyed native grasses and created the danger of regular wildfires. Also 'the widespread ringbarking that was carried out around the turn of the [20th] century was mostly of this regrowth. The landowners were attempting to re-establish the original grazing capacity.'[21] Fundamentally the pamphlet argues for the maintenance of broadacre farming, clearing, logging, and widespread burning.

While the text of the pamphlet plays the cultural and environmental card by claiming to return to Aboriginal practices, its whole approach to the past is naive. Quotations from historical texts cannot simply be taken uncritically at face value. Texts come from a context and must be understood accordingly. John Benson and Phil Redpath have already shown in a detailed and persuasive article that the quotations in the pamphlet have been wrenched out of context and do not support the conclusions attributed to them.[22] There are other historical problems with the pamphlet. The early settlers would have had little or no experience of even small fires in their home context, so that every fire in Australia would have seemed 'big' to them. The fact that they often report fires merely indicates that they thought there were a lot, but that really tells us nothing about the regularity, size or significance of these conflagrations. Also implicit in the pamphlet, as in the whole argument about Aboriginal burning, is a presumption that we can comprehend what the Aborigines were doing, and then equate that with what Europeans did and do. No one doubts that Aboriginal people lit fires for different purposes, a few of which are clear to us, most of which are not. What is lacking is an acknowledgement of the immense epistemological, cosmological and spiritual divide between traditional Aboriginal culture, and the technologically advanced, but spiritually and ethnologically blinkered eighteenth- and nineteenth-century settlers. The whites were ignorant of the ecology of Australia and their assumptions of racial superiority did not encourage them in understanding the Aborigines, let alone help them penetrate the meaning, function and significance of Aboriginal practices. Their logic, philosophy and cosmology was unique to them. The rapid and catastrophic collapse in the Aboriginal population in eastern Australia due to imported diseases also meant that settlers had no chance of coming to comprehend the meaning of Aboriginal practices. Thus reconstructing accurate pre-1788 Aboriginal views of the world and their place in it is extremely difficult. It is even more problematic for us at this distance to comprehend, on the basis of scattered, uninformed settler observations, what significance fire had and how extensively it was used by Aboriginal groups that are now extinct.

Of course they used fire, but we have no idea how frequently particular areas of the continent were burnt, to what extent, and for what purpose. We know it was used as a weapon against other tribes and whites, and to assist hunting. We also know that large tracts of eastern Australia, including dense forests, were only rarely penetrated by Aboriginal people. The alpine areas of New South Wales and Victoria were largely left alone except for moth-gathering and spiritual reasons. There was no broad-scale or frequent burning in the alpine areas prior to white settlement. To suggest that Aboriginal people wanted 'to clean up the bush' in the areas they occupied seems to be a psychological projection, almost as though they were adherents of the Victorian assumption that cleanliness is next to godliness! The notion of 'cleaning up' is an unconscious projection of white discomfort over the apparent 'untidiness' of the Australian landscape. The absurdity of the situation is further highlighted when we use technological jargon to talk about Aboriginal burning. The use of terms such as 'resource management tool' as applied to Aboriginal practices is entirely misplaced.[23] Such jargon ignores the complete disparity between postmodern managerial–technocratic rhetoric and pre-1788 Aboriginal culture. Aboriginal people did not 'manage' the landscape. Their connectedness to the land and fauna was localised, spiritual and relational. They were too civilised to see the world merely as a food source. It was this, but it also had transcendent, sacred significance, concepts that are meaningless to postmodernist secularists. Their understanding of the landscape was, to use our terms, theological, not agricultural or managerial.

Similar problems arise when it is asserted that Aboriginal people were conscious and dedicated environmentalists who lived in complete harmony with the continent. Again, the use of modern terms like 'environmentalism' to describe traditional Aboriginal relationships with the landscape is deceptive. They were not 'environmentalists' in the modern ecological sense; this is a purely modern phenomenon. The Aboriginal religio-ritual relationship to specific places in the landscape should be defined in traditionally totemic and spiritual, not in modern scientific terms. Inherent in much of this discussion of Aboriginal people as 'environmentalists' is an idealisation of traditional society and a generalisation that all Aborigines across the continent were the same over many millennia.

This argument is not only about the cessation of Aboriginal burning. What is often ignored are the fire impacts of the settlers and the introduction of a kind of European pyromania. The landscape has been further altered by settlement and clearing, and the introduction of feral floral and faunal pests. Widespread vegetation clearance has meant that the soil is unable to hold moisture. This has been compounded by the compacting effect of hard-hoofed animals, such as sheep, cattle and horses. Constant and excessive burning has led to erosion and the loss of topsoil. The loss of small animals has also been a factor in making the landscape vulnerable to fire. In his book *Heartland: The Regeneration of Rural Place*, George Main points out that, 'The widespread extinction of small mammal species worsened the [bushfire] situation. A brush-tailed bettong performs up to a hundred diggings every night. One bettong can churn six tons of earth annually. As they forage in the ground, bettongs and other creatures bury great quantities of fallen bark, leaves and other combustible materials.'[24] Several species of these small animals, such as bettongs, bandicoots and bilbies, that were once common across many parts of Australia, are now rare, endangered, or extinct.[25] As a result we have largely lost their contribution to the process of breaking down waste vegetation and making it less susceptible to fire. The historical evidence overwhelmingly indicates that the most widespread and disastrous fires in Australia began in the decades between 1830 and 1850 with settlement and squatting, reinforced by the impaction of the soil by hard-hoofed animals, the loss of fire-adapted Australian flora, and the widespread introduction of feral plants and fauna.

Drought, too, has played its role. Certainly drought was a reality prior to 1788. Michael McKernan in *Drought: The Red Marauder* records Aboriginal recollections of a terrible drought in the Murrumbidgee–Murray river system in the years around 1730–40. Have droughts increased since 1788? While some say that they have not increased, McKernan argues that settlement probably did increase drought, 'as the land's natural defences surrendered to the greed, over-grazing, inexperience and ignorance of the first settlers'. He points out that 'The Mallee was as bad as it was by the 1930s because the removal of native vegetation exposed its soils to the elements. Dust storms . . . became a pattern of most droughts of the twentieth century.'[26]

So it would be impossible to re-establish a traditional Aboriginal fire regimen, even if we knew what it was. The impact of European settlement has been so ecologically horrendous that much of the traditional landscape has been lost. William Lines is right when he says: 'Nowhere else on the earth have so few people pauperised such a large proportion of the world's surface in such a brief period of time. In under 200 years a natural world, millions of years in the making . . . vanished before the insatiable demands of a foreign invasion.'[27]

A variation on the fire-stick farming hypothesis is propounded by Professor David Bowman of Charles Darwin University.[28] He correctly asserts that the argument about prehistoric human impacts on the Australian environment is irrelevant because even if we knew the details, we cannot recreate the past. He says that our challenge is to preserve contemporary biodiversity. Attribution of destructive impacts 'to the arrival of Aborigines in Australia may be historically interesting, but such preoccupations deflect attention from pressing conservation problems'.[29]

Bowman is in no doubt that at the time of settlement skilful Aboriginal burning had created the landscape that the whites lusted after and seized. While we probably cannot recreate traditional patterns in the south-east and south-west, in places where traditional life still survives, such as the Northern Territory savanna, it is essential to involve older Aboriginal people in research on the fire ecology of the various ecosystems—something that Bowman himself has long worked to bring about. He concludes that, 'It is wrongheaded to ignore the ecological impact of a long history of Aboriginal burning. In Australia ecologists cannot retreat to the "wilderness" to study an archetype of nature because the "wilderness" has long included people.'[30]

The only place in Australia in which one might be able to make a plausible argument about the use of an Aboriginal fire regime is the Northern Territory, but even there it would rely on two suppositions: first, that there is an accurate and reliable account of traditional

Aboriginal fire practices; and second, that this burning regime is the best adapted approach for maintaining biodiversity. While there is no doubt that the Aborigines before 1788 reached a genuine equilibrium with the landscape, this does not mean that everything they did was necessarily correct. An uncritical exaltation of Aboriginal fire practice as the unchallengeable repository of all environmental wisdom flies in the face of the fact that we know that all humans are fallible. Aboriginal views on particular landscapes should be taken very seriously when accurately known, but that doesn't mean that all the evidence should be accorded equal weight.

Bowman has also attempted to evolve a theory as to why Australia is so fire-prone. He asserts that it has nothing to do with Aborigines. It all started with monsoons in northern Australia. 'This is the most lightning prone place on earth. We've got a dry season, which is characteristic of the monsoonal climate and we've got lightning. That's got to equal fire.'[31] Bowman's theory is that the eucalypts gradually took over the continent from the north using fire as their weapon. They were predator plants, full of flammable material, that spread conflagrations, destroying all flora that was not fire-adapted, especially rainforests. 'The fires which were started by the on-set of the monsoon millions and millions of year ago have ultimately come to create the whole of the Australian landscape and the vegetation, animals and plants.'

Many scientists do not agree. They argue that the key issue is climate change, with increasing aridity in Australia from about 5.2 million years ago. Everyone agrees that there is an intimate relationship between the eucalypts and fire, but they see the great dry moving from the south to the north rather than predator eucalypts moving the other way.

In regard to the south, fire scientist Jon Marsden-Smedley is investigating the traditional patterns of burning in south-west Tasmania. He faces a much more difficult task than Bowman, who has access to local Aboriginal people who are still reasonably close to traditional tribal practices. Marsden-Smedley argues that for millennia Aboriginal fire has decisively shaped the ecosystems and landscape of south-west Tasmania. From the observations of the first European explorers to penetrate the area he concludes that Aborigines occupied much of the area at least seasonally, and that they used fire to open up the country.[32] He says they would have been 'mostly low intensity fires under conditions when forested vegetation is too wet to burn . . . [They] would self-extinguish, although in some cases it appears that fires would have been beaten out using branches.'[33] Marsden-Smedley argues that, while we need to control wildfires, there is also a need to recover Aboriginal fire practices in the south-west because the vegetation is adapted to low-intensity burning and needs fire to avoid ecosystem degradation.[34]

iv

The most passionate and convincing response to the hypothesis of fire-stick farming is that of David Horton in *The Pure State of Nature* (2000). He is also the general editor of *The Encyclopaedia of Aboriginal Australia* (1994). *The Pure State of Nature* is a realistic assessment of what has happened to the Aboriginal fire debate. He says that by adopting the fire-stick farming hypothesis scientists, historians, archaeologists and Aborigines themselves are actually lining up on the side of economic rationalism and saying that 'you can conserve and protect the environment by developing and exploiting it'.[35] He correctly points out that this is a complete misreading of traditional Aboriginal attitudes, and that we already know the environmental consequences will be disastrous. Horton's view is that, 'Aboriginal society and culture and religion combine to ensure . . . that Aboriginal use of the environment results in that environment staying exactly as it is for all time—to "defeat history", in [W.E.H.] Stanner's words, by becoming "a-historical in mood, outlook and life"'.[36] Immutability rather than change was the traditional Aboriginal ideal. A European notion of linear time and a Whig notion of history as endless 'progress' was utterly alien to Aboriginal people. Traditionally they held that time was omnipresent rather than chronological, that the Dreaming existed in a kind of platonic eternity, a timeless world without duration, without a 'before' or 'after', a state of changelessness that set the pattern for life now. The Dreaming was not merely time past, but it impacted intimately and immediately on the here and now. Each individual and group was and is an incarnation of an aspect of the Dreaming in a specific place and time frame. Bill Stanner comes closest to the meaning of the Dreaming when he says that it is 'everywhen'.[37] There is no chronology nor history as Europeans understand it, for all time is accessible to time now and in fact is time now. Within this overarching context of timelessness there are the small chronological cycles of individual and clan lives and intergenerational relationships. However, these are secondary and less significant than the overarching presence of the Dreaming. Sacred places in the landscape and the seasonal round of ceremonies are essential contact points between the present and 'everywhen'.

This is in profound contrast to a western secular understanding that simply assumes change and development are signs of civilisation. And the faster the change, the better. Certainly Aboriginal society has changed over 40 or more millennia, for no society is completely static. But that change will have been slow and imperceptible to those experiencing it. As Horton says, from the Aboriginal perspective they 'not only believe that they have always been in Australia, but that they have always been in the particular place that their ancestors came from. They also say that their customs, beliefs, practices have all been handed down unchanged.'[38] The result is that many contemporary Aboriginal people live in a dual world, claiming on the one hand that they have been here in this particular place 'for ever', and on the other hand they are forced to use archaeology and anthropology to support their claims for land rights, even though their essential cultural and religious claims de facto logically exclude the use of western science.

Given these ingrained attitudes about the Dreaming and timelessness, there is no way that traditional Aboriginal people set out to manipulate the landscape in the way of modern 'managers'. They observed the world around them, understood its rhythms and developed a detailed knowledge of the way it worked. Above all they appreciated its religious significance and meaning and saw themselves as part of its dynamics. They did not interfere with it in the way of agriculturalists, they did not turn Australia into an industrial and artificial landscape like the Europeans, who have manipulated their continent since the Middle Ages. As Horton shows, Captain Arthur Phillip describes a complex ecosystem around Sydney Cove and inland from the harbour, one that had been created more by geology, soil, weather and the lie of the land than by burning.[39] This is not to say they never used fire 'but they fitted into the natural Australian fire regime and their use of fire has had little, if any, effect on vegetation'.[40] He also shows convincingly that once the fire-stick hypothesis got into popular political and cultural discourse, even intelligent people went looking for evidence of it. There is always the danger that those who start looking for evidence will unwittingly 'adjust' reality to suit their particular theory.

Nevertheless, there are contemporary Aborigines, like Professor Marcia Langton, who argue that the whole continent has been 'used [that is farmed] by indigenous people for thousands of years'.[41] While not specifically nominating the fire-stick farming hypothesis, Langton argues that the notion of 'wilderness' is as malevolent a concept as *terra nullius*. She says that the term 'wilderness' is 'a mystification of genocide', by which she means that 'where Aboriginal people had been brought to the brink of annihilation, their former territories were recast as "wilderness" '. These areas can then be turned into national parks to salve the consciences of white environmentalists 'seeking to establish a cultural security in landscapes full of colonial memories'. Langton's claim is that contemporary Aboriginal people have a 'right of stewardship of the land, its ecological systems and biodiversity'. She says that in opposition to this, environmentalists want 'to assert the supremacy of western resource management regimes'. Presumably Langton wants Aboriginal people to have powerful sway over what happens to national parks. This makes some sense in places like the Northern Territory where Aboriginal communities are still present on their land in reasonably large numbers. But it is much more problematic where this is not the case. It is also hard to see the connection between a concept of wilderness and the 'mystification of genocide'. What she seems to be arguing is that 'wilderness' is a land that has been deliberately emptied of people and then reclaimed as wilderness, whereas Horton and others would see it as an area that has not been manipulated and changed to suit human needs. Wilderness 'is landscape that has reached its potential without human interference'.[42] It doesn't necessarily presuppose the absence of people. Langton seems to be assuming that wilderness in this sense did not exist in Australia.

At the end of *The State of Pure Nature* Horton nails up 24 theses, in the style of Martin Luther's 95 theses nailed to the castle church door in Wittenberg on 31 October 1517. Horton says:

> Aboriginal use of fire changed nothing in the environment except in the sense of the short-term outcomes that follow any fire . . . If you want to practise control burning in order to protect houses or farms, then do it in the same way as you would use a bulldozer to clear a firebreak, but don't pretend you are doing anything but damaging the environment . . . The Australian environment of 1787 was not an artificial construct of human making that needed to be constantly interfered with. It was a natural construct of a long history, and the things we are losing now will not be recovered.[43]

Horton is not alone in repudiating the fire-stick hypothesis. There are other scholars who, while not going all the way with the theses of *The State of Pure Nature*, seriously question the notion of landscape 'farming'. One of Australia's most prominent scientists, Mary White, says that natural fires are inevitable in Australia, especially in the northern savannas, but she emphasises that we do not know how often these fires occurred in the past, nor do we understand the long-term ecological results of regular, deliberately lit contemporary fires. She is critical of the fire management across the tropical north and points out that there is a difference between what a traditional tribal Aboriginal on foot could do in terms of fire-lighting, and what a modern one can do dropping fire-lighters from a four-wheel drive. Her assessment of what is happening in the Northern Territory is forthright, and it applies right across the continent:

> Those who are not concerned with political correctness are increasingly admitting that Aboriginals used fire for many reasons and that conservation in the full sense of maintaining biodiversity and the sustainability of ecosystems was not within their vocabulary, and was not a deliberate intention. Today's wholesale burning of tropical savannas based on, and officially validated by, what amounts to the cultural connection between

Aboriginals and fire has to be seen as unjustifiable. It is not scientific or logical to assert that because Aboriginals burnt the land, deliberate burning must be right.[44]

v

Aboriginal fire regimes need to be questioned. It may be that they were well adapted to the environment and landscape. Perhaps they were not. Aboriginal people in the past, and now, are not infallible. Questions about their fire practices need to be asked. We will not achieve reconciliation by romanticising such practices as perfect. And we may well compound the destruction wrought by settler pyromania by using a supposed Aboriginal fire regime when the evidence is conflicting, even contradictory. Historical mistakes have a way of recurring.

There is increasing scientific evidence to support the conclusion that Aboriginal burning was far more limited than has been suggested. For instance, A. Malcolm Gill, Australia's most experienced fire researcher, speaking of central Australia, says that, 'The complete restoration of fire regimes as practised in traditional Aboriginal societies at the time of white settlement seems impossible... To achieve such a goal, traditional *mores* would have to be restored including traditional patterns of movement about the landscape and the hunting and gathering of traditional foods.'[45] A parallel conclusion has been reached through sedimentary charcoal research by Scott D. Mooney and his colleagues in the northern end of the Royal National Park, south of Sydney. It was found that there had been a much higher frequency of fires in the period after 1930 than before. 'This study suggests that Aboriginal burning... was relatively rare or consisted of small patch burning. In contrast, fire in the European period, particularly from the 1930s to *ca* 1960 was frequent. This supports the view held by Clark that "the frequency and areal extent of Aboriginal burning may well have been overestimated and European burning underestimated."[46] The researchers found that only one large fire seems to have occurred in the 1600 years prior to 1930.[47]

The most thorough repudiation of the use of explorers' and settlers' accounts to advance a theory of widespread Aboriginal burning is the work of John S. Benson and Phil A. Redpath. While conceding that some early Europeans' journals reveal 'that Aboriginal people used fire for cooking and burning the bush', they argue correctly that:

> the extent, frequency and season of their use of fire is largely unknown, particularly for southern Australia. Vegetation types such as rainforest, wet sclerophyll eucalypt forest, alpine shrub lands and herbfields, and inland chenopod shrub lands, along with a range of plant and animal species, would now be rarer or extinct if they had been burnt every few years. Furthermore . . . much evidence [has been ignored] that points to climate as being the main determinant in vegetation change over millions of years, with major changes occurring since the onset of aridity in the Miocene but continuing through the last ice age, which coincided with the occupation of Australia by Aboriginal people.[48]

Benson and Redpath conclude that modern fire management should be based on contemporary ecological understandings of species and their habitats, not on 'selective interpretations of some of the early explorers' observations'. The simple reality is that 'the main causes of change to Australia's vegetation since [1788] have been large-scale clearing and cultivation'—as well as deliberately lit fires.[49]

James L. Kohen in his *Aboriginal Environmental Impacts* (1995) attempts to find some middle ground between the two arguments about the impact of Aborigines on Australia. Kohen argues that there was a dramatic increase in the Aboriginal population from about 5000 years ago. Population increase inevitably impacts on the environment even within the context of a hunter-gatherer economy. He says the 'Aboriginal people did modify the vegetation, did have an impact on some animal species directly and did change the species they were eating and the technology they were using to exploit those species. All of these things took place while the Aboriginal population continued to grow.'[50] Kohen points out that a total collapse occurred in the Aboriginal population almost immediately after European settlement because of imported diseases. Following the research of Noel Butlin, he says that by 1800 'the vast majority of Aborigines in southeastern Australia were dead'.[51] As a result Europeans never really had a chance to observe traditional Aboriginal society. The rapid population collapse meant that traditional lifestyles were quickly abandoned. Can we discern what role burning played among traditional Aborigines? Before answering this, Kohen looks at fires ignited naturally in the landscape by lightning. He shows that many Australian animals are extremely well adapted to fire, including such apparently vulnerable creatures as lyrebirds. Observing this, Aborigines only burnt the sclerophyll forests. 'Woodlands and forests can be burnt regularly without any long-term decline in the fauna . . . It would be unwise to burn wet sclerophyll and rainforests because the animals in those vegetation associations tend to be specialists, not well-adapted to fire. If you want to protect those animals, you don't burn the rainforests.'[52]

So what does all this tell us? First, there is a real sense in which this debate about Aboriginal fire-stick farming and whether we should burn the bush or not could go on endlessly without resolution. There are two reasons: firstly, we simply don't know enough about traditional, pre-European Aboriginal fire practices. Secondly, the whole argument arises from two very different understandings of the position of humankind in the natural world and our cosmological relationship with it.[53] At one end of the spectrum are those who think that nature and the world exist to meet human needs and that in an almost Marxist sense we only become truly human by labour, by working to shape and manipulate the environment so that everything serves human purposes. Such people see everything in creation as existing to serve and support us. It was precisely because early settlers in Australia thought along these lines that they considered Aboriginal people lazy and despised them. The notion that they might not have wanted to 'farm' the continent in a British sense, and that they were content with a more spiritual and metaphysical understanding of their place in the scheme of things, remains abhorrent to those dominated by technology. Such people are determined to shape their environment, using all available means to achieve this. For them the D8 bulldozer is god. Putting it another way: if a person's *primary* concern is protecting property, stock, agriculture, forestry and other assets, and they believe, even unconsciously, in humankind's superiority over all of nature, if their classical biblical text is 'Be fruitful and multiply, and fill the earth and subdue it; and have dominance' (Genesis 1: 28), then they will want to manipulate the environment to the extent they can. Fire can be prevented by reducing fuel loads. Land must be managed for profit and the benefit of humans. Otherwise it is of no real value. Things like wilderness are meaningless, or at best secondary values. And the clinching argument has now become: 'After all, even the Aborigines *managed* the landscape'!

But if their *primary* concern is not totally focused on humankind and they think that biodiversity and the maintenance of the ecological structure of the landscape is important, then they are going to have serious questions and doubts about regular and broad-scale burning, and especially about the use of presumed Aboriginal fire regimes. They will recognise that Australia is one of the most fire-prone places on earth, and that we who live here need to adjust to its rhythms, rather than manipulating it to fit in with our needs and desires. Also, along with David Horton, they might suspect that those who want more controlled burning 'might not have the interests of the environment at heart. [These people] have always believed that the environment should be dominated and managed, and burnt constantly and frequently.'[54]

Despite these two positions there is some possibility of finding common ground, especially among people for whom the environment is a genuine priority. It is precisely this issue that will be examined in the last chapter of the book.

10

FIRE THUGS

i

One of the perennial responses to bushfires is a post-conflagration outburst of praise from politicians, community leaders and media for the generous, brave, dedicated firefighters who have laid their lives on the line. This is not to suggest that much of this praise is not deserved. But an unfortunate side effect is that adulation of this sort is highly attractive to the histrionic personality, the attentionseeker who craves recognition. What better way to become a hero than to light the fire secretly, then 'discover' and report it, then join the firefighters in the rescue?

This is precisely what Peter Cameron Burgess did. Aged twenty and unemployed, with a record for house-breaking, he heard the praise heaped on firefighters so he decided to get in on the act. He had already made several attempts to join the New South Wales Fire Brigade, but had been rejected because of lack of qualifications. He joined the RFS hoping to use that as a path to the fire brigade, but he was always a problematic member. Others started to suspect him when he was always the first on the scene of an outbreak which he himself reported. His fire-lighting career had begun in bush near the Hume Weir outside Albury. There was a break in his fire-lighting until he saw the praise heaped on New York firefighters during the 9/11 terrorist attacks. By this stage he was on the New South Wales Central Coast, and once again lighting, then reporting fires.

He moved to the Blue Mountains, joined the Glenbrook–Lapstone RFS and the same pattern occurred there. He was always the one to report a fire and then be on the spot in full uniform when the rest of his firefighting team arrived. Police were informed and Strike Force Tronto, the then New South Wales Police arson investigation unit, began monitoring Burgess. In order to get sufficient evidence to charge him, police worked with members of his RFS. 'We had our surveillance teams actually follow him from site to site. In the meantime . . . our forensic people were linking the crime scenes together . . . from the Central Coast down to Albury and then up to the Blue Mountains . . . we were able to link him into all those scenes.'[1] It is not unusual for arsonists to move around, although some prefer to stay in their own region. What was typical was the escalation in fire-lighting. Just before his arrest Burgess was lighting fires on an almost daily basis, reporting them on his mobile phone.

He was eventually sentenced in Penrith Magistrates Court to two years' jail after being charged with 25 arson offences. He pleaded guilty to lighting sixteen fires. Attorney-General Bob Debus was angry with the light sentence. Burgess should have been tried in the district court where he could have received a sentence of up to 10 years. He showed no sign of mental illness, and was fully cognisant of what he was doing. He said he was sometimes bored and fighting fires gave him a sense of achievement.

Burgess is a specific type of arsonist: his fire-lighting was the product of histrionic motives. He provoked a crisis in order to participate in its resolution. Arsonists such as Burgess tend to join volunteer units in order to participate in the excitement of fighting the fire and receiving the kudos accorded to the generous people who serve the community in the RFS.

Bushfire arson is a national problem. It costs the Australian community about $500 million per year.[2] But accurate statistics are hard to establish because arson is very hard to prosecute. An arson researcher, Richard N. Kocsis, comments 'the clearance rate for arson is remarkably poor'.[3] Nevertheless the majority of fires at the urban–bush frontier, or adjacent to accessible forests and national parks, are deliberately lit. However, there is a variation in the number of acts of arson from region to region and from time to time, explained by the fact that serial arsonists usually light many fires, and if there is one or more active in an area their fire-lighting will inflate the statistics for that area. If newspaper reports during emergencies are to be believed, deliberately lit fires constitute a very high percentage, sometimes as high as 90 per cent. This is probably an exaggeration that reflects infuriated public perceptions. On the other hand, it is hard to agree with Matthew Willis, writer on bushfire arson, when he says that 'Ultimately it is difficult to say with any certainty what proportion of fires are deliberately lit'.[4] Most natural fires begin with lightning and a thunderstorm is a pretty obvious event. Willis is critical of those who conclude that in the absence of lightning, arson is the most likely cause of fire. 'In the case of bushfires, firefighters and investigators will often develop a suspicion that a fire was deliberately lit through the absence of any other feasible explanation . . . Whether that suspicion is able to give rise to proof of deliberate ignition supported by evidence, though, is another issue.'

Certainly it is indisputable that it is hard to get evidence that will lead to a prosecution of a deliberate intention to light a fire, but that does not mean that the suspicion is unsubstantiated. This is especially true on the urban–bush fringe where arsonists are most likely to be operating. Fires originating in wilderness areas or isolated from human habitation are less likely to be deliberately lit because they do not fulfil the psychological needs to which arsonists are responding. If one includes in the statistics those who are not psychologically disturbed in the strict sense but who, for their own reasons, light fires in times of extreme fire danger, there is little doubt that up to almost two-thirds of all bushfires are deliberately lit.

Another question that Willis addresses is whether bushfire arson is on the increase. He says that, 'It is not clear whether the incidence of arson is increasing, or whether the greater attention paid to it by fire and police agencies is simply revealing the extent to which the problem always existed. Community awareness of bushfire arson has led to greater vigilance and to the community playing a significant role in the detection and apprehension of offenders.'[5] Probably in the past pathological fire-lighting tendencies were hidden in the general acceptance of communal pyromania. Another question is: how many arsonists, in the strict sense of people who light fires for malicious purposes, are actually out there? We don't know the answer to this question. Arson researcher Fabian Crowe says that, 'We have yet to develop the necessary knowledge and skill to improve the rate of detection. We learn a certain amount each time we identify the persons responsible . . . There is a very wide range of circumstances and underlying factors in the rural and forest environment that provide the stimulus for lighting fires.'[6] In other words, we still have a long way to go before we know what percentage of people are arsonists and understand their motivations.

ii

So what do we know? First, we know that there is a sharp increase in the number of fires, especially on urban–bush boundaries. Take the Blue Mountains, for example. This is one of the most fire-prone areas in the world with houses and towns sitting on the edge of steep escarpments, many of them westward facing, the direction of the prevailing winds. So fires are to be expected. The pattern used to be an outbreak about every six to seven years. But now locals expect fires much more often, sometimes every year. This increase has no natural explanation, except global warming. The only other conclusion to be drawn is that most are deliberately lit by arsonists. There are no national figures but in 1975 there were just over 1200 reports of arson; by 1995 there were almost 10,000. Some experts believe the police are lucky to catch one in 200 arsonists, and then only the stupid ones.

Second, the old-style rural pyromania is far from stamped out. There are still a lot of people who think that the bush needs to be 'cleaned up' and burnt regularly. They argue that a good burn never did any harm to the landscape. They despise 'Greenies' and environmentalists. They harbour resentment towards all ecological policies, the national parks, and any form of protection of public land, and see themselves as the true repositories of rural folk wisdom. Some will be attracted by volunteer firefighting. They are dismissive of the significance of fires that only burn the bush and do not affect property. As a result they often light fires without concern for the consequences. 'After all', they reason, 'who cares if a national park or some wilderness is burnt, as long as the fuel load is reduced and our assets protected?' Some set fire to nearby bushland to destroy 'pests' such as dingoes or native animals or birds. They hold the view that they are doing something worthwhile and are only breaking a penal law. All they will have to do is pay the fine *if* they get caught and prosecuted. And it is very hard to catch them. Some of them are powerful in their communities, and even if they can identify them, local police and fire authorities are often loath to prosecute them. Their behaviour is seen as little more than a quaint, minor breach of the law.

Nevertheless, such people, no matter what their level of respectability or their claim that they are managing the land in a 'traditional manner', are actually criminals and should be treated as such. As Fabian Crowe points out, their 'motives are not dissimilar to those of an arsonist. The only difference is that the intent does not include causing harm to life or property. To some, government land does not constitute property and in many instances there is a perception that any fire which burns forest or scrub is justifiable.' Crowe argues that this kind of atmosphere creates a breeding ground for anti-social arsonists and, 'Every fire that is not investigated is a lost opportunity to develop a better understanding of the psyche of firelighters and the triggers that cause them to light fires'.[7]

Third, there are anti-social, psychopathic arsonists like Burgess. He is typical in that he was a young, unemployed male. There are not many arsonists over 30 and very few are women. Researcher Talina Drabsch found that, 'In 1989, 85% of alleged offenders were male, 46% were juveniles and 50% were unemployed'.[8] Many arsonists are from stable but unhappy families. Nevertheless there are exceptions. On Christmas Day 2001, a man in his fifties with receding grey hair was seen lighting fires by a reliable witness in an isolated street in Valley Heights in the Blue Mountains. When challenged he jumped into his silver-blue Falcon, mumbling 'I'm trying to put it out'.[9]

Based on the work of a number of researchers, Drabsch distinguishes seven groups of people with arsonist tendencies or who suffer temptations to light fires. The first group consists of people riddled with anger and feelings of revenge: this is usually directed at an individual or, more likely, at an institution (such as a school), or the whole of society. Richard Kocsis supports the existence of this group and says 'some motivational element of anger is believed to underpin a high proportion of arson crimes'.[10] The second group comprises those who are dominated by a kind of attention-seeking which is derived from low self-esteem, or can be interpreted as a cry for help. A third group is made up of those suffering from mental illness, although most experts seriously doubt if there is such an illness as 'pyromania'. The fourth group is the many young children who simply experiment with fire out of curiosity. Adolescents with a poorly developed sense of self who light fires because of peer pressure comprise the fifth group while the sixth group is made up of those suffering intellectual impairment which results in committing arson without understanding the consequences. People using alcohol and drugs make up the seventh group: their inhibitions are loosened and other motives become more dominant. An eighth group should also be added to this list: those who act of sheer, deliberate, destructive vandalism.

The proportion of juvenile and child arsonists is much higher than for other crimes and seems to be rising. Young children between five and ten light fires out of curiosity, but 'Older fire setters usually set fires as a result of aggression, sensation seeking, social skills deficits, deviance, vandalism, covert anti-social behavior and attention-seeking behavior'.[11] Such young people seem to come from dysfunctional families where parents are often absent or abusive. However, despite attempts, it is difficult to profile juvenile arsonists. Fire-lighting might be a cry for help with physical or sexual abuse or neglect, it might be the manifestation of a young psychopath whose vandalism and lack of empathy for others is an early indication of their complete lack of conscience, it might indicate a child with cognitive difficulties or a lack of judgement, or a combination of any of these factors. Or it might be sheer bloody-mindedness.

There are other motives for arson. Some young men feel impotent and they light fires because it gives them a sense of power. Others are consumed with animosity and anger and fire-lighting gives expression to their rage. For some fire is associated with a sexual satisfaction, or the sheer thrill of seeing a wall of flame. Others are simply vandals whose one aim is to destroy, to get a malignant thrill out of fire-lighting. For some the motive is the concealment of another crime, such as burning stolen cars in the bush.

With the exception of younger children, all adolescent and young adult males know what they are doing, plan their actions carefully, cover their tracks, and often use sophisticated means to light fires so they are well away from the area before a conflagration occurs. Serial arsonists clearly understand the possible consequences for others of the fires they light, including endangering lives, homes, and other assets. Common factors motivating all of them seem to be the power they gain from their anonymity and their ability to manipulate society, and the thrill that comes from not being caught. Another common thread seems to be psychopathic tendencies, the satisfaction of anti-social behaviour, and a lack of feeling for other people, let alone for animals and the bush itself.

Yet we still don't know a lot about these people. They are very hard to catch in the act and to prosecute, and the few who are jailed and subjected to psychological assessment may not be typical. Fire researcher and consulting psychologist Rebekah Doley has interviewed many convicted arsonists. She commented in the *Sydney Morning Herald* on 7 January 2006 that the typical bushfire arsonist 'is of average to below-average intelligence'. The *Herald* continued:

> [They are loners] who find it difficult to integrate into groups successfully . . . [They are] usually unemployed or in a low-skilled occupation and have difficulties coping with day-to-day life and sustaining personal relationships. 'Fire is an effective outlet for his emotional distress or as a powerful weapon that he can wield without having to personally confront the victim', Doley says . . . 'In my exverience the common theme among them is power. Fire, inherently, is a powerful tool. Bushfire arsonists tend to be underachievers who are not getting what they want because they don't have the skills. Fire gives them

> that moment of power. It is a vicarious experience for them. Many stay to watch or they monitor the fire's progress through the media. They are extremely dangerous because they are unpredictable . . . They also tend to be serial offenders and, in my experience, most will not stop until caught. Like most arsonists, they light fires without regard for the consequences.

In fact, the most striking thing about bushfire arsonists is their malicious unconcern for others.

The Burgess story also raises the role of the media in arson. What are the consequences of 'talking up' bushfires on TV or in the newspapers? Referring to fires in the Grafton area in the 1993–4 season, Fabian Crowe says:

> The fires provided some spectacular television footage. The media coverage was widespread . . . Some of the fires were 'talked up' as being 'the worst since 1968.' In interviews firefighters and people affected by these fires provided emotional and colourful word pictures of rampaging, uncontrollable fire storms and stories of near misses and destruction. When three firefighters died on January 4, senior public figures spoke to the media using phrases such as : 'wonderful . . . heroic . . . volunteers who laid their lives on the line to fight the merciless blazes'.

These comments were followed by statements like, 'It is evident that many of these fires were deliberately lit . . . it is virtually impossible to detect people who light fires in the bush . . . The last thing we want now is a fire in the Blue Mountains'. Crowe suggests that arsonists string these disparate statements together in their minds and it triggers the impulse to go out and light more fires.[12]

Given that more than 600 wildfires occurred in New South Wales that season, Crowe's argument has persuasive force. His view is supported by Phil Koperberg: 'If we give it [bushfire] a lot of profile, such as you see in the visual electronic media, then those who from time to time are tempted to behave [as arsonists] . . . are provided with a motivation, if you like, to emulate that'.[13] Willis agrees with Crowe, but also points out that the media is also fast to scapegoat people and lay the blame for fires on either evil-minded arsonists or land management policies which are perceived as too 'soft' on green issues.[14]

Another example Crowe draws on are the 21 December 1997 Dandenong Ranges fires in Victoria which followed a fire outbreak the previous day on the Mornington Peninsula that received much media coverage. Crowe argues that this may have stimulated an arsonist to light the Dandenong Ranges fires which were ignited within two hours of each other, were lit to spread up-slope to houses on the ridge-line in heavy forest, and were in isolated locations. 'There is little doubt that the person responsible had knowledge of fire behaviour, of the local area and the potential impact.' Three people were killed that day at Ferny Creek when they made the mistake of sheltering in a cellar under a house that collapsed; 41 houses were destroyed and 45 damaged.[15] No one was ever charged. Certainly volunteer fire brigades want their work noticed in the media because good media coverage offers leverage for more funds and equipment from government. It is a sad commentary on state governments that firefighters might resort to this sort of pressure. But the danger is that this can provide a trigger for the arsonist making 'talking up' a fire event a self-fulfilling prophecy.

Kocsis also has commented on how little we know about bushfire arsonists; given the nature of Australia, 'it is surprising that research into this crime is not given far greater priority'.[16] But the police are underresourced and bushfire arson is particularly difficult to investigate. As a result police tend to give these investigations a lower priority. It is just too hard and the results in terms of prosecutions are minimal. Fire investigator Mitchell Parish says arson is not a high priority with police: 'As soon as a crime which is considered more serious—generally a crime against a person such as murder, assault, etc—the arson matter gets pushed aside. It is not unusual to be told by police . . . that when they get a fire brief, it may take anywhere between six to twelve months to have [the arson investigated]. And in most cases police have so much on their plate anyway . . .'[17]

There is no national database on arson incidents and while arsonists like to know the area in which they are operating, they do move across jurisdictions.[18] In 2005 only Victoria, Queensland and Western Australia have dedicated arson squads. New South Wales had a temporary operation, Strike Force Tronto, but it has run its course. Tracking arsonists is an unromantic, painstaking process that can take months as investigators try to find the source and cause of a fire. RFS investigator Richard Woods says: 'Very often people will light in certain patterns and we look for certain things when we're going about our investigations and try to link those back with previous fires. All that information then combines to form a profile of the fire-lighter.'[19]

So what is the community to do about bushfire arson? Is education the answer? There have been programs in schools for over a decade, including having fire brigade officers talking to children in a classroom setting. There are also child intervention programs in all states to deal with children who light fires.[20] There have been some successes. As Willis points out there is a considerable literature on child arsonists. 'The literature generally reflects the belief that child fire-setting, if addressed early enough, can be eliminated before it becomes an established behaviour or before it escalates into a more dangerous and more malicious activity.'[21]

While such programs are, no doubt, successful in many cases, we still have to deal with the adolescent and adult arsonist. We know much less about these, especially in a bushfire setting. And perhaps we take too much notice of psychological interpretations of their behaviour. Often the general public demand stiffer penalties for arsonists, especially in times of bushfire crisis. However, experts feel this is 'highly dubious'. They make the point that 'it is based on an assumption that an individual will stop and rationally consider their behaviour' before fire-lighting.[22] They emphasise the danger of profiling and targeting certain groups within the community.

Nevertheless, the sheer frustration of ordinary citizens, which the media tends to reflect, is equally understandable. Fire-lighting in such a dangerous setting as the Australian bush is malicious and destructive. It is an assault on the whole community and the environment, not just on an individual. While common law is biased towards crimes against property and the individual, people instinctively appreciate how destructive and malevolent bushfire arson is. Professionals like psychologists and social workers, who are often trapped in a therapeutic interpretation of this behaviour, miss entirely the ethical outrage that the community instinctively feels with this assault upon itself. There is a tendency by professionals to underestimate the moral responsibility of arsonists. Certainly, there may be cases of diminished responsibility, but psychologists sometimes embrace the delusory notion that disturbed people are not responsible for their actions. Actually, it is because the community perceives a considerable level of responsibility, as well as malevolence, that ordinary people feel they have the right to express outrage and demand appropriate and exemplary punishment.

So what are we to do? What about a register of known arsonists who, after serving their prison term, are placed under some form of monitoring during bushfire weather until they have convincingly demonstrated that they have dealt with or grown beyond their destructive behaviour? The parallel here is the register of pedophiles which is now accepted in many jurisdictions. It could be argued that a similar register would be justified because arsonists are actually more destructive and can affect the wider community and the environment in catastrophic ways. This approach would at least act as a restraint on convicted arsonists at dangerous times.

And then there is the problem of *suspected* arsonists who may not have been prosecuted or found guilty, although there is reasonable suspicion about their behaviour. There are many more of them than convicted arsonists. There are civil rights issues involved here and it is very hard to achieve a balance between suspects' rights to the presumption of innocence, and the equally important right of the community to protect itself and, even more importantly, to protect the environment. While a solution to this dilemma is elusive, the time has come for the community to debate it. Perhaps if police have reasonable suspicions that a certain person is a bushfire-lighter, they can apply to a judge to have that person forced to wear a GPS bracelet of some sort so that their whereabouts can be checked at all times during bushfire danger periods. In this way they are not confined, but it does place a form of genuine restraint on their behaviour. If they attempt to light fires they will easily be placed at the scene and can more easily and successfully be prosecuted.

Part of the problem is that psychologists and social workers are now so influential in thinking and decision-making about crime, punishment and restorative justice that we are all caught in their therapeutic view of reality. Their concern is with the psychologically disturbed perpetrator and his or her supposed familial, social, educational and human deprivation. The community is concerned about the broader consequences of the actions of these people. The psychological perspective is too narrow. We are beginning to recognise that a wider, non-therapeutic approach is required that involves a number of perspectives including the views of the community.

As a result we are also beginning to recognise that there is a broader context for bushfire arson. It is not just about property or threats to human lives, real as these are. It is not just an attack on the community. It is also an assault on nature itself. Embedded in this argument is a presumption that the natural world—especially the sentient animals—possesses rights. Increasing numbers of ethicists and philosophers are concerned with our insensitivity to environmental destruction. The American theologian and environmental thinker, Thomas Berry, for instance, speaks of the way we are myopically focused on 'the pathos of the human', while 'we commit biocide, the killing of the life systems of the planet . . . and we have no morality to deal with it'.[23] We have to move beyond our fixation with the psychological pathos of perpetrators and with individual and property rights, and recognise that people who, whatever their compulsions, destroy the fabric of the landscape through destructive burning must be held responsible for their behaviour and must be, if necessary, confined. Their rights are limited through interaction with other sets of rights.

In the end what we are dealing with here is the question of the rights of nature itself. There are many people out there speaking for human rights. Perhaps the time has come for a few of us to speak about the rights of other sentient beings, the landscape and the natural world itself.

11

FIREPROOFING AUSTRALIA?

i

One good result of the challenge of bushfires is that they bring Australians together. As Henry Lawson pointed out in the nineteenth century the threat of fire led men to ignore class barriers. In the twentieth century firefighting has been a force for social cohesion, transcending sectarian barriers between Catholic and Protestant, and after the Second World War gradually bringing the 'New Australian' arrivals into the multicultural mainstream through participation in volunteer bushfire brigades.[1]

But fires have also had a divisive effect. Humans are naturally and incurably manipulative; we constantly attempt to modify our environment to suit our needs. Few groups of people have had a more drastic impact on their natural surroundings than settler and migrant Australians and their descendants. In a way, it was inevitable. The landscape and biota of Australia are so different to anything Europeans had ever experienced before in the northern hemisphere. While some settlers quickly adapted to the continent's unique geography, flora and fauna, many seemed determined to create a carbon copy of the old country, a 'new Britannia' in the Great South Land.

Some people quickly connect with the landscape; others take generations to learn to 'call Australia home'. Some seem determined to be foreigners in their own land and live in a totally urbanised environment divorced from the natural world. There are also those who are imbued with a zealot-like doggedness to conquer and subdue the continent by any means available in order to promote an ideology of 'development'. Many of these people seem almost to hate nature. Developmental mania has degraded the landscape. After almost 220 years of this kind of assault the time has come for a serious rethink of how we relate to the continent. And rethinking our relationship to fire will be an integral part of that process.

Fire is endemic to Australia. Anyone who wishes to live here must come to terms with this. As long as eucalypts and native vegetation are combustible, fire-adapted species will dominate in a fire-prone landscape, one that has evolved over millions of years and is not going to adapt to us. We have to adapt to it. It seems obvious and almost silly to have to keep making this point, but the evidence is that many Australians live in denial. They think that fire can be eliminated, or confined to wilderness, that it is an 'enemy' that will only be fought and defeated by preventative burning. The image of firefighting as a paramilitary activity is revealing. It shows that some still think they are engaged in a battle against the landscape. The public memory is very short. Fires are quickly forgotten, especially by those who do not suffer their impact, or are only marginally affected. Only a few conflagrations, such as Black Friday and Ash Wednesday, have penetrated popular culture and live on in the communal memory. For most, the memory of the 2003 fires has already faded. As a result people seem surprised, even shocked, when the next fire threatens. There still seems to be no real consciousness that fire is part of the nature of Australia.

Certainly fire science has moved on from the days when broad-scale hazard reduction was seen as the only way to deal with fire. Nowadays the emphasis is on asset protection and mosaic burning dominates practical fire thinking. Several elements go to make up the mosaic:

those conducting the burn have to consider how the fire might behave once it is lit, what the vegetation community to be burnt can bear and how often it can be burnt, the impact on native animals, and the relationship of this fire to nearby individual and community assets that may be impacted by the burn.[2] Mosaic burning, especially in difficult terrain, can be dangerous, as was shown in June 2000 when a fire in the Ku-ring-gai Chase National Park killed four staff engaged in a deliberately lit burn.

Nowadays there is much more concern about loss of biodiversity and the negative impact of broad-scale burning on local ecology. Species in a given area co-exist in interdependent and complex relationships with each other and fire effects on one can have disastrous effects on others. But this does not mean that public land management agencies, such as national parks, have abandoned hazard reduction. The director of the New South Wales NPWS told a Legislative Council committee in 2000:

> In recent times we have become more sophisticated in our capacity to control hazard reduction burns, we are actually managing strategic burns at the interface where there is the greatest risk of fire moving off the park and where we need to manage that interface in order to protect life and property and be a good neighbour. We are refining our methodologies . . . We are more surgical about the way we do it, but it does not reflect in any sense a lack of commitment to appropriate hazard reduction.[3]

But while experts may have moved beyond broad-scale burning, many in the public have not. When big fires hit, as in 2003, some become almost hysterical, and there are demands for hazard reduction burns in parks and remote wilderness areas far away from property and farms. This issue has led to debate and conflict between ecologists and some groups in the community about the role of preventative burning in bushfire suppression. Yes, it might be possible for the parks and wilderness to be burnt regularly, but at what cost? Some property *might* be protected, but the impact on plants and vulnerable native animals could well be catastrophic. The simple reality is that short of eliminating all ground fuel, it is not possible to stop, or in some cases even lessen the impact of wildfire. As Rural Fire Commissioner, Phil Koperberg, commented in 2002: 'Unless you're going to keep all of NSW hazard reduced to a point where there is no fuel on the ground . . . we're going to have fires'.[4] A consistent argument in this book has been that the primary value must always be the environment. The survival of endangered species is much more important than property. Animals and plants cannot adapt quickly; human beings can, and it is about time some did.

It is worth noting, too, the particular fire danger involved in monocultural plantations which are increasing in number and size across Australia, especially as so-called 'carbon sinks'. As was seen in the massive fire in the radiata pine plantations around Canberra, single-species, same-age, closely planted trees are highly inflammable and can quickly turn into destructive wildfires with catastrophic consequences. Such fires can gain massive velocity and intensity. Again, it is not only the danger that these plantations pose to people and property that make them risky undertakings. It is also the threat they pose to the natural landscape. Essentially, they are completely unnatural, especially when the species is exotic.

ii

As discussed in Chapter 9, Aboriginal fire regimes are cited by many, including some concerned with environmental values, as examples of ecologically friendly burning. However, as has been argued, a system of repeated burning mimicking suppositions about Aboriginal fire practices is unsustainable. The time has come to ditch the so-called 'Aboriginal fire' argument—for two reasons: firstly, we simply do not know enough about traditional Aboriginal fire practices. In eastern Australia and southern Western Australia there is simply no one left with accurate knowledge of what was done. Some would say that Aboriginal people engaged precisely in 'mosaic burning'. This may or may not be so: we really do not know what they did. We cannot base our knowledge of Aboriginal fire regimes on the cursory and chance observations of Europeans. Those who attempt to make arguments based on the observations of early explorers and settlers fail to recognise the context in which these observations were made, and almost always lack the historical and critical training to make these kinds of judgements. History is one of those unfortunate disciplines that suffers from the interventions of well-intentioned amateurs from other disciplines or none, or people with a political or economic axe to grind.

Secondly, settlers have so altered the landscape that it would be impossible to reintroduce such a pattern of burning. In fact, the net result of land degradation, clearing, drought and native animal extinction has been an increase in the number and intensity of conflagrations, which in turn have led to the proliferation of fire-friendly, combustible species, which themselves have to be cleared by further broad-scale burning, thus setting up a pattern of fire breeding more fire, completely destroying the dynamic equilibrium that the landscape had reached before 1788.

As we have seen, a key element in the problem is that so many fires are not natural in origin, but are the product of human malice or foolishness. Bushfire arsonists are the real terrorists in our society, especially those who are active on the urban–bush fringe, and in areas like national parks relatively close to settlements. They don't just destroy property which, with assistance and insurance, can be replaced; they kill people. And perhaps even more importantly, because of its long-term effects, they vastly increase the incidence of fire and put the delicate balance of nature at serious risk, placing ecosystems under stress and threat. Of course Australian flora and fauna recover from bushfires; they have survived precisely because they are fire-adapted. The rebound of both animals and plants in many of the alpine areas of New South Wales and Victoria after the 2003 conflagrations indicates that our ecosystems are well adjusted to periodic, natural fire. But they are not adapted to constant and intense burning, no matter whether these fires are lit through malicious arson, or through an imposed regime of regular, broad-scale burning. The focus now must be on the preservation of our fragile and threatened environment. Built and productive assets should be protected as well as possible. But ultimately the emphasis needs to be on protecting the ecology of the landscape.

The arguments about broad-scale burning are useful in so far as they highlight the fact that the time has come to pause and rethink our whole approach to the Australian landscape and our place in it. Someone who has been talking about the 'bushfire problem' for a long time is Australia's premier bushfire scientist, Dr Malcolm Gill of the CSIRO's Centre for Plant Biodiversity Research.[5] Gill correctly says that the core question confronting us is one of values; what is most important, what asset do we most want to protect? He says 'Unless "assets" are understood, the nature of the fire problem is not understood . . . [and] what is considered to be an asset will vary within and between societies'. Gill distinguishes three types of assets: owned assets (such as land), tangible assets (for example, house, stock, pasture), and intangible assets (such as scenery, biodiversity, wilderness, the spiritual values of nature). Arguments about the priority of these assets are especially sharp in the zones that he calls 'the urban to wildland interface' and 'the farmland to wildland interface'. 'The former contrasts high-density monetary values, with low-density, abstract values; the latter contrasts public land with its low-density abstract values with the private rural land of medium-density monetary values like pastures . . . buildings, fences, machinery.'[6] In a detailed analysis Gill shows that, 'The bushfire problem is a societal problem which has many partial solutions . . . and these involve co-operation, integration, knowledge, budgets and legislative support. In some places it may be possible to adopt many of these partial solutions but otherwise there are trade-offs with other imperatives.'[7]

While acknowledging the sheer complexity of this problem, the range of conflicting interests involved and the necessity to compromise, it is important to also attempt to set some priorities, especially in light of our growing awareness of the importance of biodiversity and our greater sensitivity to the Australian environment. There is also a deepening realisation of human responsibility for almost all environmental disasters which have increased exponentially with the growth of human population on the continent. We also need to realise that the primary impact of bushfires is not on us, but on the world around us. Compared with other accidental causes of death (such as road deaths), very few people are killed, or even injured, by bushfires in Australia. Certainly the number of deaths is considerably less than 2000 since 1788, and is probably closer to about 1600 altogether. The number of road deaths vastly outnumbers that. Without doubt there has also been widespread loss of assets from bushfires, particularly in rural areas, although it is possible that the total value of material losses is inflated, particularly in the media. It can certainly be in the interests of people who have suffered losses to enlarge their deficit to highlight their situation. Inflation of losses may also be in the interest of insurance companies which can use such situations to persuade the community they need more insurance.

One of the dangers inherent in a book like this is that with an emphasis on the graphic and often tragic personal experiences of people in bushfires, the key historical lessons that can be learned from European experience since 1788 might be missed: that the real impact of fire is on the natural world, on the animals, plants and landscape itself rather than on humankind. This history is full of personal tragedies, but this can disguise the fact there are other, more important values embedded in our attitudes to fire. This book unequivocally asserts that care for the environment is the most important issue facing us and it is this that must take first priority.

As Gill says, the question of the priority of assets is central. The basic, primary asset that we have is the Australian landscape itself. What remains of it must be protected. Thus what Gill calls 'abstract values' are centrally important and far outweigh property and economic values. Without ecological integrity, without wilderness (that is, nature in its natural state, not changed, manipulated or adulterated by humankind for its own purposes), biodiversity is threatened, the land degenerates and species become extinct. This is not just an ecological problem. It is also a cultural and spiritual issue. As natural systems disintegrate and disappear and as we are forced to retreat more and more to a built, artificial environment, our imaginations are left bereft, uninspired by beauty and life that is non-human. There are profound spiritual values embedded in our relationship with nature and the landscape that far transcend material values.[8]

This, of course, involves a clear shift from previous priorities that focused on the protection of life and property. This is not to say that these priorities are not important, but they must be seen within the context of the value of maintaining ecological integrity, for it is this that provides the context for all other values. In order to achieve this the frequency of fire must be lessened. While fire is an integral part of landscape, this should occur naturally. Constant, broad-scale, deliberate lighting of fires, especially well away from settled areas, should be stopped. Only forms of mosaic burning that are sensitive to the natural values of the landscape should be permitted. There must be an unequivocal commitment to the primary value of the natural world. All other assets must be subsumed by and subordinated to our primary commitment to environmental protection.

iii

One of the most striking things about discussion of forests and bushfires is the kind of rhetoric that is often used. While writing this book I was astonished by the 'management-speak' that constantly recurs in so much of the discourse about fire. I am not speaking here of the kind of jargon every discipline has. I am referring specifically to the assumed, apparently unconscious attitude that nature needs to be 'managed'. It is apparent in common roadside signs erected by state forestry commissions: 'Managing the State's Forests Sustainably'—as though forests couldn't manage themselves sustainably and needed a government department to sort them out.

Part of this rhetoric is a result of the manipulative milieu in which we live, but it also reveals an attitude that nature is not really right unless it is managed by some human being. While I am not suggesting that we do not need to understand the world and act wisely, I am arguing that nature is not there to be 'managed', that it does not need our management. It has an integrity in itself and its own complex processes which we might gradually come to understand and respect through scientific study and observation, but not to manipulate by constant interference. But some people are dissatisfied with understanding; they feel the need to be involved in order to help, improve, or exploit. These are the 'god-botherers' of the natural world. They are dominated, albeit unconsciously, by a need to interfere. What do people addicted to the management of nature think has been going on for millions of years in the forests and bush before their arrival?

The mania to manage one's entire environment is a classical symptom of the human fear of loss of control, of an inability to stand back from nature and allow it to be itself, of a failure to show some humility toward a world that has been in evolution for millions of years and does not need us or our technology. The need to manage is driven by a highly manipulative view of reality, a sense that unless we control everything that happens we will be too vulnerable, too exposed to the vagaries of a world that is ultimately completely indifferent to us. It is a kind of extreme anthropocentrism that assumes nature exists solely for the sake of humankind, that it has no dynamc, living ability to attain its own equilibrium. For humans to embrace vulnerability we need a certain modesty and humility, combined with understanding and respect, derived from scientific study, as well as a sense of our biological connectedness to the natural world and a realisation of our complete dependence upon it. We are simply part of nature, not its 'managers'. A wildfire exposes the sheer impotence of humankind, and for the control-freaks among us that is almost unbearable.

Rather than looking for endless technological or management fixes, we should be trying to create the conditions for nature to heal and care for itself. In the area of bushfires we need to consider the possibility of reviving, at least in some areas, the kind of dynamic equilibrium that Australian forests had before the 'settler-managers' arrived in 1788. The natural ecosystem dynamics of forests and the bush are self-regulating and self-optimising. This was clearly understood from the late nineteenth century onward by the MMBW and, as we saw from the transcripts of the 1939 Stretton Royal Commission, their counsel, A.E. Kelso, argued for maintaining or trying to recover the original state of the forests and bush because they were far less fire-prone as a result of natural resistance. A forest undisturbed by fire for a reasonable period not only returns to its natural equilibrium, it protects itself by gradually closing over the canopy and preventing the growth of wattles and shrubby undergrowth that feed destructive wildfires. It retains moisture and is cooler. Kelso's arguments were supported by Australia's premier forester, C.E. Lane-Poole. Lane-Poole told Stretton that he was sure that broad-scale burning was not needed to get rid of leaf and bark litter. He maintained that the natural processes of decomposition of forest matter meant that 'after a short period of years after a fire a complete balance is established between the leaf-fall and the destruction of the leaf-fall by natural agencies . . . The usual opinion is that an enormous accumulation of inflammable fuel occurs if a forester excludes fire. That is incorrect.' He said Kelso was right when he argued that if fire was excluded the forests would return to their original stable state. This is not to say that there were no fires in the past, but that frequent fires stopped this natural process and led to an escalating build-up of inflammable material.[9]

A modern version of this approach comes from the Russian biophysicist, Victor G. Gorshkov, who argues that 'prohibiting any kind of modern land use activity is a prerequisite for ensuring the proper ecological functioning of forest ecosystems'.[10] In other words, natural forests do not need our interference, management or exploitation; they are best left alone. They are complex, interdependent, ordered realities that are intimately interactive and self-regulation occurs naturally if they are left undisturbed. They are able to recover dynamic equilibrium from intrusive, disturbing events like natural fires because their biotic regulation constantly works to protect them. That is why any form of human interference in forests, especially old-growth forests, is wrong—it interferes with natural processes. Thus all logging and broad-scale burning in such environments should cease immediately. The less human intervention the better for the recovery of the dynamic equilibrium of the forest.

Certainly, it is difficult to predict how long this process will take, as Kelso admitted to Stretton. Lane-Poole commented, 'Fires have always come in before we have been able to reach that position where the wattles have disappeared'.[11] However, perhaps selected areas in the alpine parks in Victoria and New South Wales might be ideal places to begin this experiment. If selected forest areas were set aside and protected from fire as completely as possible, it would be interesting to see how long it took for a natural stasis to return. Perhaps it might only take a decade or two? In modern Australia this approach has never really been tried on a wide scale, largely because the McArthur–Luke regimen of broad-scale burning took hold in the period after the Second World War and was dominant up until the emergence of scientific ecology. Perhaps it is now time for this kind of experiment.

iv

Having established ecological integrity as the primary principle, we still need to protect productive and other assets of those who live in bushfire-prone areas. To achieve this we need to establish 'asset protection zones', buffers around agricultural and particularly residential properties where vegetation and bushfire fuels are reduced to create a clear break between the natural environment and assets we want to protect. This can be achieved by clearing of trees and vegetation and by a carefully modulated and restrictive regime of preventative burning across this zone. While this is necessary, we should not delude ourselves: by doing this we are assuming responsibility for *destroying* part of the environment. This issue was canvassed at the 1939 Stretton Royal Commission but we still have not really come to grips with the need for a break between the natural, unspoiled world, and land exploited for agricultural or settlement purposes.

In some places, particularly across the agricultural landscape, establishing asset protection zones will be difficult to achieve simply because forest and agricultural land abut each other closely along an extended frontier, and grassfires are also part of the natural ecological rhythm. These zones should be a co-operative venture between state authorities, such as national parks, local councils and landholders. Part of the problem is that even when asset protection zones are established, the buffer may not be sufficient to protect against ember attack or spotting far ahead of a fire-front. This is where we have to admit that in some circumstances, such as in 1939 in Victoria, 1967 in southern Tasmania, or 2003 in the Alps and Canberra, probably nothing we do will protect our assets: the reality of living in Australia means living with fire and its consequences. There is no escaping it.

We also need to admit that some built assets are far too dangerous to protect, even in moderate or light fires. This is especially so along parts of the urban–bush boundary. While such places can provide a wonderful context for living in non-fire danger periods with vegetation, birds and animals coming right up to the back door, once fire danger indices rise experience shows that these residences can be deathtraps unless they are extremely well maintained. Creating an asset protection zone around them is difficult and can often be very destructive of the environment. In the end, it is a problem to be dealt with by those who choose to build there.

While not all developments along the urban–bush boundaries are bad, some developments are totally inappropriate. At the very least, owners and residents should be aware of the risk they are creating for themselves and for the firefighters who are expected to risk their lives to protect their property. Examples of the areas that fit this category are to be found particularly in bush-surrounded, isolated streets in the outer suburbs of Sydney, Melbourne, Hobart and Adelaide. The risks are obvious to anyone who buys or builds in Victoria's Dandenong Ranges and Mount Macedon, the Blue Mountains west of Sydney, or the Adelaide Hills. But some developments are particularly risky: for instance single streets along a ridge-line with bush reaching down to water on either side that can be seen from aircraft approaching Sydney from the south and the west; or single suburban streets in the Blue Mountains that reach right into the bush, again usually along a ridge-line with no escape except back along the street; or isolated roads in the Adelaide Hills; or the similar streets in the Dandenong Ranges; or on the lower slopes of Mount Wellington in Hobart. Perhaps these dangerous developments should incur higher insurance premiums because of the unsafe interface along which they are built. At the very least residents in such areas should be primarily responsible for the protection of their own homes rather than expecting firefighters to come to their rescue. Complete, realistic awareness is what is needed and a frank admission that if a person builds in a known fire-prone, isolated and risky area, they must accept the consequences of their action.

v

A question that is often asked is whether the frequency and intensity of bushfires will be exacerbated by the effects of global warming. Firstly there is no doubt that the evidence for global warming is overwhelming. Professor Stephen Schneider, a climatologist from Stanford University, points out that, 'We've long understood the basics, and the basics are when you've used the atmosphere as an unpriced sewer and there's no incentive for people to conserve and get cleaner, you're going to dump the stuff in there and it's going to build up. That's exactly what's happened.'[12] But there is debate about the consequences of global warming. This certainly affects the question of whether bushfires will increase in frequency.

The Council of Australian Governments' recent inquiry into bushfires notes that, 'Climate change is likely to increase the frequency, intensity and size of bushfires in much of Australia in the future' and refers in a footnote to the work of the CSIRO.[13] This assumption has now become widely accepted. The same certainty is expressed by the CSIRO's Climate Impact Group in Aspendale, Victoria. The group's senior research scientist, Dr Kevin Hennessy, told the ABC's *Science Show* recently that the impact of global warming on Australia indicates that 'many Australian ecosystems, regions and industries are vulnerable to climate change over the coming decades. In particular there is likely to be a reduction in water for southern and eastern Australia, with more fires and heat waves, fewer frosts, less snow and more coral bleaching. Some of these changes are already evident.'[14] This bleak scenario would certainly lead to the extinction of many species and an increase in bushfires over the coming decades and has led to a certain degree of panic among some people.

But there are many who question this. This scenario of more fires is based on a widespread popular perception that global warming inevitably means that everything will be drier, that there will be less rain, the vegetation will dry out and that there will be a resultant increase in the number and intensity of bushfires. However, many scientists are becoming aware that global warming is a complex process. While there is no doubt that it is already happening, the long-term effects on weather, a key element creating conditions for bushfires, are not entirely clear, especially at the regional and more particularly the local level.[15] Part of the problem in non-scientists getting their minds around this issue is that we instinctively tend to identify warming with drying out. But we also need to remember that global warming results from the *greenhouse* effect, and that the interior of a greenhouse is usually warm, steamy and wet. So warming does not necessarily imply less rain. In fact, most climate models predict marginally more rain at a worldwide level.

At the local level it is much harder to predict what will happen. The empirical evidence is that Australia has been getting somewhat wetter, especially in the north-west, since the 1970s, and that this will probably continue. South-eastern Australia may get marginally drier, but it is difficult if not impossible to predict precisely what will happen locally. Some parts of Australia may get wetter, some drier. Another element in this complex is that predictions are based on climate models and these differ; as a result the conclusions they reach also vary, especially if applied locally.

We also have to take into account what has recently been called 'Global Dimming'. It has been discovered that while the world's temperature has increased over the last 50 years, the rate of evaporation of water from meteorological pans (that is, the potential evaporation rate, the rate that occurs in the presence of a water supply) has actually fallen due to the fact that pollution in the atmosphere from burning fossil fuels has created a kind of polluted cloud shield that has reflected the sun's rays and prevented incoming sunlight getting through to the earth's surface.[16] While some think that this has been masking the true impact of global warming, most scientists are much more cautious. Again, we are uncertain of the consequences of this process. As a result of all the variables involved in global warming we need to be cautious about predicting any necessary increase in bushfire frequency in southern Australia. The effects of global warming should not be used as a justification for more broad-scale preventative burning. A much more cautious approach is needed.

vi

It is impossible to fireproof Australia, but what can we do? The latest in a seemingly endless procession of government reports, the Council of Australian Governments' inquiry into bushfires says that:

> Given the inevitability of bushfires, all Australians must learn how to live with them. This has been recognized at least since the Streeton [sic] Royal Commission of 1939, which identified both school and adult education as 'the best means

of fire prevention and protection'. Despite achievements ... in individual states and territories, a nationally consistent bushfire education strategy that reaches and informs all Australians is yet to be implemented.[17]

Besides the egregious error of misspelling Stretton's name, the COAG Inquiry is a fairly comprehensive treatment of the issues surrounding and arising from the 2002–3 fires. It is both balanced and predictable and is more environmentally aware than the deplorable House of Representatives Inquiry report, *A Nation Charred*. There are two main thrusts at the heart of the COAG recommendations: the first centres on educating Australians about bushfires; the second is on trying to get state governments to adopt a national approach to a reality that transcends state and regional borders. Perhaps it was this centralising tendency that held up agreement on a report that was finished in March 2004 but which was not issued until late January 2005.

The inquiry has tried to thread its way through the minefield of conflicting theories about bushfires, but ends up being all things to everybody. The heightened emphasis on education about bushfires to all Australian children as 'a basic life skill' is clearly a soft option. It is based on the belief that somehow education can achieve everything. Reading the section of the chapter on adult education, it seems the real aim is to shift responsibility for fires more and more to the community with a consequent disengagement by government. 'These programs should emphasize individual and household preparedness and survival as well as the role of fire in the Australian landscape.'[18]

Yes, but what is the role of government and fire authorities? This all seems a bit like putting the cart of survival before the horse of the bushfire. Education about fire has some importance, but it comes well down the list. The report makes the point about what it calls the 'lengthening' of the 'rural–urban interface'. With many more people retiring to coastal areas the interface zone will be much extended in the next ten years.[19] This is true, but perhaps it is the role of government to warn people honestly what they are getting into by wanting to live in such fire-prone circumstances and advise them of the consequences. Some of the most vulnerable places in Australia are small settlements strung along the coast from Queensland to Western Australia. These mainly retirement and holiday villages are surrounded by coastal scrub and forest; many of them are actually cut out of this vegetation. In a wildfire people will learn, as did the inhabitants of Snug in 1967 and Lorne in 1983, that their only escape is into the sea and that their assets are in serious danger of being destroyed. Part of the role of honest government is to confront people with this reality. Local government in coastal areas especially has onerous obligations in this matter. The terrible problem of such governments coming under the control of gung-ho, pro-development councillors who encourage the building of houses in the most inappropriate places can lead to bushfire disaster.

The other emphasis from the inquiry is on more federal government involvement in bushfires. While there are some areas where this makes sense, for example the financing and purchase of aircraft used to fight fire, one needs to be cautious. While total local control has already demonstrated its ineffectiveness, centralisation is not the answer either. There is a happy medium.

However, firefighting aircraft is one area where federal government can play a real role. We have already seen that as early as 1929 small spotter planes were used to pinpoint fires and until the end of the Second World War this was the role assigned to aircraft. After 1945, the larger planes and helicopters available could be employed for fire suppression and in North America they were used widely. But many in the firefighting community were suspicious that aircraft were an expensive waste of financial resources. Post-war thinking focused almost entirely on fuel reduction. Planes were mainly used for aerial ignition of fuel reduction burns, and they are still used for this purpose today. But the emphasis remained on groundwork, better equipment, training and integrated communications to control fires. Aircraft remained ancillary.

We have come a long way from the First World War–built Westland Wapiti bi-plane with no brakes or flaps to the Erickson Air Crane of 2003. But as with everything regarding bushfires, even the use of aircraft has been contentious. In the last two decades there has been an increase in aerial water-bombing, but operational expenses meant that this method was used mainly as a ground support. Helicopters have played an increasing role and the Erickson Air Crane has become a symbol of what such aircraft can do. But disagreements remain because the air-cranes have to be leased and imported every fire season and are very expensive to operate and maintain. There are also fixed-wing amphibious aircraft that can carry almost equivalent loads and have even better scooping ability. Probably the best of these is the Canadair CL 415 made by Bombadier in Canada.[20] This aircraft has been successfully used in Canada, Portugal, Spain, France, Italy, Greece and the former Yugoslavia as a firefighting aircraft. 'It is a purpose built amphibious firefighting aircraft which has been around for many years . . . it can land on a regular landing strip or on water. It scoops water while in flight and picks up 6 tonnes of water in ten seconds, so its return time to a fire dropping zone is very short.'[21] It has high manoeuvrability, can operate in seas up to 2 metres, and needs to refuel less often than the large air-cranes and has the added advantage of being able to be used in the non-fire season in other roles such as coastal surveillance. It can carry 27 fully armed soldiers and land them on water near an intruder, or a ship, or off-shore oil rig. The main limitation is that unlike in Canada there are not a lot of inland water sources in Australia. However, most bushfires that affect housing and settlement are reasonably close to the coast where the sea can be used as a water source.

The basic problem as to why we have never bought our own fleet of these aircraft, as have other many other countries with serious bushfire problems, seems to be that empire-building fire authorities do not want to surrender power to others, as these planes would be operated by either the navy or air force or some form of coastguard or coast watch. The individual states lease the air-cranes from the United States and in this way maintain control of them while they are in Australia. This is an area where the federal government ought to intervene and make such planes available.

vii

The Spanish called Australia the South Land of the Holy Spirit. Unconsciously they linked the continent to the Spirit of God who in the hymn 'Veni Creator' is referred to as *fons vivus ignis*, 'the living source of fire' that cleanses and renews. In biblical and Christian imagery fire and life are intimately connected and the presence of God is often symbolised by fire. This notion also finds a cosmological basis in the thought of the early Greek philosopher, Heraclitus of Ephesus, who lived around the time of the sixty-ninth Olympiad (c. 504–1 BC). What impressed Heraclitus, just as it impressed the Buddha, was the sheer instability of all phenomenal reality. 'All things are in a state of flux,' he said. 'You cannot step twice into the same river, for fresh waters are ever flowing upon you' (*Fragments*, 21, 41). Aristotle sums up Heraclitus' teaching succinctly: 'All things are in motion; nothing steadfastly is' (*De Caelo*, 298 b 30). So life is essentially change, movement, strife and tension, and Heraclitus felt that nothing symbolised this better than fire. For this reason he argued that fire is the primal stuff of life, the *Urstoff*. He saw the cosmos as 'an ever-living fire'. Fire lives by igniting and bursting out and consuming material. The result is that the world and humanity exist in a constant state of flux. Heraclitus held that as fire dies down it condenses and becomes moist and eventually coheres into water. Water then turns into earth which again produces combustible stuff. Fire then breaks out and the cycle continues. Thus reality is a constant series of fire processes going on everywhere, eternally.

Judith Wright, one of our finest poets, picks up the same theme:

> In the beginning was the fire;
> Out of the death of fire, rock and the waters;
> and out of water and rock, the single spark, the divine truth.[22]

For Wright, fire not only shapes the reality of the landscape and world around us, but in the context of the fire-shaped landscape we begin to perceive 'the divine truth', a reference to a kind of transcendent presence in the world. Similar imagery can be found in the very nineteenth-century Victorian verse of pioneer Presbyterian Rev. Dr John Dunmore Lang (1799–1878). Impressed by the sheer power of the bushfire, it reminded him of the last judgement. In *On the Conflagration of the Forest Around Sydney—November 25th, 1826* he stands fearful as:

> The forest blazed around, volumes of smoke
> Towering to Heaven obscured the face of day:

> And as the red sun shot his parting ray
> Through the dense atmosphere, the lurid sky
> Glowed with a fiercer flame—spreading dismay—
> As if the dreadful day of doom were nigh![23]

For Dunmore Lang the bushfire is the image of the day of judgement when he longs 'to stand upon the rock of ages, While all the conflagration rages'. This time fire is linked to divine judgement. The one who passes muster and is saved will be able to stand safe while 'all the conflagration rages'.

This is exactly the challenge we face in Australia today. We will only discover the truth about the nature of our continent when we realise that the ecology of the landscape makes this the most fire-prone place on earth. There is a real sense in which we will be judged by our attitude to fire. If we learn to live with it and adapt our lifestyles to it, we will be truly at home in Australia. Otherwise we will always remain mere transients here.

NOTES

CHAPTER 1 BLACK FRIDAY

1. The text is based on Melbourne newspaper reports, especially those of the *Age*, the *Argus*, the *Herald* and the *Sun*. See W.S. Noble's excellent *Ordeal By Fire: The Week the State Burned-Up*, Melbourne: Privately Published, 1979, pp. 51–2.
2. Leonard Stretton, *Royal Commission on Bushfires, 16 May 1939*. Minutes of evidence, p. 1135. Subsequently referred to as RCME in these notes.
3. There were two models of N-class steam locomotives built at the Newport Railway Workshops. I am presuming the Noojee locomotive was the lighter, earlier model. See Victorian Model Railway Society, *Victorian Railways: Rolling Stock: diagrams & particulars of locomotives, steam, electric, diesel electric, etc., 1926–1961*, Kensington, Vic.: Victorian Model Railway Society, 1986, p. 138.
4. Noble, *Ordeal*, pp. 51–2.
5. For the dimensions see John Godfrey Saxton's evidence at RCME, pp. 1061–4.
6. RCME, p. 1069.
7. For a discussion of the big trees see Bernard Mace, 'Mueller—Champion of Victoria's Giant Trees' in *The Victorian Naturalist*, Vol. 113 (4), 1996, pp. 198–207. The highest officially recorded was measured at 99.36 metres in 1880 in the Upper Latrobe Valley (see RCME, p. 1707). There is evidence that there was one over 500 feet (152.4 metres) high.
8. RCME, pp. 628, 632–4.
9. RCME, pp. 722–4.
10. F.E. Stamford, E.G. Stuckey and G.L. Maynard, *Powelltown: A History of its Timber Mills and Tramways*, Melbourne: Light Railway Research Society, 1984, pp. 51–2 and 117.
11. For 'Black Friday' in Warburton see Brian Carroll, *The Upper Yarra: An Illustrated History*, Yarra Junction: Shire of Upper Yarra, 1988, pp. 172–4.
12. RCME, pp. 482–4.
13. See Mike McCarthy, *Mountains of Ash: A History of the Sawmills and Tramways of Warburton and District*, Melbourne: Light Railway Research Society, 2001, p. 40.
14. Author interview with Basil Barnard, 1/11/04.
15. Author interview with Basil Barnard, 1/11/04.
16. ABC oral history interview with Basil Barnard at <www.abc.net.au/blackfriday/oral>. There is also an interview with Barnard at the Warburton Water-Wheel Information Centre.
17. Author interview with Basil Barnard, 16/11/04. The newspaper accounts are not entirely accurate.
18. ABC oral history interview with Murray Thompson at <www.abc.net.au/blackfriday/oral>.
19. Author interview with Basil Barnard, 1/11/04.
20. PROV 24/p, 1380, 819–821/1939, pp. 152–3.

NOTES

21. RCME, p. 436. For Rawson's coronial evidence see PROV 24/p, 1380, 819–821/193, pp. 87–8. For the Tin Hut see also Peter Evans, *Rails to Rubicon: A History of the Rubicon Forest*, Melbourne: Light Railway Research Society, 1994, p. 14.
22. PROV 24/p, 1380, 819–821/1939, pp. 7–8.
23. PROV 24/p, 1380, 819–821/1939, p. 11.
24. RCME, p. 467.
25. PROV 24/p, 1380, 819–821/1939, p. 12.
26. Other versions quoted in Noble, *Ordeal*, p. 34.
27. PROV 24/p, 1380, 819–821/1939, pp. 19–20.
28. PROV 24/p, 1380, 819–821/1939, p. 24. I have seen the Coroner's pictures from the scene and can confirm the charred and grotesque state of all the bodies. Baynes was not entirely accurate; there were eight men killed at the lowering winch.
29. Noble, *Ordeal*, pp. 34–5.
30. RCME, pp. 453–4.
31. RCME, pp. 427–626.
32. Jack Robinson's story in the ABC's oral history of Black Friday at <www.abc.net.au/blackfriday/oral>.
33. Mary Robinson's story in the ABC's oral history of Black Friday at <www.abc.net.au/blackfriday/oral>.
34. Mary Robinson, see above. See also the *Colac Herald*, 16/1/39, and the Melbourne newspapers.
35. RCME, pp. 823–4.
36. RCME, p. 829.
37. RCME, p. 799.
38. RCME, p. 784.
39. RCME, p. 845.
40. RCME, p. 827.
41. RCME, pp. 971–4.
42. RCME, pp. 742–4.
43. PROV 24/p, 1380, 819–821/1939, p. 45 and 40. For Francis' comments on the lack of leadership see, pp. 39–40. Francis also gave evidence to the royal commission at RCME, pp. 763–72.
44. PROV 24/p, 1380, 819–821/1939, p. 43.
45. PROV 24/p, 1380, 819–821/1939, pp. 73–4.
46. PROV 24/p, 1380, 819–821/1939, p. 76.
47. Ray Dafter talking about his uncle in the ABC's oral history of Black Friday at <www.abc.net.au/blackfriday/oral>.
48. RCME, p. 728.
49. Neil Ross in the ABC's oral history of Black Friday at <www.abc.net.au/blackfriday/oral>.
50. Summaries based on reports in the *Age*, 16/1/39.
51. See J.C. Foley, *A Study of the Meteorological Conditions associated with Bush and Grass Fires and Fire Protection in Australia*, Melbourne: Commonwealth of Australia, Bureau of Meteorology, Bulletin No. 38, 1947, p. 88.
52. Based on material in the *Canberra Times*, the *Sydney Morning Herald*, the *Adelaide Advertiser* and the *Mercury* for the period 1–18 January 1939.
53. R.H. Luke and A.G. McArthur, *Bushfires in Australia*, Canberra: Australian Government Publishing Service, 1978, p. 332. The Canberra fires of 2003 followed a similar pattern.
54. Luke and McArthur, *Bushfires*, p. 297.

55. This number is derived directly from newspaper reports. The EMA homepage <www.ema.gov.au/emadisasters> gives thirteen dead, but it is hard to see how they reach this figure.
56. Stephen Pyne, *Burning Bush: A Fire History of Australia*, Seattle: University of Washington Press, 1991, p. 314.

CHAPTER 2   OVER A CENTURY OF FIRES

1. Local historian, Deirdre Hawkins, helped me with information on Kinglake's history, especially the Kinglake Hotel.
2. RCME, p. 259.
3. Information comes from a personal interview with Stretton's son, Professor Hugh Stretton, in Adelaide on 28 July 2004, and from 'Judge Stretton's reminiscences' in the *La Trobe Library Journal*, Vol. 5, No. 17, April 1976, pp. 1–21. See also *Australian Dictionary of Biography* (ADB), Vol. 16, pp. 336–7.
4. RCME, p. 1361.
5. Leonard Stretton, *Report of the Royal Commission on Bushfires*, 16 May 1939, p. 7. Subsequently referred to as RCR in these notes.
6. RCME, p. 1986.
7. RCME, pp. 252–6.
8. RCME, p. 1985.
9. Pyne, *Burning Bush*, pp. 249, 270.
10. RCME, p. 1260.
11. RCME, p. 2382.
12. RCME, p. 1004.
13. RCME, pp. 988–9.
14. RCME, pp. 1020, 1021.
15. For Barrett see ADB, Vol. 7, pp. 186–9 and for his evidence see RCME, pp. 1541–65. For Hardy see RCME, pp. 1699–1712.
16. Grant W. Laver, *Official tourist map of Kinglake showing the new Kinglake National Park with descriptive notes of the locality and how to get there*, Melbourne: Ad-Art Studios, 1930, pp. 22–4.
17. RCME, p. 139.
18. RCME, pp. 140, 150.
19. Reported in the *Age*, 4/2/39.
20. RCME, p. 185.
21. RCME, pp. 187–9.
22. RCME, pp. 263–5.
23. RCME, pp. 275–9.
24. RCME, pp. 280–2.
25. Reported in the *Age*, 4/2/39.
26. RCR, p. 5.
27. RCME, p. 1796. See also Noble, *Ordeal By Fire*, p. 12.
28. David Collins, *An Account of the English Colony in New South Wales*, London: T. Cadell & W. Davies, 1802, Vol II, pp. 12, 17.
29. Hunter to the Duke of Portland, 10 June 1797. Historical Records of New South Wales (HRNSW), Vol. III, pp. 219, 220.
30. Government and General Order, 24 November 1797. HRNSW, Vol. III, p. 309.
31. Collins, op. cit., pp. 100, 103, 140, 143.
32. *Diary* quoted in F.W. and J.M. Nichols, *Charles Darwin in Australia*, Melbourne: Cambridge University Press, 1989, pp. 47 and 61.

33. Baron Charles von Hügel, *New Holland Journal, November 1833–October 1834*, Dymphna Clark (ed.), Melbourne: Melbourne University Press, 1994.
34. Von Hügel, *New Holland Journal*, pp. 122, 123, 124, 129–31.
35. See Henry Reynolds' books *The Other Side of the Frontier: Aboriginal Resistance to the European Invasion of Australia* (Melbourne: Penguin, 1982) and *Frontier: Aborigines, Settlers and Land* (Sydney: Allen & Unwin, 1987).
36. The pioneering book on the environmental impact of land settlement is W.K. Hancock's, *Discovering Monaro: A Study of Man's Impact on His Environment*, Cambridge: Cambridge University Press, 1972. Other books are Michael Cannon's two books, *The Land Boomers* (Melbourne: Melbourne University Press, 1966), and *Life in the Country* (Vol. 2 of *Australia in the Victorian Age*, Melbourne: Thomas Nelson, 1973. See especially pp. 214–36); C.G. Bolton, *Spoils and Spoilers: Australians Make their Environment*, Sydney: Allen & Unwin, 1981; William Lines, *Taming the Great South Land*, Sydney: Allen & Unwin, 1991, and Tom Griffiths, *Forests of Ash: An Environmental History*, Cambridge: Cambridge University Press, 2001.
37. Stephen H. Roberts, *The Squatting Age in Australia 1835–1847*, Melbourne: Melbourne University Press, 1964 edition, especially pp. 277ff.
38. Henry Gyles Turner, *A History of the Colony of Victoria from its Discovery to its Absorption into the Commonwealth of Australia*, London: Longmans, Green, 1904, Vol. I, pp. 204–6 for statistics, pp. 310–22 for settlement of the colony, and pp. 331–5 for Black Thursday.
39. There is really no comprehensive modern account of Black Thursday. However, see Pyne, *Burning Bush*, pp. 221–4, and Robert Murray and Kate White, *State of Fire: A History of Volunteer Firefighting and the Country Fire Authority in Victoria*, Melbourne: Hargreen Publishing Company, 1995, pp. 13–24.
40. Most of the material on Black Thursday comes from the newspapers of the time. The account is based on the *Argus*, the *Melbourne Morning Herald* (forerunner of the present *Herald-Sun*), the *Port Phillip Gazette*, the *Geelong Advertiser*, and the *Portland Guardian*. Diane H. Edwards, *The Diamond Valley Story*, Greensborough, Vic: Shire of Diamond Valley, 1979, p. 97. See also William Strutt, *The Australian Journal of William Strutt, ARA, 1850–1862*, Sydney: edited and published by George Mackaness, Privately printed, 1958. For Strutt see also *National Library of Australia News*, 14/11(2004), pp. 3–6.
41. Edwards, *Diamond Valley*, p. 19.
42. *The Australian Journal of William Strutt*, pp. 19–20.
43. James Fenton, *Bush Life in Tasmania: Fifty Years Ago*, London: Hazel, Watson & Viney, 1891. Reprint by Regal [Launceston, 1970 and 1989], pp. 79–81.
44. J.H. Kerr, *Glimpses of Life in Victoria by a Resident*, Edinburgh: Edmonton & Douglas, 1872, pp. 104–5.
45. William Howitt, 'Black Thursday: the great bush fire of Victoria' in *Cassell's Illustrated Family Paper*, Vol. I, nos 6,8,9 (February 5, 18, 25, 1854), pp. 46–7, 59, 67.
46. 'Black Thursday' was completed in England in the early 1860s and is now part of the La Trobe Collection in the State Library of Victoria.
47. Strutt quoted in Peter Quartermaine and Jonathan Watkins, *A Pictorial History of Australian Painting*, London: Bison Books, 1989, p. 63.
48. For a detailed analysis of Strutt's 'Black Thursday' see Heather Curnow, *William Strutt*, Sydney: Australian Gallery Directors' Council, 1980, pp. 46–8 and p. 101 for a black-and-white reproduction of the picture with a history of its provenance.
49. Pyne, *Burning Bush*, p. 229.

50. Pyne, *Burning Bush*, pp. 200–1.
51. Quoted by Gipps in a report to the Colonial Secretary, 28 September 1840, in Hancock, *Discovering Monaro*, p. 57.
52. Cannon, *Life in the Country*, p. 217.
53. *The Golden Age* (Queanbeyan), 6 February 1861.
54. J.C.G. Banks, 'A History of Forest Fire in the Australian Alps' in Roger Good, *The Scientific Significance of the Australian Alps: The Proceedings of the First Fenner Conference*, Australian Alps National Parks Liaison Committee, 1989, p. 271.
55. Frank Dodd, 'Recollections and Experiences' in South Gippsland Pioneers Association, *The Land of the Lyre Bird: A Story of Early Settlement in the Great Forest of South Gippsland*, Melbourne: Gordon & Gotch, 1920. It has been reprinted four times, most recently by the Korumburra and District Historical Society, 1998, p. 132.
56. Joseph Jenkins, *Diary of a Welsh Swagman, 1869–1894*, abridged and annotated by William Jenkins, Melbourne: Macmillan, 1975, pp. 104, 120, 139, 181, 187, 188, 203.
57. Jon Marsden-Smedley, 'Changes in Southwestern Tasmanian Fire Regimes since the Early 1800s', *Papers and Proceedings of the Royal Society of Tasmania*, Vol. 132, 1998.
58. Marsden-Smedley, art. cit., p. 23.
59. J.B. Walker quoted in Marsden-Smedley, art. cit., p. 20.
60. I have discussed in some detail these rainforests in my book *Hell's Gates: The Terrible Journey of Alexander Pearce, Van Diemen's Land Cannibal*, Melbourne: Hardie Grant Books, 2002, pp. 81–2, 122–4.
61. John Bradshaw, *The Early Settlement of South Gippsland: A Pictorial History*, Korumburra, Vic.: Coal Creek Heritage Village, 1999.
62. Anonymous, *Land of the Lyre Bird*, Preface.
63. W.H.C. Holmes, 'Picking Up', *Land of the Lyre Bird*, p. 67.
64. I am grateful to Brian Walters, SC for alerting me to Bernard Mace's unpublished paper 'Mountain Ash Forests of Victoria: Past Glory of the World's Tallest Trees', pp. 4–5 where Mace mentions 'hunting' trees.
65. A.W. Elms, 'A Fiery Summer', *Land of the Lyre Bird*, pp. 306, 307.
66. Quoted in W.S. Noble, *The Strzeleckis: A New Future for the Heartbreak Hills*, Melbourne: Forests Commission Victoria, No date but 1986, p. 22.
67. A.W. Elms, 'A Fiery Summer', *Land of the Lyre Bird*, pp. 307–8.
68. A.W. Elms, 'A Fiery Summer', *Land of the Lyre Bird*, pp. 308–10.
69. Pyne, *Burning Bush*, p. 243.
70. Robert Martin, *Under Mount Lofty: A History of the Stirling District in South Australia*, District Council of Stirling, second edition, 1996, p. 57.
71. Luke and McArthur, *Bushfires*, p. 279.
72. Luke and McArthur, *Bushfires*, p. 242.
73. For fire in Western Australia see B. Dell, J.J. Havel and N. Malajczuk, *The Jarrah Forest: A Complex Mediterranean Ecosystem*, Dordrecht, Netherlands: Kluwer Academic Publishers, 1989; Department of Conservation and Land Management, *Fire in the Ecosystems of South-West Western Australia: Impacts and Management*. Volume 2 *Community Perspectives About Fire*, Proceedings of April 2002 Symposium; Ian Abbott and Neil Burrows (eds), *Fire in the Ecosystems of South-west Western Australia: Impacts and Management*, Leiden, Netherlands: Backhuys Publishers, 2003. This is the published volume of the proceedings of the 2002 conference; P.E.S. Christensen, *The Karri Forest: Its Conservation Significance and Management*, Como, WA: Department of Conservation and Land Management.
74. Von Hügel, *New Holland Journal*, pp. 30, 40.

NOTES

75. W.L. McCaw and N.D. Burrows, 'Fire Management' in Dell et al., *The Jarrah Forest*, p. 318.
76. The term 'brickfielder' originated in Sydney to denote a southerly gale which blew dust and dirt into central Sydney from Brickfields Hill just outside the city. The term migrated to Victoria. See Sidney J. Baker, *The Australian Language: The Meanings, Origins and Usage; from Convict Days to the Present*, Sydney: Currawong Press, 1978, p. 210.
77. Figures supplied by the Australian War Memorial, Canberra.
78. See *Year Book Australia 2002*, Population Distribution, ABS.
79. *Year Book Australia, 1925*, Special Article, 'Settlement of Returned Soldiers 1914–18', ABS.
80. Marilyn Lake, *Limits of Hope: Soldier Settlement in Victoria, 1915–38*, Melbourne: Oxford University Press, 1987, p. 238.
81. Luke and McArthur, *Bushfires*, p. 271.
82. For Richardson see ADB, Vol. 11, pp. 384–5.
83. Quoted in Graeme Butler, *Buln Buln: A History of the Buln Buln Shire*, Drouin, Buln Buln Shire Council, 1979, pp. 714–15.
84. See F.E. Stamford, G.L. Stuckey and G.L. Maynard, *Powelltown*, p. 52.
85. This description is based on reports in the *Sydney Morning Herald* in the days following the fires.
86. RCME, pp. 2281–5, 2074.

CHAPTER 3   'BURN, BURN, BURN'

1. Vito Fumagalli, *Landscapes of Fear: Perceptions of Nature and the City in the Middle Ages*, Cambridge: Polity Press, English trans., 1994, pp. 126, 15.
2. RCME, p. 963.
3. RCME, p. 1985.
4. RCME, pp. 486–8.
5. RCME, pp. 1133–4.
6. For an outline history of the Creswick school see <www.sfc.edu.au/pages/history>.
7. RCME, pp. 2202–3.
8. RCME, pp. 251–2.
9. RCME, pp. 1071, 1074.
10. For Galbraith's evidence, RCME, pp. 2136–279.
11. RCR, p. 7.
12. RCR, p. 11.
13. Quoted in RCME, p. 2136.
14. RCME, p. 2149.
15. RCME, pp. 2174–5, 2138.
16. RCME, pp. 2193–5.
17. RCME, pp. 2205–6.
18. RCME, p. 2209.
19. RCME, p. 2276. For the whole argument see pp. 2274–6.
20. RCME, p. 2277.
21. RCME, p. 2213.
22. RCME, pp. 304–5.
23. RCME, pp. 365–6.
24. RCME, pp. 375–80, 385–6, 390.
25. RCME, p. 396.
26. I was assisted in researching Kelso's background and career by Dr Carolyn

Rasmussen, both through discussion and through the book she wrote with Tony Dingle, *Vital Connections: Melbourne and its Board of Works, 1891–1991*, Melbourne: McPhee Gribble, 1991. See also her *Increasing Momentum: Engineering at the University of Melbourne*, Melbourne: Melbourne University Press, 2004, p. 74. Dr Rasmussen searched out and obtained for me copies of *The Journal of the Institution of Engineers, Australia*, Vol. 14 (August 1942), No. 8, p. 189 and Vol. 15 (May 1943), No. 5, p. 121 where Kelso's appointment to the Department of the Army and death are reported.

27. Information from the 'MMBW Federation Journal' (November 1934) given to me by Dr Rasmussen. Kelso's war record is held at the Australian War Memorial, Canberra.
28. RCME, p. 425.
29. Kelso's evidence can be found at RCME, pp. 89–129 and at pp. 1778–1882, and his summing-up is at pp. 2520–55.
30. See William J. Robertson, 'Kelso's Contribution to Dam Safety', *Cranks and Nuts*, 1963. Here I am indebted again to Dr Rasmussen.
31. See Dingle and Rasmussen, *Vital Connections*, pp. 114–19 and 143–5, and Tom Griffiths, *Forests of Ash*, pp. 90–101.
32. RCME, pp. 110–11.
33. RCME, p. 111.
34. RCME, pp. 1780–1.
35. RCME, p. 112.
36. RCME, p. 111.
37. RCME, p. 111.
38. RCME, pp. 112–13.
39. RCME, pp. 1783–4.
40. RCME, p. 1786.
41. RCME, p. 1787.
42. RCME, p. 1807.
43. RCME, p. 1810.
44. RCME, p. 1814.
45. RCME, p. 1821.
46. For biographical details see ADB, Vol. 9, pp. 660–1. See also Jenny Mills, 'The impact of man on the northern jarrah forest from settlement in 1829 to the Forests Act 1918', and J.J. Havel, 'Land use conflicts and the emergence of multiple land use' and W.L. McCaw and N.D. Burrows, 'Fire management', all found in B. Dell et. al., *The Jarrah Forest*, pp. 274–6, 282 and pp. 318–20.
47. For Kessell see ADB, Vol. 15, pp. 14–15.
48. *The Forest Resources of the Territories of Papua and New Guinea* available in typescript copy, NLA, MS 496.
49. RCME, p. 2377.
50. RCME, pp. 2381–2.
51. RCME, p. 2383.
52. RCME, p. 2395.
53. For biographical details see ADB, Vol. 7, pp. 186–9.
54. RCME, p. 1545.
55. RCME, p. 1547.
56. RCME, p. 1543.
57. RCME, pp. 1563–5.
58. RCME, p. 1551.
59. Griffiths, *Forests of Ash*, pp. 186–7.
60. RCME, p. 1498.

61. RCME, p. 1533.
62. RCME, p. 1538.
63. RCME, pp. 1538–9.
64. See Bernard Mace, 'Mueller–Champion of Victoria's Great Trees', *The Victorian Naturalist*, 1996, pp. 198–205.
65. RCME, p. 1708.
66. RCME, pp. 1708–9.
67. Pyne, *Burning Bush*, p. 309.
68. All quotes except where stipulated from RCR, pp. 5–7.
69. RCR, p. 8.
70. RCR, p. 9.
71. RCR, p. 18.
72. RCME, p. 1568.
73. RCR, p. 10.
74. RCR, p. 10.
75. RCR, p. 11.
76. RCR, p. 11.
77. RCR, pp. 11–12.
78. RCR, p. 13.
79. RCME, pp. 789–90.
80. RCR, p. 17.
81. RCR, p. 14.
82. RCR, p. 18.
83. For Galbraith's evidence see RCME, pp. 2276–7.
84. RCR, p. 20.
85. RCME, pp. 1597–9.
86. RCR, p. 21.
87. RCME, p. 419.
88. RCR, p. 27.
89. Murray and White, *State of Fire*, pp. 113–20.
90. RCR, p. 24.
91. RCR, p. 31.
92. See *Victoria. Parliamentary Debates*, Session 1939, Vol. 207, pp. 52–69. Subsequently referred to as VPD in these notes.
93. VPD, 207, p. 59.
94. VPD, 207, p. 60.
95. VPD, 207, p. 61.
96. RCR, pp. 19–20. For Lind see ADB, Vol. 10, pp. 102–3.
97. RCR, p. 32.
98. VPD, 207, p. 64.
99. VPD, 207, p. 575.
100. VPD, 207, p. 461.

CHAPTER 4   BLACK DAYS

1. Murray and White, *State of Fire*, pp. 113–20.
2. Luke and McArthur, *Bushfires*, p. 309.
3. G.J. Barrow, 'A survey of houses affected in the Beaumaris fire, January 14, 1944' in the *Journal of the CSIR*, 18 (1945), pp. 27ff. See also Justin E. Leonard and Neville A McArthur, 'A History of Research into Building Performance in Australian Bushfires' in *Conference Proceedings: Australian Bushfire Conference, July 1999*, Charles Sturt University, 1999.

4. Leonard and McArthur, 'Building Performance', p. 4.
5. There seems to be some confusion about the exact number killed in the 1943–4 fires. The *Age* (3/2/44) gives the number 25 in an editorial; I have accepted this.
6. John D. Cash and Martin D. Cash, *Producer Gas for Motor Vehicles*, Sydney: Angus & Robertson, 1940 and 1942 describes how to make these engines. This book has been recently reprinted in the United States.
7. See his homepage at <www.freeman.powerup.com.au/jsfstory/gasprod.htm>.
8. J.C. Foley, 'Study', p. 102. See also Luke and McArthur, *Bushfires*, p. 309.
9. Leonard Stretton, *Report of the Royal Commission to inquire into The Place of Origin and the Causes of the Fires which commenced at Yallourn on the 14th Day of February 1944; The Adequacy of the Measures which had been taken to Prevent Damage; and The Measures to Protect the Undertaking and Township of Yallourn*, in *Victoria. Legislative Assembly. Votes and Proceedings 1945–1947*, Melbourne: Government Printer, 1945, Vol I, pp. 1769–81. Subsequently referred to as RCYF in these notes.
10. All quotations from RCYF, p. 3.
11. RCYF, p. 5.
12. RCYF, p. 5.
13. RCYF, p. 5.
14. RCYF, p. 7.
15. RCYF, p. 8.
16. RCYF, p. 8.
17. RCYF, p. 10. Later these buildings were decommissioned and dismantled by explosives in 1989. Clouds of asbestos were released by the explosion. According to a 2001 *Four Corners* investigation, 'Power Without Glory' the SEC knew all about the dangers of asbestos from the 1940s onwards.
18. RCYF, p. 12.
19. Murray and White, *State of Fire*, pp. 116–20.
20. Murray and White, *State of Fire*, p. 121.
21. Leonard Stretton, *Report of Royal Commission to inquire into Forest Grazing in Victoria. Legislative Assembly. Votes and Proceedings 1945–1947*, Melbourne: Government Printer, 1947, Vol. II, pp. 427–56. Subsequently referred to as RCFG in these notes. For John Cain (1882–1957) see ADB, Vol. 13, pp. 335–7.
22. These figures are derived from various sources and brought together by J.C. Foley, 'Study', pp. 9–10 and the Victorian Forests Commission, *Annual Report*, 30 June 1944. Interestingly, there was an 'outstanding' decrease in the number of fires started by graziers in the 1945 Forests Commission report: they had dropped to 10.6 per cent of known causes.
23. RCFG, p. 6.
24. RCFG, p. 6.
25. RCFG, p. 7.
26. RCFG, p. 11.
27. RCFG, p. 12.
28. RCFG, p. 12.
29. RCFG, p. 13.
30. RCFG, p. 14.
31. RCFG, p. 15.
32. RCFG, p. 15.
33. Pyne, *Burning Bush*, p. 325.
34. Information on the Act from *The Soil Conservation Authority*, self-published, 1951. Quotations from pp. 10–11.
35. Pyne, *Burning Bush*, p. 323.

NOTES

36. Libby Robin, *Building a Forest Conscience: A Historical Portrait of the Natural Resources Conservation League*, Springvale South, Vic.: Natural Resources Conservation League of Victoria, 1991. See chapters 1 and 2.
37. John Edwards, *Curtin's Gift: Reinterpreting Australia's Greatest Prime Minister*, Sydney: Allen & Unwin, 2005.
38. Pyne, *Burning Bush*, p. 326.
39. William J. Lines, *Patriots: Defending Australia's Natural Heritage, 1946–2004*, Brisbane: University of Queensland Press, 2006.
40. First published in 1978 by the Australian Government Publishing Service, Canberra.
41. Reprinted in 1953.
42. R.H. Luke, *Bush Fire Control in Australia*, Melbourne: Hodder & Stoughton, 1961.
43. Pyne, *Burning Bush*, pp. 360–3.
44. Quoted in Pyne, *Burning Bush*, p. 361.
45. The bibliographical details are already noted. It is the Commonwealth Bureau of Meteorology's Bulletin No. 38.
46. I have found Foley to be a reliable source, although there are fires he mentions that I have not been able to trace. This does not mean that he is wrong, but he never cites sources.
47. *Courier Mail*, 15/1/51 to 25/1/51.
48. McArthur and Luke, *Bushfires*, p. 260.
49. Peter Brookhouse and Don Nicholson, 'Large Pilliga fires and the development of fire management and fire suppression strategies', *Australian Bushfire Conference, Albury, July 1999. Conference Proceedings*.
50. Australian Bureau of Meteorology's paper, 'El Niño—Detailed Australian Analysis' at <www.bom.gov.au/climate/enso/australia_detail.shtml>.
51. Wagga's *Daily Advertiser*, 12/1/52 to 25/1/52.
52. Sandra Florance (ed.), *The Bega Bushfires of 1952: A Fiftieth-Anniversary Commemoration*, Bega: Bega Pioneers Museum, 2002.
53. *Bega District News* quotations from Florance, *Bega Bushfires*, pp. 29–33.
54. Quoted in Florance, *Bega Bushfires*, pp. 31–2.
55. All quotations from *Sydney Morning Herald*, 2/2/52.
56. The description is based on reports in the Wagga *Daily Advertiser*, 1/2/52.
57. This story is taken from the *Benalla Ensign*, 14/2/52.
58. 'The Fires of 1951/52' from <www.esb.act.gov.au/firebreak/1952fire.html>.
59. For Wilfred Kent Hughes, see ADB, Vol. 15, pp. 6–7.
60. *Canberra Times*, 7/2/52. See also Doris Hogg, '1952 Mt Stromlo Fire Remembered' at <www.mso.anu.edu.au/info/fire/1952_fire.php>.
61. Minard Fannie Crommelin (1881–1972) was born in Sydney. Postmistress at Woy Woy from 1906–10, she donated a block near Pearl Beach to Sydney University and gave money to the Australian Academy of Science. The Crommelin Conservation Fund is named after her. See Ruth Teale, 'Minard Crommelin—Conservationist' in Heather Radi (ed.), *Two Hundred Australian Women: A Redress Anthology*, Sydney: Women's Redress Press, 1988.
62. 'From Wandilo to Linton: Lessons learned from an in-depth analysis of 40 years of Australian Bushfire Tanker Burnovers' at <www.esb.act.gov.au/firebreak/paix-sydney99>.
63. Luke and McArthur, *Bushfires*, p. 282.
64. Luke and McArthur, *Bushfires*, pp. 245–6.
65. Murray and White, *State of Fire*, pp. 175–187.
66. Australian Bureau of Meteorology, 'El Niño—Detailed Australian Analysis'.

67. *Age*, 25/1/2003.
68. See W.L. William's interesting essay 'Snowfalls in New South Wales 1957–1979' found at <www.theweather.com.au>.

CHAPTER 5   ABLAZE

1. This is the number given in the *Mercury* (9/2/67). The figure of 200 students is also mentioned.
2. R.L. Wettenhall, *Bushfire Disaster: An Australian Community in Crisis*, Sydney, Angus & Robertson, 1975, p. 82.
3. Statistics from EMA at <www.ema.gov.au/ema/emadisasters>.
4. Figures from the *Saturday Evening Mercury*, 4/3/67, quoted in Wettenhall, p. 182. However, the Hobart Catholic newspaper, *The Standard*, (17/2/67) gives the statistic of seventeen churches. I have followed it rather than the Wettenhall statistics. The Anglican figures are confirmed in the *Mercury* (11/2/67).
5. Tony Coleman, 'The Impact of Climate Change on Insurance against Catastrophes', Institute of Actuaries of Australia, 2003 Biennial Convention, p. 7.
6. The main sources for the fires are D.M. Chambers and C.G. Brettingham-Moore, *The Bush Fire Disaster of 7th February, 1967. Report and Summary of Evidence*, Hobart, Parliamentary Paper, 16/1967. Subsequently cited as 'TBFD'. Also A.G. McArthur, 'The Tasmanian Bushfires of 7th February, 1967, and Associated Fire Behavior Characteristics', *Second Australian Conference on Fire—Conference Papers*, Melbourne, 1968. Subsequently cited as 'McArthur, TTBA'. These two papers provide the main sources for this chapter, together with Wettenhall's book *Bushfire Disaster*, as well as newspaper reports. Other sources are cited in the notes.
7. McArthur, TTBA, p. 29.
8. D.M. Chambers, G.G. Sinclair, A.G. McArthur and D.L. Burbury, *Fire Protection and Suppression*, Hobart, Parliamentary Paper 28/1967, p. 8.
9. TBFD, p. 6.
10. TBFD, p. 7.
11. Luke and McArthur, *Bushfires*, pp. 114–15.
12. McArthur, TTBA, p. 29.
13. McArthur, TTBA, p. 33. See also TBFD, p. 17.
14. TBFD, p. 16 and Appendix A.
15. TBFD, p. 16.
16. TBFD, p. 17.
17. TBFD, pp. 20–30.
18. TBFD, p. 21
19. 'Wellington Park Fire Management Strategy' (28 February 2000), p. 57.
20. McArthur, TTBA, p. 33 says it was 37,800 acres in area.
21. Brian Andrews, *Creating a Gothic Paradise: Pugin in the Antipodes*, Hobart: Tasmanian Museum and Art Gallery, 2002, pp. 129–32.
22. Wettenhall, *Bushfire Disaster*, p. 82.
23. TBFD, p. 26.
24. TBFD, p. 30.
25. TBFD, pp. 27–9.
26. TBFD, p. 28.
27. TBFD, p. 28.
28. Julie Gardem, *Peppermint Bay: A History of the Woodbridge Area from Settlement to the 1967 Bushfires*, Snug: Self-published, 1992, p. 64.
29. TBFD, pp. 21–2.

NOTES   391

30. Wettenhall, *Bushfire Disaster*, pp. 82–3 deals with the response of schools to the fires and the panic that can occur among parents if they do not know where their children are.
31. TBFD, p. 22.
32. TBFD, p. 23.
33. TBFD, p. 23.
34. I am following the time line set out in TBFD, pp. 23–5. It outlines the fire's progress at quarter or half hour intervals.
35. *Mercury*, 17/6/67.
36. TBFD, p. 24.
37. TBFD, p. 24.
38. TBFD, p. 25.
39. All statistics on house losses from McArthur, TTBA, 38.
40. TBFD, p. 25.
41. Wettenhall, *Bushfire Disaster*, pp. 92–3.
42. McArthur, TTBA, p. 38.
43. Wettenhall, *Bushfire Disaster*, p. 81.
44. McArthur, TTBA, p. 32.
45. TBFD, pp. 34–6.
46. TBFD, p. 35.
47. Wettenhall, *Bushfire Disaster*, p. 21.
48. Wettenhall, *Bushfire Disaster*, pp. 83, 106.
49. Wettenhall, *Bushfire Disaster*, p. 269.
50. Pyne, *Burning Bush*, p. 366.

CHAPTER 6   ON THE URBAN FRONTIER

1. Pyne, *Burning Bush*, p. 410.
2. Quoted by South Australian Premier Mike Rann in a speech at the Ash Wednesday Memorial Service, 16 February 2003.
3. For details see Martin, *Under Mount Lofty*, p. 240.
4. Rev. Dr. Gordon Powell, Diary, 20 February 1983. Kindly provided by his daughter, Jenny Goldie.
5. *Sydney Morning Herald*, 1/1/69.
6. N. Krusel and S.N. Petris, 'A Study of Civilian Deaths in the 1983 Ash Wednesday Bushfires Victoria, Australia', CFA Occasional Paper No. 1, pp. 8–9.
7. Statistics from newspapers and EMA. See <www.ema.gov.au/ema/emadisasters>.
8. Murray and White, *State of Fire*, pp. 208–9.
9. Based on Bryan Power's article, 'The 1973 Bushfire in Lysterfield and Rowville' in *Rowville-Lysterfield Community News*, March 2003.
10. The EMA website seems to contain an error. It gives the date of these fires as 12/18/76—EMA has the annoying habit of giving dates in the American fashion with the month first—whereas the actual date as recorded in the *Sydney Morning Herald* is 3 December 1976. Also EMA states that three people were killed in these fires and this statistic is repeated by the Council of Australian Governments' *National Inquiry on Bushfire Mitigation and Management* (p. 341). I can only find verification of one death in the contemporary newspapers.
11. See ACT ESB, 'From Wandilo to Linton—Lesson learned from an indepth analysis of 40 years of Australian Bushfire Tanker Burnovers', p. 1 on the Australian Capital Territory ESB website. See <www.esb.act.gov.au/firebreak>. Subsequently referred to as 'Tanker Burnovers'.
12. Martin, *Under Mount Lofty*, pp. 238–40.

13. Martin, *Under Mount Lofty*, pp. 246–51.
14. Bureau of Meteorology, 'El Niño—Detailed Australian Analysis'. See also the BOM's 1984 Report, 'Severe Fire Weather. A Case Study of Ash Wednesday, 16 February 1983'.
15. Bureau of Meteorology, 'Report on the meteorological aspects of the Ash Wednesday fires, 16 February 1983', p. 12.
16. See J. Oliver, N.R. Britton and M.K. James, *Disaster Investigation Report No. 7. The Ash Wednesday Bushfires in Victoria*, Townsville, Centre for Disaster Studies, James Cook University, 1984, p. 6. Subsequently referred to as *Disaster Investigation, Ash Wednesday*. See also the book published by the *Age* and *Adelaide Advertiser* newspapers, *Ash Wednesday. Wednesday February 16, 1983*. Subsequently referred to as *Age/Advertiser* (not paginated). John Baxter, *Who Burned Australia?* London, New English Library 1984.
17. *Disaster Investigation, Ash Wednesday*, p. 6.
18. *Disaster Investigation, Ash Wednesday*, p. 9.
19. *Disaster Investigation, Ash Wednesday*, pp. 21–2.
20. Quotations from the *Standard* (Warrnambool), 17/2/8.
21. See claims in the *Age*, 18/2/83.
22. Quoted in Murray and White, *State of Fire*, p. 234.
23. *Age/Advertiser*.
24. *Disaster Investigation, Ash Wednesday*, p. 41.
25. Statistics based on *Age/Advertiser* which gives the number of dead at eight in the Macedon fires. *Disaster Investigation, Ash Wednesday* (p. 42) gives the number at seven. I have followed this.
26. The book, published in Sydney by Thomson Law Book Company, has gone through three editions, 1979, 1986 and the most recent in 2004.
27. *Age/Advertiser*.
28. Quoted in Ian F. McLaren, *Aireys Inlet: From Anglesea to Cinema Point*, Anglesea and District Historical Society, 1988, p. 115.
29. Warren Bebbington (ed.), *A Dictionary of Australian Music*, Melbourne: Oxford University Press, 1998, p. 142.
30. *Age/Advertiser*. See also the Joan Hammond papers in the National Library, Canberra, at NLA MS8648.
31. McLaren, *Aireys Inlet*, p. 116.
32. See 'Remembering Ash Wednesday' at <www.anglesea-online.com.au/AshWednesday>.
33. *Age/Advertiser*.
34. *Age/Advertiser*.
35. Statistics drawn from Murray and White, *State of Fire*, p. 228.
36. *Disaster Investigation, Ash Wednesday*, p. 28.
37. *Disaster Investigation, Ash Wednesday*, p. 28.
38. Quoted in Murray and White, *State of Fire*, p. 240.
39. Quoted in Murray and White, *State of Fire*, p. 240.
40. All quotations from Murray and White, *State of Fire*, pp. 226–7.
41. *Disaster Investigation, Ash Wednesday*, p. 28.
42. Based on CFA statistics.
43. Sometimes the number of dead is given at 26. I have followed the statistics given by the Security and Emergency Management Office of South Australia.
44. Murray Nichol talking to George Negus, ABC TV, *New Dimensions*, 18/11/2002.
45. See 'Tanker Burnovers', p. 2.
46. The most detailed treatment of these fires is in Pam and Brian O'Connor, *Out of the Ashes: The Ash Wednesday Bushfires in the South East of SA*, Privately

published, Mount Gambier, 1993.
47. O'Connor, *Ashes,* p. 19.
48. O'Connor, *Ashes,* p. 20.
49. O'Connor, *Ashes,* p. 32.
50. O'Connor, *Ashes,* pp. 27–8.
51. See *Border Watch,* 21/2/83.
52. O'Connor, *Ashes,* p. 80.
53. See Pyne, *Burning Bush,* p. 413. House of Representatives Standing Committee on Environment and Conservation, *Bushfires and the Australian Environment: Report, August 1984,* Canberra, Government Printer, 1984.
54. New South Wales Rural Fire Service, *A Guide to the Rural Fires Act, 1997.*
55. There is a difference between the EMA figure for the area burnt and that of the New South Wales RFS. EMA says 650,000 hectares and RFS 733,342. I have followed the RFS.
56. Council of Australian Governments, *National Inquiry on Bushfire Mitigation and Management,* 2004, p. 40
57. Luke and McArthur, *Bushfires,* p. 271.
58. Luke and McArthur, *Bushfires,* p. 283.
59. Luke and McArthur, *Bushfires,* p. 271.
60. Luke and McArthur, *Bushfires,* p. 272.
61. I have followed the dates in Luke and McArthur, *Bushfires,* p. 296. According to the EMA, drawing on the New South Wales RFS, the Roto fires lasted from New Year's Day 1970 until 21 January.
62. Luke and McArthur, *Bushfires,* p. 296.
63. Luke and McArthur, *Bushfires,* pp. 339–44.
64. See *Sydney Morning Herald,* 19, 20, 21, 24, 27, 31/12/74 and *Sunday Sun-Herald,* 22/12/74.
65. EMA website. However, in newspaper reports I can only find references to one man being critically injured in New South Wales. See *Sydney Morning Herald,* 20/12/74. Luke and McArthur, *Bushfires* (pp. 339–44) make no reference to anyone being killed.
66. Statistics from EMA homepage.
67. Luke and McArthur, *Bushfires,* p. 271.

CHAPTER 7 'STINKING HOT AND WINDY'

1. The 1:50,000: 'Numbla Vale' map sheet 8624-N (second edition) clearly show it as 'Milligans Mountain' whereas other maps such as the RACV's 'East Gippsland and South East NSW' and George Seddon (*Searching for the Snowy: An Environmental History,* Sydney, Allen & Unwin, 1994, p. 49) call it 'Mulligans (or Minnigans) Mountain'.
2. 'Coronial Inquiry into the Fires Referred to as "Jagungal Wilderness Area, Kosciuszko National Park Complex of Fires" ', File No. 43/2003—Westmead Coroners Court. Subsequently referred to as 'COR INQ Kos' in these notes. There is no pagination in the Coronial Report.
3. Klaus Hueneke, *Huts of the High Country,* Canberra, Tabletop Press, 1999 reprint, p. 165.
4. Seddon, *Searching,* pp. 45–6.
5. COR INQ Kos.
6. 'Submission to the Coronial Inquiry into the December 2002–March 2003 Bushfires', Department of Environment and Conservation (New South Wales),

subsequently referred to as 'DEC SUB'. The 'Coronial Brief' prepared by the New South Wales Police was kindly made available to me by Detective Sergeant Paul Barclay and Detective Senior Constable Nicole Mulready. Subsequently referred to as 'POL BRIEF' with page reference.

7. For Victoria see Bruce Esplin, Malcolm Gill and Neal Enright, *Report of the Inquiry into the 2002–2003 Victorian Bushfires*, Melbourne: Dept. of Premier and Cabinet, 2003. Subsequently referred to as 'ESPLIN' with page reference. Kevin Wareing and David Flinn prepared the DSE's *The Victorian Alpine Fires January–March 2003* including maps and a daily narrative. Subsequently referred to as 'DSEVAF' with page reference. For the House of Representatives Select Committee see Garry Nairn (Chairman), *A Nation Charred: Inquiry into the Recent Australian Bushfires*, Canberra: Commonwealth of Australia, 2003. Subsequently referred to as 'NAIRN' with page reference.
8. NAIRN, 'Additional Comments', p. 3.
9. Bureau of Meteorology, *Annual Australian Climate Summary*, p. 1.
10. DEC SUB, p. 18.
11. DEC SUB, p. 19.
12. DEC SUB, pp. 19–20.
13. DEC SUB, p. 20. According to DSEVAF (p. 187) 'a total of 86 and 42 fires from lightning strikes were reported in Victoria and NSW respectively'.
14. NPWS, Media Release, 24/2/05.
15. COR INQ Kos.
16. COR INQ Kos.
17. New South Wales does not have a readily accessible narrative of the fires. Mine is constructed from briefs of evidence given to the New South Wales Coroner. The most helpful chronology is that of Sergeant Paul Batista in POL BRIEF.
18. DEC SUB, p. 23.
19. DEC SUB, p. 25.
20. POL BRIEF, p. 149.
21. DEC SUB, p. 23.
22. POL BRIEF, pp. 200–5.
23. Nick Goldie, personal communication, March 2005.
24. Information from the CD *Spirit and Survival: Stories from the 2003 Snowy Mountains Bushfires* recorded and edited by Louise Darmody of Sound Memories for the Snowy Bushfire Recovery Taskforce, track 6.
25. POL BRIEF, p. 150
26. POL BRIEF, pp. 341–2.
27. POL BRIEF, pp. 151–2.
28. POL BRIEF, p. 301.
29. DEC SUB, p. 24.
30. POL BRIEF, p. 153.
31. POL BRIEF, p. 154.
32. For the Wellsmore/Troha narrative, Darmody, *Spirit and Survival*, tracks 2–4. Material also from personal talks with Kerry Wellsmore.
33. DEC SUB, p. 24.
34. POL BRIEF, p. 154.
35. COR INQ Kos.
36. ESPLIN, p. 30.
37. ESPLIN, p. 30.
38. ESPLIN, p. 30.
39. ESPLIN, p. 173.
40. ESPLIN, p. 172.

NOTES

41. ESPLIN, pp. 172–9.
42. *Age*, 15/1/03. This article incorrectly asserts the air-crane carries 40,000 litres of water.
43. DSEVAF, p. 66.
44. DSEVAF, pp. 66, 68.
45. Here the road is called 'Gelantipy Road' and becomes 'Snowy River Road' north of the junction with the McKillops Bridge Road (see Victorian 1:50,000 map sheet 'Deception-Deddick'). I have called it the 'Snowy River Road' to avoid confusion. In New South Wales it becomes Barry Way.
46. ESPLIN, p. 33.
47. John Cook in Leanne Appleby and others (eds), *Flames Across the Mountains: Personal Accounts of the Bogong, Razorback and Pinnibar Fires*, Benambra: Leanne Appleby, 2004, p. v.
48. *Age*, 22/1/03.
49. ESPLIN, p. 33.
50. DSEVAF, p. 81.
51. DSEVAF, p. 87.
52. Mark Reeves in Appleby, *Flames*, pp. 49–53, 47–8.
53. Mark Reeves in Appleby, *Flames*, p. 51.
54. Georgina Williams in Appleby, *Flames*, p. 31
55. DSEVAF, pp. 95, 98.
56. DSEVAF, pp. 103, 118.
57. DSEVAF, p. 98 and *Age*, 29/1/03.
58. ESPLIN, p. 36.
59. DSEVAF, pp. 103, 105.
60. T. Bannister quoted in DSEVAF, p. 108.
61. Picture in DSEVAF, p. 198.
62. Beth Allen in Appleby, *Flames*, pp. 124–5.
63. DSEVAF, p. 111.
64. Clive Richardson in Appleby, *Flames*, p. 153.
65. ESPLIN, p. 36.
66. DSEVAF, p. 182.
67. *Canberra Times*, 6/2/03.
68. *Age*, 24/1/03.
69. POL BRIEF, pp. 214–16, 227–9.
70. POL BRIEF, p. 269.
71. POL BRIEF, pp. 691, 693.
72. Quoted in *Age*, 27/1/03.
73. DSEVAF, p. 12.
74. Mark Robinson, 'Bushfires, 2003—a rural GP's perspective', *Australian Family Physician*, Vol. 31, No. 12, pp. 987–8.
75. Brad van Wely, 'Hell's angels: Aerial firefighters play a vital role in Australia's constant battle against bushfires', *Flight Safety Australia*, March–April, 2003, p. 24.
76. van Wely, *Flight Safety Australia*, p. 26.
77. van Weky, *Flight Safety Australia*, pp. 30–1.
78. NAIRN, pp. 218–29.
79. NAIRN, p. 229.
80. Media Release of Prime Minister, 24/1/05.
81. Institute of Public Affairs, *IPA Review*, March 2003, p. 4.
82. A.J. Myers, 'In the matter of a Report of the Inquiry of the 2002–2003 Victorian Bushfires. Memorandum'. 15 July 2004.
83. ESPLIN, p. 86.

84. My thanks to Dr Malcolm Gill of CSIRO and Dr Brendan Mackey of ANU for information on the cypress pine. See Mary E. White, *Australia's Prehistoric Plants*, North Ryde, NSW: Methuen, 1984, pp. 22, 28.
85. 'Joey in the Snowy', *Quantum*, ABC TV, 1 March 2001.
86. 'Bushfire Recovery Reports', *The Shadow*, No. 14, Winter 2005, p. 11.
87. Personal communication and the *Age*, 2/6/03.
88. For Dr Ken Green see Darmody, *Spirit and Survival*, track 9.
89. DSEVAF, p. 187.
90. Ian Lunt, 'Vegetation Management' in *New Times*, Charles Sturt University e-magazine, March 2003 at <www.csu.edu.au/division/marketing/tms>.
91. NAIRN, p. 14.
92. Council of Australian Governments, *National Inquiry on Bushfire Mitigation and Management*, 31 March 2004, pp. 15–27, and ABS *Year Book Australia 2004*.

CHAPTER 8   A PERFECT FIRE DAY

1. Ron McLeod, *Inquiry into the Operational Response to the January 2003 Bushfires in the ACT*, 1 August, 2003, p. 42. Subsequently referred to as MCLR with page reference.
2. I watched the landing of the aircraft from Mount Ainslie to the west of the airport. A westerly downwind approach is somewhat unusual for Canberra.
3. Quotations from Des Fooks' account of his experiences recorded just after the fires. Later published in Stephen Matthews (ed.), *How Did the Fire Know We Lived Here?* Canberra: Ginninderra Press, 2003, pp. 41–4.
4. The text was partly constructed from the *Canberra Times* and *Sydney Morning Herald* for January 2003, and Peter Clack, *Firestorm: Trial by Fire*, Milton, Qld: John Wiley, 2003.
5. This account of the ACT fires is based on Ron McLeod's report (MCLR) for the ACT Government, pp. 16–20. The New South Wales narrative is based on the report of the Coroner, Carl Milovanovich, 'Coronial Inquiry into the Circumstances of the Fire(s) in the Brindabella Range in January 2003' (18 September 2003), subsequently 'COR INQ NSW' with page reference.
6. MCLR (p. 16) gives 'about 3.30pm' as the time. COR INQ NSW (p. 13) says that the Mount Coree fire-spotter observed 'a flash of lightning around 2.30pm in the direction of McIntyre's Hut and further lightning strikes in the direction of Corin Dam and Bendora'.
7. ACT Volunteer Brigades Association, *What You Wouldn't Believe . . . The January 2003 Bushfires in the ACT as seen by Bushfire and Emergency Services Personnel*, Canberra: ACT Volunteer Brigades Association, 2003, p. 2. With splendid photographs.
8. MCLR, p. 20.
9. *What You Wouldn't Believe . . .*, p. 8.
10. Quoted in MCLR, p. 22.
11. Quoted in MCLR, p. 22.
12. Clack, *Firestorm*, pp. 2–5. ABC TV, *Stateline*, 'Where There's Smoke', broadcast 9/5/03.
13. MCLR, pp. 57, 60–1, 62.
14. COR INQ NSW, p. 15.
15. MCLR, pp. 25, 26, 28, 29, 31.
16. MCLR, p. 35.
17. Quoted in MCLR, p. 36.
18. Matthews, *How Did the Fire*, p. 12.

NOTES

19. Quoted in MCLR, p. 43.
20. Figures from MCLR, p. 42.
21. MCLR, p. 44.
22. Matthews, *How Did the Fire*, pp. 51–3.
23. *Canberra Times*, 21/1/03.
24. Matthews, *How Did the Fire*, pp. 28–31.
25. Figures from MCLR, p. 47.
26. Tim Borough, Mount Stromlo Stories at <www.mso.anu.edu.au/info/fire/stories>.
27. Clack, *Firestorm*, pp. 101–6.
28. Clack, *Firestorm*, p. 102.
29. Media Release Actew AGL, 3 September 2003 commenting on Clack's book *Firestorm*.
30. MCLR, p. 45.
31. Matthews, *How Did the Fire*, p. 32.
32. For an account of the fire see Epilogue, Tom Frame and Don Faulkner, *Stromlo: An Australian Observatory*, Sydney: Allen & Unwin, 2003. For more accounts of the fire see Mount Stromlo Stories at <www.mso.anu.edu.au/info/fire/stories>.
33. Andrew Thompson, Mount Stromlo Stories. See notes 26, 32.
34. Matthews, *How Did the Fire*, p. 101.
35. Matthews, *How Did the Fire*, p. 120.
36. For ACT deaths see interim findings of ACT Coroner, October 2003.
37. Eda and Tony McGloughlin, personal interview, 4/1/05.
38. Based on *Canberra Times*, 26/1/01 to 10/1/02.
39. Matthews, *How Did the Fire*, pp. 89, 90–1.
40. Tritia Evans, personal communication, 27/1/05.
41. From program record quoted in MCLR, pp. 178–180, and from a copy provided by the ABC.
42. MCLR, pp. 180–1.
43. MCLR, p. 1. Statistics on the fires vary. For instance Emergency Management Australia says that 530 homes were destroyed and the total cost was 1 billion dollars.
44. Matthews, *How Did the Fire*, p. 141.
45. Phil Chaney, ABC TV program *Stateline*, 16 May 2003.
46. From author's interview with Eda and Tony McGloughlin, 4/1/05.
47. Clack, *Firestorm*, pp. 47–8.
48. Clack, *Firestorm*, p. 66.
49. MCLR, p. 57.
50. MCLR, p. 231.
51. MCLR, p. 63.
52. RCME, pp. 2386, 2390.
53. MCLR, p. 239.
54. MCLR, p. 11. The 'history' is on pp. 11–12.
55. MCLR, p. 239.
56. Joan Webster, *The Complete Bushfire Safety Book*, Sydney: Random House, 2000.
57. MCLR, pp. 250–1. For what is the most detailed but still incomplete history of fire in the ACT see ACT ESB. See <www.esb.act.gov/firebreak>.
58. MCLR, p. 240.
59. MCLR, p. 22.
60. Quoted in *What You Wouldn't Believe*, p. 87.
61. MCLR, p. 241.

62. MCLR, p. 241.
63. Quoted in MCLR, p. 53.
64. See J.C.G. Banks' article, 'A History of Forest Fire in the Australian Alps' in Roger Good (ed.), *The Scientific Significance of the Australian Alps. The Proceedings of the First Fenner Conference*, Australian Alps National Parks Liaison Committee, 1989, pp. 265–80.
65. Banks, 'Forest Fire', p. 278.

CHAPTER 9   TO BURN OR NOT TO BURN?

1. Dean Yibarbuk in D. Schulz (ed.), *Fire in the Savannas: Voices from the Landscape*, Darwin: Co-operative Research Centre for the Sustainable Development of Tropical Savannas.
2. John Mulvaney and Johan Kamminga, *Prehistory of Australia*, Sydney: Allen & Unwin, 1999, pp. 143–5.
3. James L. Kohen, *Aboriginal Environmental Impacts*, Sydney: UNSW Press, 1995, p. 16, and David Horton, *The Pure State of Nature: Sacred Cows, Destructive Myths and the Environment*, Sydney: Allen & Unwin, 2000, p. 8. See also the *Encyclopaedia of Aboriginal Australia*, Canberra: Aboriginal Studies Press, 1994.
4. Norman Tindale, *The Aboriginal Tribes of Australia*, Canberra: ANU Press, 1974.
5. Rhys Jones, 'Fire-stick Farming' in *Australian Natural History*, 16(1969), pp. 224–8. David Horton, 'Fire and Australian Society' in WA Department of Conservation and Land Management, *Fire in Eco-Systems of South-West Western Australia*, Perth: Proceedings of April 2002 Symposium, Vol. 2, p. 8.
6. Sylvia Hallam, *Fire and Hearth: A study of Aboriginal Usage and European Usurpation in South-Western Australia*, Canberra: Institute of Aboriginal Studies, 1975.
7. Eric Rolls, *A Million Wild Acres*, Ringwood, Vic: Penguin, 1984, p. 248.
8. Thomas Livingstone Mitchell, *Journal of an Expedition into the Interior of Tropical Australia* (1848) quoted by Rolls, *Wild Acres*, p. 249.
9. Horton, 'Fire', p. 9.
10. Josephine Flood, *The Archaeology of the Dreamtime: The Story of Prehistoric Australia and its People*, Sydney: Angus & Robertson, 1983, p. 170. See pp. 157–70.
11. Tim Flannery, *The Future Eaters: An Ecological History of the Australasian Lands and People*, Sydney: Reed Books, 1994.
12. Flannery, *Future Eaters*, p. 223.
13. Tim Flannery, 'Beautiful Lies. Population and Environment in Australia' in *Quarterly Essay*, No. 9, Melbourne: Black Ink, 2003.
14. Flannery, *Beautiful Lies*, p. 20.
15. Flannery, *Beautiful Lies*, p. 41
16. Bob Debus' letter in *Quarterly Essay*, No. 11, pp. 112–13.
17. Lines, *Great South Land*, p. 12.
18. Horton, 'Fire', p. 11.
19. D.G. and J.R. Ryan and B. J. Starr, *The Australian Landscape: Observations of Explorers and Early Settlers*, Wagga Wagga: The Murrumbidgee Catchment Management Committee, 1995.
20. *The Australian Landscape*, p. 2.
21. *The Australian Landscape*, p. 15.
22. John S. Benson and Phil A. Redpath, 'The nature of pre-European vegetation in south-eastern Australia: A critique of Ryan, D.G., J.R. and Starr, B.J.', *Cunninghamia*, 5/2 (December 1997), pp. 285–328.
23. This phrase is used in the abstract of a paper by Beth Gott, 'Fire as an Aboriginal

NOTES

management tool in South-Eastern Australia' in the Conference Proceedings of the Australian Bushfire Conference, Albury, July 1999.

24. George Main, *Heartland: The Regeneration of Rural Place*, Sydney: UNSW Press, 2005, p. 210.
25. Ronald Strahan, *The Australian Museum Complete Book of Australian Mammals*, North Ryde: Cornstork Publishing, 1991, pp. 184–5. See pp. 93–109 for bandicoots and bilbies.
26. Michael McKernan, *Drought: The Red Marauder*, Sydney: Allen & Unwin, 2005, pp. 8–9.
27. Lines, *Great South Land*, pp. 12–13.
28. David Bowman, Tansley Review No. 101 paper, 'The impact of Aboriginal landscape burning on the Australian biota', in *New Phytologist*, 140 (1998), pp. 385–410. See also his *Australian Rainforests: Islands of Green in the Land of Fire*, Cambridge: Cambridge University Press, 2000, and especially 'Bushfires: A Darwinian Perspective' in Geoffrey Cary et al., *Australia Burning: Fire Ecology, Policy and Management Issues*, Melbourne: CSIRO Publishing, 2003, pp. 3–14.
29. Bowman, art. cit., p. 404.
30. Bowman, art. cit., p. 405.
31. Bowman in 'Monsoon Fire' on the ABC TV program *Catalyst*, 29 May 2003.
32. Jon Marsden-Smedley, 'Changes in Southwestern Tasmanian fire regimes since the early 1800s' in *Papers and Proceedings of the Royal Society of Tasmania*, Vol. 132, 1998, p. 17. See also W.D. Jackson, 'The Tasmanian legacy of man and fire', *Papers and Proceedings of the Royal Society of Tasmania*, Vol. 133/1 (1999), pp. 1–14.
33. Marsden-Smedley, art. cit., p. 19.
34. Marsden-Smedley, Jon 'Fire management in Tasmania's Wilderness World Heritage Area: Ecosystem restoration using Indigenous-style fire regimes?' in *Ecological Management and Restoration*, 1/3 (2000), pp. 195–203.
35. Horton, *State of Nature*, p. 19.
36. Horton, *State of Nature*, p. 7.
37. W.E.H. Stanner, 'The Dreaming' in W.H. Edwards (ed.) *Traditional Aboriginal Society: A Reader*, Melbourne: Macmillan, 1987, p. 225.
38. Horton, *State of Nature*, p. 33.
39. Horton, *State of Nature*, pp. 81–5.
40. Horton, *State of Nature*, p. 17.
41. Marcia Langton, 'The European construction of wilderness' in *Wilderness News*, No. 143, Summer 1995/96, pp. 16–17.
42. Horton, *State of Nature*, p. 100.
43. Horton, *State of Nature*, pp. 171–2.
44. Mary E. White, *Listen . . . Our Land is Crying. Australia's Environment: Problems and Solutions*, Sydney: Kangaroo Press, 1997, pp. 220–1.
45. A. Malcolm Gill, *Fire Pulses in the Heart of Australia: Fire Regimes and Fire Management in Central Australia*. Report to Environment Australia, Research paper, Chapter One, Conclusion.
46. Scott D. Mooney, Kate L. Radford and Gary Hancock, 'Clues to the "burning question": Pre-European fire in the Sydney coastal region from sedimentary charcoal and palynology' in *Ecological Management & Restoration*, Vol. 2, No. 3, December 2001, p. 211. The 'Clark' referred to is R.L. Clark, 'Pollen and charcoal evidence for the effects of Aboriginal burning on the vegetation of Australia' in *Archaeology in Oceania*, Vol. 18, pp. 32–7.
47. See abstract of Mooney, et al., in the 'Clues' article, art. cit., p. 203.
48. Benson and Redpath in *Cunninghamia*, art. cit., p. 286.

49. Benson and Redpath in *Cunninghamia*, art. cit., pp. 322, 323.
50. Kohen, *Impacts*, p. 95.
51. Kohen, *Impacts*, p. 95. For Noel Butlin see *Our Original Aggression*, Sydney: George Allen & Unwin, 1983.
52. Kohen, *Impacts*, p. 132.
53. I am grateful to John Benson of the Royal Botanical Gardens, Sydney, and to Bill Lines for provocative questions and discussions on these issues.
54. Horton, 'Fire', p. 11.

CHAPTER 10   FIRE THUGS

1. *Four Corners*, 24/2/03, ABC TV, interview with Assistant Police Commissioner John Laycock.
2. Rebekah Doley, Submission to Nairn, Garry (Chairman), House of Representatives Select Committee, *A Nation Charred*, 2003.
3. Richard. N. Kocsis, 'Arson: Exploring Motives and Possible Solutions', Australian Institute of Criminology, Research Paper 236, August 2002, p. 2.
4. Matthew Willis, *Bushfire Arson: A Review of the Literature*, Canberra: Australian Institute of Criminology, Research and Public Policy Series, No. 61, p. 77.
5. Willis, *Bushfire Arson*, p. 78.
6. Fabian Crowe, 'The Arsonist's Mind', National Academies Forum, *Fire! The Australian Experience*, Seminar at University of Adelaide, 30 September–1 October 1999, p. 45.
7. Crowe, 'The Arsonist's Mind', pp. 46–7.
8. Talina Drabsch, 'Arson', Briefing Paper 2/2003. The Parliament of NSW, p. 11.
9. *Sydney Morning Herald*, 27/12/01.
10. Kocsis, 'Arson', p. 2.
11. Drabsch, 'Arson', p. 14.
12. Crowe, 'The Arsonist's Mind', p. 47.
13. Commissioner Phil Koperberg, *Lateline*, ABC TV, 26/11/2002.
14. Willis, *Bushfire Arson*, p. 127
15. For the fatalities see Joan Webster, *The Complete Bushfire Safety Book*, Sydney: Random House Australia, p. 135.
16. Kocsis, 'Arson', p. 6.
17. Reported on ABC TV, *Lateline*, 25/11/02.
18. Drabsch, 'Arson', p. 18 and ABC TV, *Lateline*, 25/11/02.
19. ABC TV, *Lateline*, 25/11/02.
20. Drabsch, 'Arson', p. 16. Drabsch is speaking only about New South Wales. She refers to similar programs in other states (pp. 18–20).
21. Willis, *Bushfire Arson*, p. 107.
22. Willis, *Bushfire Arson*, p. 107.
23. Thomas Berry, *Insights*, ABC Radio National, 27 January 1991.

CHAPTER 11   FIREPROOFING AUSTRALIA?

1. It was Frank Purcell of Shepparton, Victoria who reminded me of the cohesive power of the bushfire threat in bringing people together in the twentieth century.
2. See R. Rose et al., 'The importance and application of spatial patterns in the management of fire regimes for the protection of life and property and the conservation of biodiversity', Australian Bushfire Conference, Albury, July 1999. See also Stewart Smith, 'Bushfires', Briefing Paper, 5/2002, New South

NOTES

       Wales Parliamentary Library, p. 10.
3. Quoted in Smith, op. cit., 'Bushfires', p. 10.
4. ABC TV *7.30 Report*, 9/12/2002. See also Ian Brandes, 'Burning Questions. Australia's Bushfire Policy', 7 February 2003 at <www.econ.usyd.edu.au/drawingboard/digest/ 0302/brandes>.
5. A. Malcolm Gill, 'Landscape fires as social disasters: An overview of the "bushfire problem" ', *Environmental Hazards*, 6(2005), pp. 65–80.
6. Gill, 'Landscape Fires', p. 66.
7. Gill, 'Landscape fires', p. 78.
8. See Collins, *God's Earth: Religion as if Matter really Mattered*, North Blackburn, Vic.: HarperCollins Religious, 1995. See pp. 171–230. See also my discussion of imagination in *Between the Rock and a Hard Place: Being Catholic Today*, Sydney: ABC Books, 2004, pp. 72–108.
9. RCME, pp. 2381–2.
10. Brendan Mackey (ed.), *Wildfire, Fire and Future Climate: A Forest Ecosystem Analysis*, Melbourne: CSIRO Publishing, 2002, p. 3. See also V.G. Gorshkov's own book, *Physical and Biological Bases of Life Stability: Man, Biota, Environment*, Berlin: Springer-Verlag, 1995.
11. RCME, p. 2383.
12. Professor Stephen H. Schneider, ABC Radio National, *The Science Show*, 18/6/05.
13. *Council of Australian Governments, National Inquiry on Bushfire Mitigation and Management*, 31 March 2004, p. 3.
14. Dr Kevin Hennessy, ABC Radio National, *The Science Show*, 13/12/03.
15. In this part of the text I was very much helped by Dr Michael L. Roderick of the Research School of Biological Studies and Professor Graham Farquhar of the Environment Biology Group at the ANU, Canberra.
16. Michael L. Roderick and Graham Farquhar, 'The cause of decreased pan evaporation over the last 50 years', *Science*, Vol. 298, No. 5597 15 November 2002.
17. *Council of Australian Governments, National Inquiry on Bushfire Mitigation and Management*, 31 March 2004, p. xii, but not released until the last week of January 2005. Subsequently referred to as COAGBMM with page reference.
18. COAGBMM, Recommendation 3.1, p. xxi.
19. COAGBMM, p. 40.
20. My thanks to former Democrat Senator Dr. John Coulter for information on the CL 415.
21. John Coulter in an email to the author, 28/10/05.
22. Judith Wright, 'The two fires', *Collected Poems 1942–1985*, Sydney: Angus & Robertson, 1994.
23. Lang, John Dunmore, *Poems Sacred and Secular—Written Chiefly at Sea Within the Last Half Century*, Sydney: William Maddock, 1873. My thanks to Dr Jeff Brownrigg, Head of Research at Screensound and Professor of Cultural Histroy at the University of Canberra for bringing this Dunmore Lang poem to my attention.

BIBLIOGRAPHY

REPORTS

Barber, Esler, *Report of the Board of Inquiry into the Occurrence of Bush and Grass Fires in Victoria*, 1977.

Chambers, D.M. and Brettingham-Moore, C.G., *The Bush Fire Disaster of 7th February, 1967. Report and Summary of Evidence*, Hobart, Parliamentary Paper, 16/1967.

Council of Australian Governments, *National Inquiry on Bushfire Mitigation and Management*, Canberra: Commonwealth of Australia, 2004.

Department of Sustainability and Environment, *The Victorian Alpine Fires January–March 2003*, Melbourne: Department of Sustainability and Environment, 2003.

Esplin, Bruce, Gill, Malcolm and Enright, Neal, *Report of the Inquiry into the 2002–2003 Victorian Bushfires*, [the Esplin Report] Melbourne: Department of Premier and Cabinet, 2003.

Foley, J.C., *A Study of the Meteorological Conditions associated with Bush and Grass Fires and Fire Protection in Australia*, Melbourne: Commonwealth of Australia, Bureau of Meteorology, Bulletin No. 38, 1947.

House of Representatives Standing Committee on Environment and Conservation, *Bushfires and the Australian Environment: Report, August 1984*, Canberra, Government Printer, 1984.

McArthur, A.G., 'The Tasmanian Bushfires of 7th February, 1967, and Associated Fire Behavior Characteristics', *Second Australian Conference on Fire—Conference Papers*, Melbourne, 1968.

McLeod, Ron, *Inquiry into the Operational Response to the January 2003 Bushfires in the ACT*, [the McLeod Report], Canberra: Chief Minister's Department, 2003.

Nairn, Garry (Chairman), House of Representatives Select Committee, *A Nation Charred: Inquiry into the Recent Australian Bushfires*, Canberra: Commonwealth of Australia, 2003.

Oliver, J., Britton, N.R. and James, M.K., *Disaster Investigation Report No. 7. The Ash Wednesday Bushfires in Victoria*, Townsville, Centre for Disaster Studies, James Cook University, 1984.

Stretton, Leonard, *Report of the Royal Commission into the causes and measures taken to prevent the bushfires of January 1939, and to protect life and property*, Melbourne: Government Printer, 16 May, 1939.

———, *Royal Commission on Bushfires, 16 May 1939. Minutes of Evidence*. Typescript now available on a set of CDs at NLA.

———, *Report of the Royal Commission to inquire into The Place of Origin and the Causes of the Fires which commenced at Yallourn on the 14th Day of February 1944; The Adequacy of the Measures which had been taken to Prevent Damage, and The Measures to Protect the Undertaking and Township of Yallourn, in Victoria. Legislative Assembly, Votes and Proceedings 1945–1947*, Melbourne: Government Printer, 1945.

———, 'Report of Royal Commission to inquire into Forest Grazing', *Victoria. Legislative Assembly. Votes and Proceedings 1945–1947*, Melbourne: Government Printer, 1947.

BOOKS

Abbott, Ian and Burrows, Neil (eds), *Fire in the Ecosystems of South-west Western Australia: Impacts and Management*, Leiden, Netherlands: Backhuys Publishers, 2003.

ACT Volunteer Brigades Association, *What You Wouldn't Believe . . . . The January 2003 Busfires in the ACT as seen by Bushfire and Emergency Services Personnel*, Canberra: ACT Volunteer Brigades Association, 2003.

Age/Adelaide Advertiser, *Ash Wednesday: Wednesday February 16*, 1983, Melbourne: *Age*, 1983.

Andrews, Brian, *Creating a Gothic Paradise: Pugin in the Antipodes*, Hobart: Tasmanian Museum and Art Gallery, 2002.

Appleby, Leanne et al. (eds), *Flames Across the Mountains: Personal Accounts of the Bogong, Razorback and Pinnibar Fires*, Benambra: Leanne Appleby, 2004.

Baker, Sidney J., *The Australian Language: The Meanings, Origins and Usage; From Convict Days to the Present*, Sydney: Currawong Press, 1978.

Banks, J.C.G., 'A History of Forest Fire in the Australian Alps' in Roger Good (ed.), *The Scientific Significance of the Australian Alps: The Proceedings of the First Fenner Conference*, Australian Alps National Parks Liaison Committee, 1989.

Baxter, John, *Who Burned Australia?* London: New English Library, 1984.

Bebbington, Warren (ed.), *A Dictionary of Australian Music*, Melbourne: Oxford University Press, 1998.

Bolton, C.G., *Spoils and Spoilers: Australians Make their Environment*, Sydney: Allen & Unwin, 1981.

Bowman, David, *Australian Rainforests: Islands of Green in the Land of Fire*, Cambridge: Cambridge University Press, 2000.

———, 'Bushfires. A Darwinian Perspective' in Geoffrey Cary et al., *Australia Burning: Fire Ecology, Policy and Management Issues*, Melbourne: CSIRO Publishing, 2003.

Bradshaw, John, *The Early Settlement of South Gippsland: A Pictorial History*, Korumburra, Vic: Coal Creek Heritage Village, 1999.

Britton, Neil R., *Disaster Investigation Report No. 6. The Bushfires in Tasmania, February 1982*, Townsville, Centre for Disaster Studies, James Cook University, 1984.

Butler, Graeme, *Buln Buln: A History of the Buln Buln Shire*, Drouin, Buln Buln Shire Council, 1979.

Butlin, Noel, *Our Original Aggression*, Sydney: George Allen & Unwin, 1983.

Cannon, Michael, *The Land Boomers*, Melbourne: Melbourne University Press, 1966.

———, *Life in the Country*, Melbourne: Thomas Nelson, 1973.

Carroll, Brian, *The Upper Yarra: An Illustrated History*, Yarra Junction: Shire of Upper Yarra, 1988.

Cash, John D, and Cash, Martin D., *Producer Gas for Motor Vehicles*, Sydney: Angus & Robertson, 1940.

Christensen, P.E.S., *The Karri Forest: Its Conservation Significance and Management*, Como, WA: Department of Conservation and Land Management, 1992.

Clack, Peter, *Firestorm: Trial by Fire*, Milton, Qld: John Wiley, 2003.

Collins, David, *An Account of the English Colony of New South Wales*, London: T. Cadell & W. Davies, Vol. II, 1802.

Collins, Paul, *God's Earth: Religion as if Matter really Mattered*, North Blackburn, Vic:, HarperCollins Religious, 1995.

———, *Hell's Gates: The Terrible Journey of Alexander Pearce, Van Diemen's Land Cannibal*, Melbourne: Hardie Grant Books, 2002.

———, *Between the Rock and a Hard Place: Being Catholic Today*, Sydney: ABC Books, 2004.

Curnow, Heather, *William Strutt*, Sydney: Australian Gallery Directors' Council, 1980.

Dell, B., Havel, J.J. and Malajczuk, N., *The Jarrah Forest: A Complex Mediterranean Ecosystem*, Dordrecht, Netherlands: Kluwer Academic Publishers, 1989.

Department of Conservation and Land Management, *Fire in the Ecosystems of South-West Western Australia: Impacts and Management*. Volume 2, *Community Perspectives About Fire*, Proceedings of April 2002 Symposium (available on Internet).

Dingle, Tony and Rasmussen, Carolyn, *Vital Connections: Melbourne and its Board of Works, 1891–1991*, Melbourne: McPhee Gribble, 1991.

Edwards, Diane H., *The Diamond Valley Story*, Greensborough: Shire of Diamond Valley, 1979.

Edwards, John, *Curtin's Gift: Reinterpreting Australia's Greatest Prime Minister*, Sydney: Allen & Unwin, 2005.

Evans, Peter, *Rails to Rubicon: A History of the Rubicon Forest*, Melbourne: Light Railway Research Society, 1994.

Fenton, James, *Bush Life in Tasmania: Fifty Years Ago*, Launceston: Regal Publications, 1989 (reprint).

Flannery, Tim, *The Future Eaters: An Ecological History of the Australasian Lands and People*, Sydney: Reed Books, 1994.

Flood, Josephine, *The Archaeology of the Dreamtime: The Story of Prehistoric Australia and its People*, Sydney: Angus & Robertson, 1983.

Florance, Sandra, (ed.), *The Bega Bushfires of 1952: A Fiftieth-Anniversary Commemoration*, Bega: Bega Pioneers Museum, 2002.

Forshaw, Joseph M., *Australian Parrots*, Melbourne: Lansdowne Press, 1969.

Frame, Tom and Faulkner, Don, *Stromlo: An Australian Observatory*, Sydney: Allen & Unwin, 2003.

Fumagalli, Vito, *Landscapes of Fear: Perceptions of Nature and the City in the Middle Ages*, Cambridge: Polity Press, English trans., 1994.

Gardem, Julie, *Peppermint Bay: A History of the Woodbridge Area from Settlement to the 1967 Bushfires*, Snug: Self published, 1992.

Gill, A. Malcolm, *Fire Pulses in the Heart of Australia: Fire Regimes and Fire Management in Central Australia*. Report to Environment Australia, 2000.

Gorshkov, Victor G., *Physical and Biological Bases of Life Stability: Man, Biota, Environment*, Berlin: Springer-Verlag, 1995.

Griffiths, Tom, *Forests of Ash: An Environmental History*, Cambridge: Cambridge University Press, 2001.

Hallam, Sylvia, *Fire and Hearth: A Study of Aboriginal Usage and European Usurpation in South-Western Australia*, Canberra: Institute of Aboriginal Studies, 1975.

Hancock, W.K., *Discovering Monaro: A Study of Man's Impact on His Environment*, Cambridge: Cambridge University Press, 1972.

Horton, David, *The Pure State of Nature: Sacred Cows, Destructive Myths and the Environment*, Sydney: Allen & Unwin, 2000.

Hueneke, Klaus, *Huts of the High Country*, Canberra: Tabletop Press, 1999 reprint.

Jenkins, Joseph, *Diary of a Welsh Swagman, 1869–1894*, abridged and annotated by William Jenkins, Melbourne: Macmillan, 1975.

Kerr, J.H., *Glimpses of Life in Victoria by a Resident*, Edinburgh: Edmonton and Douglas, 1872.

Kocsis, Richard. N., 'Arson: Exploring Motives and Possible Solutions', Australian Institute of Criminology, Research Paper 236, August 2002.

Kohen, James L., *Aboriginal Environmental Impacts*, Sydney: UNSW Press, 1995.

Lake, Marilyn, *Limits of Hope: Soldier Settlement in Victoria, 1915–38*, Melbourne: Oxford University Press, 1987.

Lang, John Dunmore, *Poems Sacred and Secular—Written Chiefly at Sea Within the Last Half Century*, Sydney: William Maddock, 1873.

Lines, William J., *Taming the Great South Land*, Sydney: Allen & Unwin, 1991.
——, *Patriots: Defending Australia's Natural Heritage, 1946–2004*, Brisbane: University of Queensland Press, 2006.
Luke, R.H., *An Outline of Forest Fire Control Principles for the Information of NSW Foresters*, Sydney: Forestry Commission of NSW, 1947.
——, *Bush Fire Control in Australia*, Melbourne: Hodder & Stroughton, 1961.
Luke, R.H. and McArthur, A.G., *Bushfires in Australia*, Canberra: Australian Government Publishing Service, 1978.
Mackey, Brendan (ed.), *Wildfire, Fire and Future Climate: A Forest Ecosystem Analysis*, Melbourne: CSIRO Publishing, 2002.
Main, George, *Heartland: The Regeneration of Rural Place*, Sydney: UNSW Press, 2005.
Martin, Robert, *Under Mount Lofty: A History of the Stirling District in South Australia*, Stirling: District Council of Stirling, second edition, 1996.
Matthews, Stephen (ed.), *How Did the Fire Know We Lived Here?* Canberra: Ginninderra Press, 2003.
McCarthy, Mike, *Mountains of Ash: A History of the Sawmills and Tramways of Warburton and District*, Melbourne: Light Railway Research Society, 2001.
McKernan, Michael, *Drought: The Red Marauder*, Sydney: Allen & Unwin, 2005.
McLaren, Ian F., *Aireys Inlet: From Anglesea to Cinema Point*, Anglesea: Anglesea and District Historical Society, 1988.
Mulvaney, John and Kamminga, Johan, *Prehistory of Australia*, Sydney: Allen & Unwin, 1999.
Murray, Robert and White, Kate, *State of Fire: A History of Volunteer Firefighting and the Country Fire Authority in Victoria*, Melbourne: Hargreen Publishing Company, 1995.
Nichols, F.W. and J.M., *Charles Darwin in Australia*, Cambridge: Cambridge University Press, 1989.
Noble, W.S., *The Strzeleckis: A New Future for the Heartbreak Hills*, Melbourne: Forests Commission Victoria, No date but 1986.
——, *Ordeal By Fire: The Week the State Burned-Up*, Melbourne: Privately Published, 1979.
O'Connor, Pam and Brian, *Out of the Ashes: The Ash Wednesday Bushfires in the South East of SA*, Mount Gambier: Privately published, 1993.
Pannam, Clifford, *The Horse and the Law*, Sydney: Thomson Law Book Company, 1979.
Pyne, Stephen, *Burning Bush: A Fire History of Australia*, Seattle: University of Washington Press, 1991.
——, *The Still-Burning Bush*, Melbourne: Scribe, 2006.
Quartermaine, Peter and Watkins, Jonathan, *A Pictorial History of Australian Painting*, London: Bison Books, 1989.
Rasmussen, Carolyn, *Increasing Momentum: Engineering at the University of Melbourne*, Melbourne: Melbourne University Press, 2004.
Reynolds, Henry, *The Other Side of the Frontier: Aboriginal Resistance to the European Invasion of Australia*, Ringwood: Penguin, 1982.
——, *Frontier: Aborigines, Settlers and Land*, Sydney: Allen & Unwin, 1987.
Roberts, Stephen H., *The Squatting Age in Australia 1835–1847*, Melbourne: Melbourne University Press, 1964.
Robin, Libby, *Building a Forest Conscience: A Historical Portrait of the Natural Resources Conservation League*, Springvale South: Natural Resources Conservation League of Victoria, 1991.
Rolls, Eric, *A Million Wild Acres*, Ringwood: Penguin, 1984.

Seddon, George, *Searching for the Snowy: An Environmental History*, Sydney: Allen & Unwin, 1994.
South Gippsland Pioneers Association, *The Land of the Lyre Bird: A Story of Early Settlement in the Great Forest of South Gippsland*, Korumburra: Korumburra and District Historical Society, 1998.
Stamford, F.E., Stuckey, G.L. and Maynard, G.L., *Powelltown: A History of its Timber Mills and Tramways*, Melbourne: Light Railway Research Society, 1984.
Stanner, W.E.H., 'The Dreaming' in W.H. Edwards (ed.) *Traditional Aboriginal Society: A Reader*, Melbourne: Macmillan, 1987.
Strahan, Ronald, *The Australian Museum Complete Book of Australian Mammals*, North Ryde: Cornstalk Publishing, 1991.
Strutt, William, *The Australian Journal of William Strutt, ARA, 1850–1862*, Sydney: edited and published by George Mackaness, Privately printed, 1958.
Teale, Ruth, 'Minard Crommelin—Conservationist' in Heather Radi (ed.) *Two Hundred Australian Women: A Redress Anthology*, Sydney: Women's Redress Press, 1988.
Tindale, Norman, *The Aboriginal Tribes of Australia*, Canberra: ANU Press, 1974.
Turner, Henry Gyles, *A History of the Colony of Victoria from its Discovery to its Absorption into the Commonwealth of Australia*, London: Longmans, Green, 1904.
Victorian Model Railway Society, *Victorian Railways: Rolling Stock: Diagrams & Particulars of Locomotives, Steam, Electric, Diesel Electric, etc., 1926–1961*, Kensington: Victorian Model Railway Society, 1986.
von Hügel, Baron Charles, *New Holland Journal November 1833–October 1834*, Dymphna Clark (ed.), Melbourne: Melbourne University Press, 1994.
Webster, Joan, *The Complete Bushfire Safety Book*, Sydney: Random House, 2000.
Wettenhall, R.W., *Bushfire Disaster: An Australian Community in Crisis*, Sydney, Angus & Robertson, 1975.
White, Mary E, *Australia's Prehistoric Plants*, North Ryde: Methuen, 1984.
———, *Listen . . . Our Land is Crying: Australia's Environment: Problems and Solutions*, Sydney: Kangaroo Press, 1997.
Willis, Matthew, *Bushfire Arson: A Review of the Literature*, Canberra: Australian Institute of Criminology, Research and Public Policy Series, No. 61.
Wright, Judith, *Collected Poems 1942–1985*, Sydney: Angus & Robertson, 1994.
Yibarbuk, Dean in D. Schulz (ed.), *Fire on the Savannas: Voices from the Landscape*, Darwin: Co-operative Research Centre for the Sustainable Development of Tropical Savannas, 1998.

ARTICLES

ACT ESB, 'From Wandilo to Linton—Lesson learned from an indepth analysis of 40 years of Australian Bushfire Tanker Burnovers', p. 1 on the ACT ESB website. <www.esb.act.gov.au/firebreak>
Barrow, G.J., 'A survey of houses affected in the Beaumaris fire, January 14, 1944', *Journal of the CSIR*, 18 (1945), pp. 27ff.
Benson, John S. and Redpath, Phil A., 'The nature of pre-European vegetation in south-eastern Australia: A critique of Ryan, D.G., J.R. and Starr, B.J.', *Cunninghamia*, 5/2 (December 1997).
Bowman, David, Tansley Review No. 101 paper, 'The impact of Aboriginal landscape burning on the Australian biota', in *New Phytologist*, 140(1998).
Brandes, Ian, 'Burning Questions. Australia's Bushfire Policy', 7 February 2003 at <www.econ.usyd.edu.au/drawingboard/digest/0302/brandes>.
Brookhouse, Peter and Nicholson, Don, 'Large Pilliga fires and the development of

fire management and fire suppression strategies', *Conference Proceedings, Australian Bushfire Conference, Albury, July 1999.*

Chambers, D.M. et al., 'Fire Protection and Suppression', Hobart, Parliamentary Paper 28/1967.

Coleman, Tony, 'The Impact of Climate Change on Insurance against Catastrophes', Institute of Actuaries of Australia, 2003 Biennial Convention.

Crowe, Fabian, 'The Arsonist's Mind', National Academies Forum, *Fire! The Australian Experience,* Seminar at University of Adelaide, 30 September–1 October 1999.

Drabsch, Talina, 'Arson', Briefing Paper 2/2003. The Parliament of New South Wales.

Flannery, Tim, 'Beautiful Lies: Population and Environment in Australia' in *Quarterly Essay,* No. 9, Melbourne: Black Ink, 2003.

Gill, A. Malcolm, 'Landscape fires as social disasters: An overview of the "bushfire problem" ', *Environmental Hazards,* 6(2005), pp. 65–80.

Gott, Beth, 'Fire as an Aboriginbal management tool in South-Eastern Australia', *Conference Proceedings,* Australian Bushfire Conference, Albury, July 1999.

Horton, David, 'Fire and Australian Society', WA Department of Conservation and Land Management, *Fire in Eco-Systems of South-West Western Australia,* Perth: Proceedings of April 2002 Symposium.

Howitt, William, 'Black Thursday: the great bush fire of Victoria' in *Cassell's Illustrated Family Paper,* Vol. I, Nos 6,8,9 (February 5, 18, 25, 1854).

Jackson, W.D., 'The Tasmanian legacy of man and fire', *Papers and Proceedings of the Royal Society of Tasmania,* Vol. 133/1, (1999).

Jones, Rhys, 'Fire-stick farming', *Australian Natural History,* 16(1969), pp. 224–8.

Krusel, N. and Petris, S.N., 'A Study of Civilian Deaths in the 1983 Ash Wednesday Bushfires Victoria, Australia', CFA Occasional Paper No. 1.

Langton, Marcia, 'The European construction of wilderness', *Wilderness News,* No. 143, Summer 1995/96.

Leonard, Justin E. and McArthur, Neville A., 'A History of Research into Building Performance in Australian Bushfires', *Conference Proceedings,* Australian Bushfire Conference, July 1999, Charles Sturt University, 1999.

Lunt, Ian, 'Vegetation Management' in *New Times,* Charles Sturt University e-magazine, March 2003 at <www.csu.edu.au/division/marketing/tms>.

Mace, Bernard, 'Mueller—Champion of Victoria's Giant Trees', *The Victorian Naturalist,* Vol. 113 (4), 1996.

——, 'Mountain Ash Forests of Victoria, Past Glory of the World's Tallest Trees', unpublished paper.

Marsden-Smedley, Jon, 'Changes in Southwestern Tasmanian fire regimes since the early 1800s', *Papers and Proceedings of the Royal Society of Tasmania,* Vol. 132, 1998.

——, 'Fire management in Tasmania's Wilderness World Heritage Area: Ecosystem restoration using Indigenous-style fire regimes?' in *Ecological Management and Restoration,* 1/3, 2000.

Mooney, Scott D., Radford, Kate L. and Hancock, Gary, 'Clues to the "burning question": Pre-European fire in the Sydney coastal region from sedimentary charcoal and palynology', *Ecological Management & Restoration,* Vol. 2, No. 3, December 2001.

Power, Bryan, 'The 1973 Bushfire in Lysterfield and Rowville', *Rowville-Lysterfield Community News,* March 2003.

Robertson, William J., 'Kelso's Contribution to Dam Safety', *Cranks and Nuts,* 1963.

Robinson, Mark, 'Bushfires, 2003—a rural GP's perspective', *Australian Family Physician,* Vol. 31, No. 12.

Roderick, Michael L. and Farquhar, Graham, 'The cause of decreased pan evaporation over the last 50 years', *Science,* Vol. 298, No. 5597, 15 November 2002.

Rose, R. et al., 'The importance and application of spatial patterns in the management of fire regimes for the protection of life and property and the conservation of biodiversity', *Conference Proceedings*, Australian Bushfire Conference, Albury, July 1999.

Smith, Stewart, 'Bushfires', Briefing Paper, 5/2002, New South Wales Parliamentary Library.

Stretton, Leonard, 'Judge Stretton's reminiscences, *La Trobe Library Journal*, Vol. 5, No. 17, April 1976, pp. 1–21.

van Wely, Brad, 'Hell's angels: Aerial firefighters play a vital role in Australia's constant battle against bushfires', *Flight Safety Australia*, March–April, 2003.

OTHER SOURCES

Darmody, Louise (ed.), *Spirit and Survival: Stories from the 2003 Snowy Mountains Bushfires*, CD recorded by Sound Memories for the Snowy Bushfire Recovery Taskforce.

Historical Records of New South Wales (HRNSW).

Ryan, D.G. and J.R. and Starr, B.J., *The Australian Landscape: Observations of Explorers and Early Settlers*, Wagga Wagga: Murrumbidgee Catchment Management Committee, 1995.

Victoria. Parliamentary Debates (VPD), Session 1939, Vol. 207.

INDEX

Some personal and place names have been deliberately omitted for brevity.

ABC Radio xix, 39, 181, 288, 295, 313–16, 373
ABC TV 292, 319
Aberfeldy (Vic) 40, 61
Aboriginal fire 61, 62, 66, 67, 80, 84, 85, 242, 244, 279, 282, 329–48, 364
Acheron River (Vic) 20, 23
Acheron Way (Vic) 15, 20, 21, 22–3, 27, 52, 59
ACT parks service 320
Adam, David 225
Adaminaby (NSW) 253, 256, 257
Adelaide (SA) 105, 156, 165, 227–8, 372
aerial firefighting xix, 102, 122, 174, 235–6, 247, 253, 257, 263, 264, 265, 269, 277–8, 290, 291, 321, 376–7
Aireys Inlet (Vic) 206, 221–2
Aitchison, Barry 260
Albany (WA) 84, 103
Albury (NSW) 87, 92, 94, 95, 101, 242, 350
Aldgate (SA) 166, 229
Alexandra (Vic) 21, 26, 27, 53
Alice Springs (NT) 240, 243
Allen, Beth 272
Allen, Craig 276
alpine ash (*Eucalyptus delegatensis*) 253
Alpine National Park (Vic) 139, 146, 174, 262, 264–6, 273
Alpine Way (NSW) 247, 251, 253, 254, 256, 257, 275
Anakie (Vic) 208
Anglesea (Vic) 206, 222–3
ANZAAS Conference (1939) 42–3
Apollo Bay (Vic) 12, 34, 54, 60, 104, 109, 121
Appin (NSW) 161, 234, 237
Ararat (Vic) 41, 95, 99
arson xvii, 12, 34, 132, 145, 170, 172, 176, 194, 211, 213, 215, 223, 227, 234, 235, 236, 238, 243, 311, 349–60, 365
Arthurs Seat (Vic) 13
Arve Valley (Tas) 190
attitudes to fire 49, 54, 61, 91, 107, 110, 112, 122, 135, 153, 232, 236, 329, 341, 367, 379
Augathella (Qld) 105
Australian Forestry School 119, 154
Avoca (Vic) 95
Avonside (NSW) 261

Babinda (Qld) 11
Bacchus Marsh (Vic) 163
back-burn 132, 191, 211, 215, 256–7, 274, 298, 320, 321
Bacon, Mark 303
Bairnsdale (Vic) 13, 56, 95, 106, 134, 175, 264, 269
Ballan (Vic) 26, 218
Ballarat (Vic) 26, 54, 72, 95, 99, 106, 141, 143
Ballina South (NSW) 94
Banks, J.C.G. 324
Baradine (NSW) 105
Barcaldine (Qld) 92
Barclay, Paul xix, 394
Barling, James Hartley 14–15
Barnard, Basil xix, 20–2, 24
Barnawatha (Vic) 162
Barongarook (Vic) 31
Barrenjoey Peninsula (NSW) 45
Barrett, Sir James William 58, 121–2, 123
Barrow, G.J. 142–3
Barry Way (NSW) 256–8
Basin, The (Vic) 171
Bass Strait 16, 72, 91, 105
Batchelor (NT) 243
Batemans Bay (NSW) 46, 161
Bathurst (NSW) 45, 75, 101
Batlow (NSW) 95, 251–2
Baulkham Hills (NSW) 65, 234
Baynes, Reginald 29, 381
Beaumaris (Vic) 142–3
Beech Forest (Vic) 104, 128
beech trees (*Nothofagus*) 80, 330
Beechworth (Vic) 87, 142, 266
Bega (NSW) 46, 158–60, 161, 234, 242
Belair (SA) 105
Belgrave (Vic) 26, 54, 223, 225
Bellarine Peninsula (Vic) 72, 74
Bellerive (Tas) 188, 197
Bellingen (NSW) 92
Bells Line of Road (NSW) 215, 235, 243
Belrose (NSW) 215
Bemboka (NSW) 159
Benalla (Vic) 12, 156
Benambra (Vic) 174, 263, 265, 267–9, 270
Bendigo (Vic) 95, 144, 174
Bendora Dam (ACT) 291, 293–4
Benson, John S. xix, 336, 346
Bermagui (NSW) 161
Berowra (NSW) 173, 239
Berriedale (Tas) 194–5

Berrigan (NSW) 103
Berrybank (Vic) 143
Berwick (Vic) 224
Big Desert Wilderness Park (Vic) 243, 262
Bilpin (NSW) 235
Birregurra (Vic) 87
Black Mountain Reserve 311
Black Rock (Vic) 142
Blackall (Qld) 92
Blackheath (NSW) 156, 161, 162, 166, 235
Blackmans Bay (Tas) 192
Blackmore, John 58–9
Blacks Spur, The (Vic) 14, 21, 22
Blackwood (SA) 105
Blaxland (NSW) 207, 215
Blue Gum Forest (NSW) 166–7
Blue Mountains (NSW) 45, 64, 66, 92, 101–2, 105, 156, 161, 166, 167–8, 176, 177, 207–8, 210, 211, 212, 213, 215, 234, 235, 237, 239, 350, 352, 353, 356, 372
Bogong High Plains (Vic) 13, 26, 40, 264–5, 269
Bolaro (NSW) 256
Bolt, Andrew 279
Bolton, Leslie 191–2
Bombala (NSW) 101, 261
Bond, Geoffrey 192
Bonython, Kym and Julie 205
Bordertown (SA) 242
Borough, Tim 298
Bothwell (Tas) 184
Boulia (Qld) 92, 241
Bourke (NSW) 44, 100, 241
Bowenfels (NSW) 64
Bowman, Helen 273
Bowman, David 339–40
Boyer (Tas) 185
Boyup Brook (WA) 104
Bradman, Sir Donald 17
Braidwood (NSW) 44, 352
Branxholme (Vic) 219
Briagolong (Vic) 175
Bridgetown (WA) 104, 105
Bridgewater (SA) 216, 229
Bridgewater (Tas) 185, 200
Bright (Vic) 13, 26, 41, 95, 99, 264, 266, 269, 275
Brindabella Ranges (ACT/NSW) 43, 78, 92, 94, 120, 251, 290, 292, 293, 298, 320, 324, 325
Brisbane (Qld) 100, 101, 140
Broomehill (WA) 105
Brown Mountain (Tas) 187
Brown, Ralph 111–12
Bruny Island (Tas) 189
brush-tailed rock wallaby 176, 181–2, 296, 297
Bruthen (Vic) 175
Buchan (Vic) 274
Buchan River Valley (Vic) 176, 270

Buckland (Tas) 189
Buckland Valley (Vic) 40, 41, 175
Bugg, Alex 192
Bullaburra (NSW) 212
Bullengarook (Vic) 219–20
Bulls Head (ACT) 294
Bullumwaal (Vic) 175
Bunbury (WA) 104, 105
Bundanoon (NSW) 46, 173
Bundeena (NSW) 211
Burgess, Peter Cameron 349–50, 353, 356
Burra (NSW) 164
Burrill Lake (NSW) 161
Bush Users Group 279
Bush Fire Brigades Association 130, 140
Bush Fires Board (WA) 149
bushfires
   Black Thursday (1851) 67–75, 79, 86, 103
   Red Tuesday (1898) 178–204
   Black Sunday (1926) 6, 95–100
   1931–2 fires 103–4
   Black Friday (1939) xviii, xix, 3–50, 51–133, 142, 157, 171, 206, 244, 250, 260, 266, 309, 362
   1943–4 fires 140, 141–5, 148, 149, 388
   1951–2 fires 155–65
   Ash Wednesday (1981) xix, 205–6, 216–26, 226–32, 244, 250, 265, 362
   1993–4 fires 232, 234–6, 356
   2001–2 fires 237–8, 323
   Alpine fires (2003) 247–86
   Canberra fires (2003) 287–325
bushwalking 51, 102
Busselton (WA) 104
Butcher, Laurie 225
Butlin, Noel 346–7
Buxton (Vic) 21, 27
Byadbo Wilderness (NSW) 248, 251, 252, 282

Cabramurra (NSW) 173, 253
Cain, John 149, 151, 218
Cairns (Qld) 106
Calder Highway (Vic) 220
Calga (NSW) 214
Caloundra (Qld) 106
Campania (Tas) 187
campers and fires 12, 31, 59, 96, 111, 149
Canada 102, 122, 153, 377
Canadair CL 415 377
Canberra (ACT) xvii, xviii, xix, 42, 43, 78, 92, 94, 120, 154, 164, 165, 173, 237, 250, 251, 252, 254, 255, 264, 269, 286, 287–325, 371
Canberra Nature Park 299, 309, 312
Cann River (Vic) 218
Cape Otway (Vic) 12, 34
Cape York (Qld) 241
Carabost (NSW) 157

INDEX

Carnarvon (WA) 105
Carnell, Kate 318
Carr, Bob 238, 278
Carter, Barry 166–7
Carter, Colin 214
Cascade Brewery 197, 199
Cascades, The (Tas) 193, 197, 198
Casterton (Vic) 41, 170
Castle Hill (NSW) 45, 65
Castlemaine (Vic) 190
Castles, Mike 295
Catling, Peter 285
Cessnock (NSW) 173, 273
Chaney, Phil 204, 232, 318–19
Channel district (Tas) 105, 182, 189–93
Chapman (ACT) 288, 309, 312, 316
Charleville (Qld) 92, 105, 155
Charlotte Pass (NSW) 253
Chigwell (Tas) 194–6
Chiltern (Vic) 266
Chisholm, Alec H. 173
Christensen, Paul 12–13
Christmas Hills (Vic) 171
Chum Creek (Vic) 15, 95
Clack, Peter 292, 300, 320
Clare Valley (SA) 227, 230
Claremont (Tas) 193–4
Clarendon (SA) 47
Claridge, Andrew 282
Clark, Constable 195
Clark, Marie 201–2
Clunes (Vic) 141
Coal River (Tas) 187
Cobar (NSW) 44, 103, 241–2
Cobargo (NSW) 158, 161
Cobbannah (Vic) 175
Cobden (Vic) 29, 219
Cobungra Station/Estate (Vic) 40, 267–8
Cockatoo (Vic) 13, 26, 95, 223–4, 225–6
Cockle Creek (Tas) 190
Code, Robert 54, 55, 110
Coffs Harbour (NSW) 105, 106
Colac (Vic) 31, 34, 53, 86, 87, 95, 143
Colebrook (Tas) 186–7
Coles, Anthony 307–8
Collarenebri (NSW) 44
Collie (WA) 104
Collins Cap (Tas) 186
Collins, David 61, 382
Collinsvale (Tas) 186
Colo (NSW) 65, 212
Como (NSW) 235
Condingup (WA) 243
Condobolin (NSW) 240
Conningham (Tas) 191
controlled burning 56, 119–20, 129, 135, 154, 170, 208, 243, 284, 319, 321, 348
Cook (SA) 217
Cook, John 265, 395
Cooleman Ridge (ACT) 288, 309, 312

Cooma (NSW) 234, 254, 255
Coonabarabran (NSW) 105, 155, 172
Coonamble (NSW) 101–2
Coopracambra National Park 263–4
Cordwell, David 'Les' 195
Corin Dam (ACT) 396
Corio Bay (Vic) 69, 72
Corowa (NSW) 16
corroboree frog 281, 282, 283
Corryong (Vic) 41, 158, 298
Cotter Dam (ACT) 242, 298
Cotter, Kevin 299
Coulter, John xix
Council of Australian Government (COAG) inquiry on 2003 fires 374, 375
Country Fire Authority (CFA) 149, 211, 221, 223, 224, 225, 226, 239, 264, 265, 267, 270, 276
Country Fire Brigades Board 130
Cowes (Vic) 103
Cowra (NSW) 100
Crafers (SA) 47, 166, 205, 229
Cravensville (Vic) 263
Cremorne (Tas) 188
Cressy (Vic) 143–4
Creswick (Vic) 106, 141
Creswick Forestry School 110, 111
Crommelin, Minard Fannie 168
Crowe, Fabian 252, 353, 356–7
CSIRO 142, 232, 285, 311, 318, 366, 373
Cudgewa (Vic) 158
Culcairn (NSW) 103
Culgoa River (Qld) 141
Curdie Vale (Vic) 219
Curtin (ACT) 195, 312
Curtin, John 152, 389, 404
Cygnet (Tas) 189, 192
cypress pine (*Callitris endlicheri*) 280–1, 396

Dafter, Bill 38, 381
Dalgety (NSW) 255, 261
Dandenong Ranges (Vic) 10, 13, 26, 72, 94, 95, 96, 99, 102, 104, 106, 123, 170, 171, 172, 211, 223, 357, 372
Danggali Conservation Park (SA) 242
Dare, Tim 213
Dargo High Plains (Vic) 174
Dargo River (Vic) 41, 271
Darling Ranges (WA) 105
Darling River (NSW) 241
Darlington, Dave 251
Darmody, Louise xix
Dartmouth (Vic) 263, 266
Darwin, Charles, Vernon 122–3
Davis, Cliff 201
Davis, Lynda and Trevor 256–7
Daylesford (Vic) 26, 106
Deans Marsh (Vic) 221, 223
Debus, Bob 317, 334, 350
Delegate (NSW) 14

Demby, Charles Isaac 14–15
Denmark (WA) 103
Department of Environment and
  Conservation (DEC) (NSW) 249, 252
Derrett, Julie 314
Derrinallum (Vic) 143
Derwent River (Tas) 65, 185, 193, 194
Derwent Valley (Tas) 100, 105, 182
Devonport (Tas) 48, 70, 91
Diamond Creek (Vic) 72, 163
Dimboola (Vic) 170
Dinner Plain (Vic) 265, 267
Dirranbandi (Qld) 144
Dismal Swamp (SA) 231
Dodd, Frank 78
Doherty, Megan 317
Doley, Rebekah 355
Doogan, Maria 325
Dorrigo (NSW) 92
Douglas Park (NSW) 161
Dover (Tas) 190
Drabsch, Talina 353, 354
Dromana (Vic) 13, 54, 95
Dromedary (Tas) 185
Drouin (Vic) 82
Department of Sustainability and
  Environment (DSE) (Vic) 249, 262, 263,
  264, 265, 266, 267, 273, 274, 283, 284,
  294–6
Dubbo (NSW) 44, 100, 106, 261
Duffy (ACT) 289, 302, 307–9, 310, 314, 315
Duffys Forest (NSW) 215
dugouts 8, 9, 16, 18–19, 21, 24, 25–6, 27,
  28, 29–30, 31, 35, 40, 113, 114, 130,
  133, 142
Dunkeld (Vic) 143
Dunlop (ACT) 298
Dunmarra (NT) 240
Dunstan, Albert 135, 141, 146, 149
Dural (NSW) 65
Dutkiewicz, Matt 301–2
Dwellingup (WA) 169–70
Dynnyrne (Tas) 198–9

Eastern View (Vic) 221
Echunga (SA) 47
Eden (NSW) 101
Edwards, Frank 20–3, 26, 52
El Niño event 155, 217
Elanora Heights (NSW) 215
Elder, Malcolm 210–11
Eldorado (Vic) 141, 266
Electrona (Tas) 180, 193
Eltham (Vic) 54, 72, 110, 176
Emu Plains (NSW) 207
Ensay (Vic) 176
Environment ACT 320, 321
Erickson Air-Crane 237–8, 263, 270, 376
erosion 55, 77, 116, 120, 149, 150, 152,
  248, 274, 280, 338

Emergency Services Bureau (ESB) (ACT)
  290, 291, 293, 295, 312, 314, 315, 317,
  320, 321, 324
Esk (Qld) 101
Eskdale (Vic) 270
Esperance (WA) 84, 243
Esplin Report (Vic, 2003) 249, 262, 270,
  279–80
Eucla (WA) 217
Eucumbene Drive (ACT) 289, 307, 308, 315
Euroa (Vic) 14, 79, 95, 174
Evans, Tritia xix, 313–14
Everton (Vic) 140, 142
Eyre Peninsula (SA) 105, 156

Fairhaven (Vic) 221
Farquhar, Graham xix
Faulconbridge (NSW) 156, 207, 211, 212
Feiglins' mills (Vic) 19–21, 24, 25, 114
Fenton, James 70
Fern Tree (Tas) 193, 197, 199, 200–1, 202
Fern Tree Gully (Vic) 26, 171
Ferny Creek (Vic) 104, 357
fire science 153–4
fire-stick farming 331–2
Fitzpatrick's mill (Vic) 35–8, 39, 52, 113,
  114, 125, 130
Fixter, Paul 307–8
Flannery, Tim 236, 333–5
Flinders Ranges (SA) 243
Flood, Josephine 332–3
Flowerpot (Tas) 191
Foley, James Charles 154
Fooks, Des xix, 288
Forbes (NSW) 100, 101
Ford, Ann 312–13
forest conscience 122, 152
Forest Fire Danger Index (FFDI) 183
Forestry Commission (Tasmania) 183
Forestry Commission (NSW) 153
Forests Commission (Vic) 14, 19, 21, 26,
  29, 31, 38, 54–5, 56, 58, 59, 102, 109–10,
  111–12, 113–14, 123, 125, 126–7, 128,
  129–30, 131, 132, 134, 174, 175, 218
Forests League (Victoria) 58, 123, 128
Forrest (Vic) 53
Foster (Vic) 88, 90, 104
Francis, William 35–6
Frankland (WA) 104
Franklin (Tas) 190
Franklin River (Tas) 79
Franklin-Browne, Graham 294
freckled ducks 296
Freeburgh (Vic) 41, 266
Freeman, Anna, Jessie (WA) 85
Frenchs Forest (NSW) 45, 157
Fumina district (Vic) 96–7

Galbraith, Alfred Vernon 14, 26, 110,
  111–14, 130

INDEX

Garden Island Creek (Tas) 191
Garigal National Park (NSW) 208, 215
Garvoc (Vic) 219
Geelong (Vic) 33, 43, 69, 72, 87, 143, 144, 208–9
Geeveston (Tas) 189, 190
Gelantipy (Vic) 265, 270–1, 272, 273
Gellibrand River (Vic) 34, 104
Gembrook (Vic) 13, 26
Georges River (NSW) 44, 235
Geraldton (WA) 84, 169
Gerraty, Finton 113
Gilderoy (Vic) 97–8, 103
Gill, A. Malcolm xix, 345, 366, 367
Gingin (WA) 85
Gipps, Sir George 77
Gippsland (Vic) 13, 18, 41, 54, 56–7, 72, 81–2, 91, 92, 96, 99, 103–4, 146, 175–6, 262, 274, 330
Gippsland East (Vic) 31, 40, 128, 134, 263, 265, 272, 282–1
Gippsland South (Vic) 69, 70, 78–84, 85, 90, 92, 94, 117, 263
Glen Innes (NSW) 100, 105, 156
Glenbrook (NSW) 105, 350
Glencoe (SA) 230
Glenlyon (Vic) 144
Glenorchy (Tas) 184, 193, 194–5, 196–7
Glenorie (NSW) 239
Glenrowan (Vic) 163
Gnowangerup (WA) 105
Goldie, Jenny xix
Goldie, Nick xix, 254
Goodradigbee River (NSW) 92, 290
Goondiwindi (Qld) 105, 106
Gordon (Tas) 191
Gorrie Station (NT) 239
Gorshkov, Victor G. 370
Gosford (NSW) 239
Goulburn (NSW) 46, 173, 250
Goulburn River (Vic) 39, 69, 72
Gould, Jim 318–19
Government House ('Yarralumla') 43, 311
Gowans, Gregory 52, 53, 108, 113, 117
Grafton (NSW) 92, 156, 234, 237, 356
Grainger, Tim 316–17
Grampians Range (Vic) 41, 103, 170
Grand Ridge Road (Vic) 146
Granton (Tas) 195
grassfires 62, 64, 75, 86, 88, 92, 94, 103, 104, 106, 139–40, 141, 143–4, 145–6, 154, 155, 156, 157, 160–1, 170, 173, 174, 182, 194, 195, 208, 240, 241–3, 286, 332, 371
graziers 12–13, 31, 34, 56, 58, 93, 112, 121, 122, 128, 129, 148, 149, 150, 151, 155
Great Alpine Road (Vic) 140, 264, 267, 271
Great Ocean Road (Vic) 221
Great Victoria Desert (SA) 241
Great Western Highway (NSW) 207, 235

Green, Ken 283
green pick 34, 58, 68, 75, 96, 107, 121
Greenhill Road (SA) 227, 228–9
Greta (Vic) 163
Griffith (NSW) 242
Griffiths, Tom 122
Grose Valley (NSW) 166, 207, 215, 235
Gulf of Carpentaria (Qld/NT) 241
Gundagai (NSW) 252
Gunter, Betty 167–8
Gunyah (Vic) 104
Guthega (NSW) 253
Gympie (Qld) 101, 105, 106

Hahndorf (SA) 217, 229
Hallam, Sylvia 332, 335
Hall's Gap (Vic) 41
Hamilton (Vic) 143–4, 219
Hamilton, Derek 306
Hammond, Dame Joan 222
Harden (NSW) 144, 242
Hardy, Alfred Douglas 58, 123
Harrietville (Vic) 13, 40, 41, 175, 264, 266, 270, 275
Hartz Mountains (Tas) 79
Hastings (Tas) 189–90
Hawkesbury (NSW) 64, 235, 237
Hawkesbury River (NSW) 63, 65, 101, 235
Hawkins, Deirdre xix
Hay (NSW) 241, 242
Hazelbrook (NSW) 212
heads, treatment of 30–1, 36, 60, 85, 114, 129, 133
Healesville (Vic) 14–15, 20, 21, 26, 31, 53, 54, 55, 59, 94–5, 104, 114, 131, 170, 171
Heathcote (NSW) 214
Heathcote (Vic) 144
Heathfield (SA) 216
Helensburgh (NSW) 213, 236, 237
helicopters, fire-fighting 206, 214, 238, 252, 257, 277–8, 289, 291, 376–7
Hennessy, Kevin 373
Hepburn Springs (Vic) 104
Hillston (NSW) 44, 240
Hobart (Tas) 65, 105, 154, 177, 178, 181, 183–6, 188–202, 203–4, 208, 272
Hoddle Ranges (Vic) 88
Holbrook (NSW) 92, 242
Holder (ACT) 307, 309, 310, 317
Holey Plains State Park (Vic) 263
Holmes, W.H.C. 81
Holsworthy (NSW) 234, 239
Holyoake (WA) 169
Hopetoun (Vic) 121
Hornsby (NSW) 161, 207, 239
Horsham (Vic) 99
Horton, David 331, 323, 341–4, 348
Howard government 278, 317
Howitt, William 71–2
Howlong (NSW) 92, 162

Howrah (Tas) 188
Hueneke, Klaus 248
Hügel, Charles von 65, 85
Hughenden (Qld) 92, 106
Hume Weir (NSW) 150, 350
Hume Highway (NSW/Vic) 46, 139, 157, 162–3, 173, 174
humidity 12, 34, 44–5, 83, 94, 142, 154, 155, 157, 169, 183, 208, 217, 230, 233, 238, 265–7, 290, 294, 295, 296
Humula (NSW) 157, 161
Hunter, Governor John 62
Hunter district (NSW) 234, 237
Huon district (Tas) 66, 91, 100, 105, 180, 189, 190, 202
Huonville (Tas) 189, 190, 192
Hurstbridge (Vic) 72

Illawarra district (NSW) 65
Ingebyra (NSW) 256, 258, 259, 276
Ingleside (NSW) 215
Inglewood (Qld) 105
Inglewood (Vic) 174
Institute of Foresters Australia 119, 285
Institute of Public Affairs 279
Inverell (NSW) 172–3
irrigation 49, 273
Ivanhoe (NSW) 44, 103, 240, 241

Jannali (NSW) 235, 236
jarrah (*Eucalyptus marginata*) 84
Jeffery, C.C. 'Val' 319–20
Jenkins, Joseph 78
Jervis Bay (NSW) 237
Jindabyne (NSW) 153, 247, 252, 253, 255, 256, 257–8, 275
Jingellic (NSW) 101
Jones, Inigo 42–3, 44
Jones, Owen 112
Jones, Rhys 331–3, 335
Judbury (Tas) 190
Jugungal Wilderness (NSW) 251
Julia Creek (Qld) 140
Junee (NSW) 144

Kalangadoo (SA) 231
Kalgoorlie (WA) 241
Kallista (Vic) 171
Kalorama (Vic) 26, 106, 171
Kambah (ACT) 299, 301, 312
Kangaroo Ground (Vic) 172, 210
kangaroos 46, 71, 72, 74, 146, 259, 282, 285, 297, 331, 332
Kaoota (Tas) 189
karri (*Eucalyptus diversicolor*) 84
Katanning (WA) 103
Katherine (NT) 239–40, 243
Katoomba (NSW) 45
Keiran, Rodger 215
Keith (SA) 242

Kellerberrin (WA) 104
Kellner, Arthur 34
Kelly, Matthew xviii
Kelso, Alexander Edward xix, 57, 60, 114, 115–18, 119, 120, 123, 129, 151, 168, 369–70, 285
Kempsey (NSW) 105, 237
Kempton (Tas) 184
Kent Hughes, Sir Wilfred 164
Kenthurst (NSW) 234
Kerslake family 20–4, 26, 27, 28, 52, 171
Kessell, Stephen 86, 93, 119
Kettering (Tas) 189, 191
Keysbrook (WA) 169
Khancoban (NSW) 251, 253
Kiandra (NSW) 173
Kiewa Valley (Vic) 26
Killarney-Top Spring fire (NT) 239
Kilmore (Vic) 68, 69, 73, 75
Kimberley (WA) 243
King, Harry 97
Kingaroy (Qld) 106
Kinglake (Vic) xix, 3, 14, 20, 27, 51–3, 58–60, 99, 106, 171, 382
Kingston (Tas) 180, 184, 189, 190, 192
Kissane, John 270
Kitson, Ione 306
koalas 80, 173, 296, 297
Kocsis, Richard N. 351, 354, 357, 400, 404
Kohen, James L. 331, 346–7, 398, 400, 404
Kojonup (WA) 105
Koperberg, Phil 212, 213, 215, 252, 257, 294, 356, 364, 400
Koppio (SA) 156
Kosciuszko National Park (NSW) 173, 242, 247, 248, 249–58, 261, 264, 276, 280, 283, 290, 318, 393
Kotz, Gillian 229
Kununurra (WA) 240
Ku-ring-gai (NSW) 174, 207
Ku-ring-gai Chase National Park (NSW) 215, 235, 363
Kurrajong (NSW) 64, 235, 243
Kyneton (Vic) 41, 69, 95, 144

Lake Bolac (Vic) 143
Lake Burley Griffin (ACT) 311
Lake Eucumbene (NSW) 253, 257
Lake Grace (WA) 104
Lake Macquarie (NSW) 101, 235
Lake, Marilyn 93, 385, 404
Lake Mundi (Vic) 41
Lancefield (Vic) 171
land clearing 49, 84, 93, 335
land rights (Aboriginal) 332, 342
Lands Department (Vic) 126–7, 148
Landsborough (Qld) 106
Lane Cove National Park (NSW) 235
Lane-Poole, Charles Edward 56, 58, 86, 93, 118–20, 128, 322, 369, 370

INDEX

Lang, John Dunmore 379, 401, 404
Lang Lang (Vic) 224
Langton, Marcia 343, 399, 407
Lara (Vic) 209
Latrobe River (Vic) 4, 6, 16, 96, 147
Lauderdale (Tas) 188
Launceston (Tas) 181, 186, 200, 383, 404
Lawson (NSW) 212
Lawson, Henry 77, 361
Lemke, Andrew 231
Lenah Valley (Tas) 196–7, 198, 199
Leonard, Mike 284
Leura (NSW) 167
Levendale (Tas) 189
Lilydale (Vic) 16, 99
Limekiln Gully Reserve fire (Tas) 196
Lind, Albert 31, 133–5, 141, 142, 387
Linden (NSW) 212
Lindenmayer, David 322–3
Lines, William A. xix, 334, 339
Lismore (NSW) 156, 160
Lismore (Vic) 143
Lithgow (NSW) 45, 64, 92, 94, 101, 156, 160, 168, 215
Little Desert (Vic) 170
Little River Gorge (Vic) 281
Little Yarra Valley (Vic) 16, 97
Liverpool Ranges (NSW) 176
Livingstone, Bill 273
local government/shires 130, 131, 376
Londonderry (NSW) 156
Longreach (Qld) 92, 155
Longwood (SA) 216
Longwood (Vic) 174
Lorne (Vic) xix, 54, 176, 206, 221, 223, 276
Louth (NSW) 241
Lucas Heights (NSW) 215
Lucas-Smith, Peter 290, 293–4, 295, 315
Lugarno (NSW) 44, 106
Luke, R.H. 'Harry' 44, 153–4, 160, 169, 170, 233, 240, 323, 370
Lunt, Ian 284
Lyell Highway (Tas) 79
Lymington (Tas) 189, 192
Lyons (ACT) 312
Lyons, Joseph 17, 48
Lysterfield (Vic) 211

Macarthur, Ian 324
McArthur, A.G. 'Alan' 44, 143, 153, 154, 169, 170, 182, 183, 201, 204, 218, 233, 240, 323, 370
Macclesfield (SA) 47
Mace, Bernard 81
Macedon (Vic) 220
McGloughlin, Eda and Tony xix, 309–10, 320
McGrath, Dorothy 306
Macgregor (ACT) 298

McGuinness, family 97
McIlroy, William 126
McIntyres Hut fire (NSW) 290–1, 292–3, 294, 300
Mackay, John 300
McKenzie, Euan 293–4
McKernan, Michael 338
Mackey, Brendan xix
McKillops Bridge (Vic) 274
Macksville (NSW) 92
McLaren Flat (SA) 230
McLelland family 72–3, 163
McMahon, James and John 59–60
McMahons Creek (Vic) 226
Maffra (Vic) 54, 106, 174
Magra (Tas) 185
Main, George 338
Maitland (SA) 156
Maitland, Geoffrey 229
Maldon (Vic) 78
Mallacoota (Vic) 121
Mallee district (Vic) 72, 93, 121, 163, 241, 243, 280, 339
Maltby, Thomas 133–4, 135
Mangoplah (NSW) 157
Manly (NSW) 45
Mansfield (Vic) 12, 39, 53
Margaret River (WA) 103
Maroondah Highway (Vic) 14, 20, 21
Marradong (WA) 105
Marsden-Smedley, Jon 79, 340–1
Martin, Raz 381–2
Martin, Sarah 297–8, 319
Marulan (NSW) 173
Maryborough (Qld) 105, 141
Marysville (Vic) 20, 53
Matlock (Vic) 27, 35–8, 52, 114, 130
Maydena (Tas) 79
Meadows (SA) 47, 230
megafauna 332–3
Megalong Valley (NSW) 168
Melbourne (Vic) 5, 11, 13, 15, 16, 17, 18, 31, 36, 39, 44, 51, 53, 57, 67, 68, 69–74, 78, 80, 91, 94–5, 96, 97, 103, 121, 142, 146, 157, 162–3, 170, 172, 193, 208, 210, 211, 218, 265, 266, 272
Melbourne and Metropolitan Board of Works (MMBW) 110, 114, 115, 116, 123, 126–30, 132, 151, 369
Menai (NSW) 239
Menindee (NSW) 103
Merredin (WA) 104
Michelago (NSW) xix, 254–5
Middleton (Tas) 191
Milawa (Vic) 95
Mildura (Vic) 11, 87
Millgrove (Vic) 226
Millhouse, Lavinia 178–9
Millicent (SA) 231, 243
Milligans Mountain (NSW) 248

Milovanovich, Michael 249, 292–3
Milton (NSW) 161
Mintaro (SA) 230
Mirboo (Vic) 81
Mirboo North (Vic) 95, 111, 146
Misner, Ludwig and Leila 299
Mitchell River National Park (Vic) 175, 176
Mitchell River Wilderness (Vic) 174
Mitchell, Thomas Livingstone 68, 332
Mitta Mitta (Vic) 265, 270
Mittagong (NSW) 46, 101
Moe (Vic) 13, 15, 54, 65, 95, 99, 146
Moggs Creek (Vic) 221
Molesworth (Tas) 186
Molong (NSW) 106
Molonglo River (ACT/NSW) 42, 164, 300–1, 311, 331
Monaro Highway (NSW) 254, 255, 256
Montrose (Vic) 13
Moonah (Tas) 193, 195
Moonbah (NSW) 256
Mooney, Scott D. 345
Moree (NSW) 242
Mornington (Vic) 16, 26, 99, 170, 357
Morphett Vale (SA) 47
Mortlake (Vic) 143
Morton National Park (NSW) 173
Morwell (Vic) 82, 146
mosaic burning 233, 262–3, 264, 268
Mosman (NSW) 45
Moss Vale (NSW) 173
Mount Barker (SA) 47, 103, 229
Mount Best (Vic) 88–9
Mount Bogong (Vic) 264
Mount Bonython (SA) 229
Mount Buffalo (Vic) 106, 264, 266, 267, 269
Mount Burr (SA) 231
Mount Bute (Vic) 143
Mount Coree (ACT) 43, 290, 291
Mount Donna Buang (Vic) 218
Mount Druitt (NSW) 45
Mount Eliza (Vic) 141
Mount Faulkner (Tas) 194
Mount Feathertop 264
Mount Franklin (ACT) 43
Mount Franklin Road (ACT) 290–1
Mount Gambier (SA) 87, 105, 106, 168, 227, 230
Mount Hotham (Vic) 40, 41, 265, 267, 269, 270, 275
Mount Kosciuszko (NSW) 95, 150, 247
Mount Ku-ring-gai (NSW) 147, 207
Mount Lofty Ranges (SA) 84, 100, 165, 205, 216, 227
Mount Macedon (Vic) 12, 26, 73, 75, 106, 163, 218, 220–1
Mount Mittamatite (Vic) 263
Mount Nelson (Tas) 184, 193, 197, 198, 199, 200

Mount Osmond (SA) 229
Mount Pleasant (SA) 100
Mount Riddell (Vic) 95
Mount Stromlo (ACT) 43, 164–5, 303–5, 307, 316
Mount Stuart (Tas) 193, 197, 198
Mount Taylor (ACT) 212–3
Mount Tennant (ACT) 297, 319
Mount Tomah (NSW) 215
Mount Torrens (SA) 47
Mount Victoria (NSW) 45
Mount Wellington (Tas) 65–6, 91, 105, 181, 193, 194, 199, 200, 201, 372
Mount White (NSW) 214
Mount Wilson (NSW) 215, 235
mountain ash (*Eucalyptus regnans*) 3, 7, 10, 13, 18, 22, 23, 52, 56, 79, 80, 81, 88, 95, 99, 109, 115, 116–18, 123, 141, 226
Moyston West (Vic) 41
Mudgee (NSW) 237, 242
Mueller, Ferdinand von 10, 12–13
Mulready, Nicole xix
Mulvaney, John 331
Murphy, Kerry 220
Murray Gates (NSW) 256
Murray River 92, 94, 95, 140, 150, 152, 158, 161, 162, 243, 256, 263, 338
Murray, Robert 149
Murrumbidgee Catchment Management Committee 335
Murrumbidgee River (NSW/ACT) 43, 92, 164, 254–5, 297, 299, 301, 312, 313, 319, 322, 338
Murrumburrah (NSW) 144
Murwillumbah (NSW) 92, 100
Muttaburra (Qld) 155
Myers, Allen 279–80
Mylor (SA) 105, 216, 229
Myrtleford (Vic) 41, 264
Mystery Lane (Vic) 263, 264

Naas Road (ACT) 296, 297
Nairn Report (House of Reps, 2003) 249–50, 278, 285–6
Namadgi National Park (ACT) 254, 256, 290, 293, 295, 319
Nambour (Qld) 101
Nanga Brook (WA) 169
Nangwarry (SA) 231
Narbethong (Vic) 14, 20, 21–3
Narembeen (WA) 104
Narooma (NSW) 161
Narrabeen (NSW) 45
Narrabri (NSW) 105
Narrandera (NSW) 44
Narre Warren (Vic) 224
Narromine (NSW) 237
National Aerial Fire-Fighting Centre 278
National Association of Forest Industries 318
National Bushfire Research Unit 232

INDEX

National Equestrian Centre 311
National Parks and Wildlife Service (NPWS) 248, 251, 252, 256–7, 276, 282, 283, 290, 291, 293, 363
native animals and fires 48, 108, 121, 124, 173, 210, 238, 281, 353, 363
Nayook West (Vic) 16
Needham, Reginald 56
Neerim (Vic) 81
Neerim Junction (Vic) 4–6
Nelligen (NSW) 46
Nelson Bay (NSW) 214
Nepean River (NSW) 46
New Norfolk (Tas) 91, 185, 186, 194
New Zealand 99, 110, 176, 237, 264, 270
Newcastle (NSW) 94, 101234, 235, 240, 243
Newcastle Waters (NT) 240
Ngarkat Conservation Park (SA) 242, 243
Nhill (Vic) 41
Nichol, Murray 205, 227–8
Nicholls Rivulet (Tas) 192
Noble, W.S. 29–30
Noojee (Vic), Charles 27, 59
Northam (WA) 85, 103, 169
Northern Territory 86, 94, 155, 232, 239–40, 241, 242, 243, 286, 329–30, 339, 343, 344
Norton Summit (SA) 227
Nowra (NSW) 46, 161, 173
Nullarbor Plain (SA/WA) 241
Numbla Vale (NSW) 255, 258, 261
Nymagee (NSW) 242

Oberon (NSW) 45, 237
Ockwell family 171
Officer (Vic) 224
Olinda (Vic) 26, 170, 171
Olsen family 96
Omeo (Vic) xix, 26, 40, 42, 54, 151, 174, 176, 263, 265, 267, 269, 270, 271–2, 290
Orange (NSW) 45
Orbost (Vic) 14, 151, 269
Oregon, USA 118
Orielton (Tas) 187
Orroral Valley (ACT) 294–5
Otton family 158–60
Otway Ranges (Vic) 12, 17, 26, 31, 33, 34, 54, 60, 86, 87, 96, 104, 109, 128, 221
outback/inland fires 86, 103, 239, 240, 241, 242–3, 248
Ouyen (Vic) 217, 243
Ovens Valley (Vic) 40, 264
Oxenbury family 174
Oyster Cove (Tas) 189, 191, 192, 201

Paddys River road (ACT) 296
Pakenham (Vic) 26, 224
Palmer family 191
Pambula (NSW) 101
Pannam, Clifford 221, 405
Panton Hill (Vic) 172, 224

Parish, Mitchell 357–8
Parkes (NSW) 100
Parramatta (NSW) 45, 62, 63
Paupong (NSW) xix, 248, 258, 260, 261
Peachtown (SA) 229
Peakhurst (NSW) 45
Pearl Beach (NSW) 168
Penna (Tas) 187
Pennant Hills (NSW) 45, 211
Penola (SA) 231
Penrose (NSW) 46
Penshurst (Vic) 87
Perisher (NSW) 256
Perry, Russell 229
Perth (WA) 84, 85, 103, 104, 140, 169
Petrenko, Anna 198–9
Phillip Island (Vic) 103, 121
Picnic Point (NSW) 239
Picton (NSW) 46
Pieman River (Tas) 79
Pierces Creek (ACT) 298
Pilliga Scrub (NSW) 105, 155, 389, 406
pine plantations (ACT) 291, 294, 364
Pinjarra (WA) 104, 105, 169, 170
Pinnibar Fire complex (Vic) 253, 256–7, 263, 264, 266, 267, 269
Pitt Town (NSW) 65
Pitt Water (Tas) 187
Pittwater (NSW) 45
Playford, John 227
Plenty Ranges (Vic) 69, 106
Pope, Matthew 252
Porepunkah (Vic) 264, 266
Port Albert (Vic) 69
Port Augusta (SA) 156
Port Fairy (Vic) 69, 72
Port Hacking (NSW) 45
Port Lincoln (SA) 105, 156
Port Macquarie (NSW) 160
Port Phillip Bay/District 13, 18, 66, 68, 71, 142, 209
Port Stephens (NSW) 65
Port Welshpool (Vic) 88, 90
Portland (Vic) 69, 72, 104, 219
post-war reconstruction 152, 376
Powell, Reverend Gordon 206, 391
Powelltown (Vic) 13, 15, 16, 97, 99, 103, 144, 226
Poynton family 222
Princes Highway (Vic/NSW) 13, 158, 159, 175, 209, 219, 224
prohibited season (for fires) 55, 132, 149
prospectors 12, 18, 52, 59, 79
Pryor, Geoff 316
Pryor, L.G. 324–5
Puen Buen (NSW) 159
Pugin, Augustus Welby 186
Purnululu National Park (WA) 243
Putty Road (NSW) 239, 243

Pyne, Stephen xviii, 49, 55, 76, 77, 83, 124, 151, 152, 153, 204, 205
pyromania xviii, 50, 55, 75–6, 102, 107, 111, 145, 204, 233, 338, 345, 352, 354

Queanbeyan (NSW) 43, 44, 164, 288, 290
quolls 282

rainforests 78, 80, 123, 330, 340, 347
Ranelagh (Tas) 190
Rankin, John 96–7
Rasmussen, Carolyn xix
Red Hill (ACT) 164
Redpath, Phil A. 336, 346, 398
Reeves, Mark 267–8
regent honey eaters 296
Remmele, Fred 229
remote area fire-fighting 247, 248, 253, 270, 292
Renmark (SA) 242
Rennoldson, Norm 229
Richardson, Clive 274
Richardson, Horace 96
Richmond (Tas) 187
Richmond (NSW) 64, 156
Riverina District 12, 78, 86, 87, 92, 94, 101, 103, 158, 161
Rivett (ACT) 307, 309
Robinson family 31–2
Robinson, Mark 277
Rochester (Vic) 95
Rocky Plains Brigade (NSW) 255
Roderick, Michael xix
Rogers, Gavin 231–2
Rogerville (Tas) 185
Rokeby (Tas) 188
Rokewood (Vic) 143
Rolls, Eric 332, 335
Rosetta (Tas) 194–5
Roto Fire (1969–70, NSW) 240
Rowe family 98
Rowley family 9–10
Royal National Park (NSW) 173, 211, 212, 213, 214, 235, 236, 238, 345
Royalla (NSW) 164
Rubicon Forest (Vic) 19, 26, 27, 30, 110
Runnymede (Tas) 189
Rural Fire Service (RFS) 229, 249, 251, 252, 257, 276, 290, 291, 293, 294, 299, 300, 321, 349, 358
Rural Fires Act (NSW, 1997) 184, 236, 252
Rural Fires Board Tasmania 183

St Andrews (Vic) 171
St Ives (NSW) 207
St Marys (NSW) 156, 211
Sale (Vic) 60, 95, 146, 175, 176, 263
Sandy Bay (Tas) 193, 199, 200
Sassafras (Vic) 26, 104, 170–1
Saxton's mill 7–9, 10, 52
Schatzle, Christine 275
Schneider, Stephen 272, 401
Scone (NSW) 177
Scorpion Springs Conservation Park (SA) 243
Scrivener Dam (ACT) 311
Sea Lake (Vic) 11
Seabourne, Warren 190
Seddon, George 248
Selby (Vic) 99
Seldom Seen (Vic) 272, 273
Sellers, George 37
Serpentine (WA) 169
Seven Hills (NSW) 65
Sevenhill (SA) 230
Seymour (Vic) 68
Shannons Flat (NSW) 254, 255, 256
Shepparton (Vic) 20
Sherbrooke (Vic) 123, 223
Shoalhaven district (NSW) 234, 237
Simmonds, Angus 140
Simpson Desert (NT/SA/Qld) 241
Skipton (Vic) 143, 144
Smiggins Holes (NSW) 253
Smith, Ivan 224
Smith, Stuart 152
Snowy Mountains (NSW) xx, 46, 153, 247, 251, 261
Snowy Mountains Highway (NSW) 159, 252, 253, 254
Snowy Plains (NSW) 257
Snowy River (NSW/Vic) 153, 247, 248, 272, 280
Snowy River National Park (Vic) 265, 271, 272, 273, 281
Snug (Tas) 178–80, 189, 191, 192, 193, 276
soil conservation 151
soil humus 55, 116, 168
Solaaris, Chris 23–4
Somerville (Vic) 170
Sorell (Tas) 187
Sorell, John 186–7
South Hobart (Tas) 193, 197, 198, 199
Southport (Tas) 189, 190
Spark, Elly 251
Spencer, Allan 22–3
Springwood (NSW) 156, 207, 211, 212–13, 235
squatters 66–9, 74, 76, 77
Stanhope, Jon 293–4, 315, 321, 322, 323
Stanley (Vic) 266
Stanner, W.E.H 341
Stanwell Tops (NSW) 237
State Electricity Commission (Vic) 146–8
Stern, 'Butch' 221–2
Sternbauer, Brigid 212
Stirling (SA) 47, 216, 229, 384, 405
Stockyard Spur (ACT) 291
Strathalbyn (SA) 47

INDEX

Strathblane (Tas) 190
Strathfieldsaye (Vic) 87
Stretton Group 279–81
Stretton, Hugh xix
Stretton, Leonard 50, 51–61, 86, 102, 107, 108–35, 141, 146–8, 149–52, 183, 204, 210, 322, 369, 370, 371, 375
Strickland Avenue (Tas) 193, 194, 197, 199, 202
Strike Force Tronto 350, 358
Stromlo Forest (ACT) 301, 303, 306, 310, 311
Stromlo Forestry Settlement 294, 306
Strutt, William 69, 74
Strzelecki, Paul Edmund 77, 80
Strzelecki Ranges (Vic) 79, 80, 83, 104, 146
Stuart Highway (NT/SA) 240
Suggan Buggan (Vic) 274
Sussex Inlet (NSW) 237
swagmen 68, 74, 76, 78, 152
Swain, E.H.F. 31
Swifts Creek (Vic) 176, 263, 264, 269, 272
Sydlowski, Lydia 199
Sydney (NSW) 44–5, 46, 61, 63, 68, 78, 101, 140, 155, 156, 157, 160, 161, 166, 173, 193, 207, 208, 211, 212, 214, 215–16, 234, 235, 237–8, 238, 240, 243, 250, 251, 258, 311, 332, 342, 345, 372, 378
Sydney Harbour National Park (NSW) 233
Sylvania (NSW) 45

Taggerty (Vic) 27
Talbot (Vic) 141
Tallangatta (Vic) 162, 263
Tallong (NSW) 173
Tambo Crossing (Vic) 176
Tanami Desert (NT) 240
Tanjil Bren (Vic) 7, 10, 309
Tanjil Loop (ACT) 309
Tarcutta (NSW) 157
Taree (NSW) 92, 105
Taroona (Tas) 184, 193, 194, 199–200
Tarpeena (SA) 231
Tarrawingee (Vic) 139–41
Ten Mile (Vic) 40
Tennant Creek (NT) 240
Terang (Vic) 219
terra nullius, theory of 332, 333, 334
Terry Hills (NSW) 215
Tharwa (ACT) 92, 296, 297–8, 319, 320
Thirlmere (NSW) 46
Thompson, Andrew 303–5
Thoona (Vic) 156
Thorpdale (Vic) 82
Thredbo (NSW) 247
Thredbo River (NSW) 247, 253–6, 257
Tiaro (Qld) 105
Tibooburra (NSW) 241
Tidbinbilla (ACT) 44, 296

Tidbinbilla Nature Reserve (ACT) 294–5, 296, 297, 298, 299
Tilley, Iola 225
Tilpa (NSW) 103, 241
Timboon (Vic) 219
Tindale, Norman 331
Tingaringy (NSW) 261
Tingate, A.C. (Victorian Coroner) 27, 29
Tintaldra (Vic) 158
Tolosa Street (Tas) 193, 196–7
Tom Groggin (NSW) 251, 256–7, 263
Toodyay (WA) 85
Toogoolawah (Qld) 101
Toolangi (Vic) 14, 58, 59
Toolern Vale (Vic) 122, 176
Tooma (NSW) 253
Toombullup (Vic) 12, 13, 26
Toongabbie 62
Toora (Vic) 13, 88, 90
Torquay (Vic) 222
Torrens (ACT) 312
Townrow, Patricia 163
Trafalgar (Vic) 56
Traralgon 60, 95, 146, 263
Trentham (Vic) 219
Trinca, Helen 236–7
Troha, Mark 258–61
Trunkey (NSW) 45
Tubbut (Vic) 265, 272
Tuckey, Wilson 317
Tuggeranong Parkway (ACT) 310, 311, 312
Tumbarumba (NSW) 95, 161, 251, 252
Tumby Bay (SA) 156
Tumut (NSW) 101, 161, 251, 252, 253, 256
Tumut River Gorge (NSW) 253
Turramurra (NSW) 211
Tynong (Vic) 95

Uhlman, Chris 292
Ulladulla (NSW) 46
Ulverstone (Tas) 91
Underwood, Geoff 296–7
Unger, George 25
United States 42, 153, 319, 377
Unkles, Jim 276
Upper Beaconsfield (Vic) 223–4
Upper Brogo (NSW) 158–9
Upper Murray (Vic/NSW) 41, 95, 99, 158, 263, 264
Upper Sturt (SA) 166
Urambi Hills (ACT) 312
Urandangi (Qld) 241
Uriarra Forestry Settlement (ACT) 299
Uriarra pine plantation (ACT) 295
Uriarra Road (ACT) 163, 301
Uriarra Station (ACT) 43

Valley Heights (NSW) 45, 207, 353
Victoria Highway (NT) 240

Victorian Railways 5
Violet Town (Vic) 94

Wadbilliga National Park (NSW) 158
Wagga Wagga (NSW) 11, 86, 94, 101, 261
Walgett (NSW) 155
Walhalla (Vic) 104
Wallacia (NSW) 46
Wallangarra (Qld) 105
Walters, Brian xix
Walwa (Vic) 106, 158
Wandiligong (Vic) 266
Wandilo (SA) 168
Wangaratta (Vic) 139, 140, 156, 162
Warburton (Vic) xix, 6, 15, 16, 20, 21, 25, 35, 38, 40, 95, 98, 99, 103, 104, 226
Warby, Peter 166–7
Warenda Station (Qld) 92
Warragamba (NSW) 46, 166, 237
Warragamba Avenue (ACT) 307–8
Warragul (Vic) 4, 5, 7, 82, 103
Warrandyte (Vic) 16, 41, 54, 171, 176, 210
Warrane (Tas) 188
Warren, Stan 271–2
Warrimoo (NSW) 205, 215
Warrnambool (Vic) 157, 219
Waterfall (NSW) 216, 272
Waterworks Road (Tas) 198–9
Wave Hill Station (NT) 240
Weather Bureau/Bureau of Meteorology 142, 217, 226, 248, 250, 266, 270, 295
Webb family 197
Webster, Joan 323
Wee Jasper (NSW) 164
Weetangera (ACT) 43
Wellington (NSW) 172, 176
Wellington Range (Tas) 186, 189
Wellsmore, Don 260
Wellsmore, Kerry xix, 258–61
Wely, Brad van 277
Wentworth (NSW) 44
Wentworth Falls (NSW) 167
Werribee (Vic) 209
West Coast Range (Tas) 79
West Hobart (Tas) 197, 198, 203
West Wyalong (NSW) 101
Western District (Vic) 12, 26, 72, 79, 86, 94, 99, 143, 144, 151, 157, 219
Weston Creek (ACT) 301, 302, 312
White, Mary E. 344–5
Wilcannia (NSW) 44, 242
Williams family 231–2
Williams, Georgina 268
Williams, Ronald 195

Willis (NSW/Vic) 280
Willis, Matthew 351, 357, 358
Wilmington (SA) 156
Wilsons Promontory (Vic) 88, 121, 122
Wilton (NSW) 161
Wimmera district (Vic) 72, 140
Windsor (NSW) 45, 65, 239
Wingello (NSW) 173
Wingham (NSW) 92
Winmalee (NSW) 235
Wodonga (Vic) 158, 162
Wollemi National Park (NSW) 215
Wollondilly Valley (NSW) 101
Wollongong (NSW) 46, 119, 214, 237
Wombat State Forest (Vic) 219
wombats 80, 128, 260, 282, 332
Wonthaggi (Vic) 95
Woodbridge (Tas) 189, 191
Woodburn, David 273
Woodend (Vic) 12, 14, 41, 163, 220
Woods Point (Vic) 35, 38, 39, 40, 41, 52, 54, 99, 108
Woods, Richard 358
Woolley, Richard van der Riet 165
Woori Yallock (Vic) 171
Worlley's mill (Vic) 97–8
Woronora (NSW) 211
Woy Woy (NSW) 168, 235
Wright, Kate 149
Wulgulmerang (Vic) 270, 272, 273
Wyandra (Qld) 155
Wycheproof (Vic) 163
Wyperfield National Park (Vic) 121, 262

Yallourn (Vic) 13, 26, 146–8, 171
Yamba (NSW) 160
Yaouk (NSW) 254, 256
Yarra Glen (Vic) 110, 171
Yarra Junction (Vic) 15, 16, 97, 99, 103, 226
Yarra River (Vic) 16, 98, 226
Yarrabee Road (SA) 227–8
Yarralumla (ACT) 164
Yarram (Vic) 90, 92, 146, 263
Yarramundi (NSW) 235
Yarrangobilly (NSW) 252, 253
Yass (NSW) 43, 46, 73, 242, 252
Yea River (Vic) 58
Yelland's mill (Matlock) 35–7
Yibarbuk, Dean 329–30
Yorke Peninsula (SA) 156
You Yangs (Vic) 209

Zeehan (Tas) 91